More praise for *The Universe Below*

"We stand, says Pulitzer Prize-winning science writer William J. Broad, on the edge of an age of discovery that will open the deep seas to us in the same way that we have recently explored near space and that 15th-and-16th-century explorers mapped the world. . . . Broad reports his adventures with wit and insight . . . a genial, knowledgeable guide."—Willard St. John, *Detroit Free Press*

"*The Universe Below* reminds us that the deep sea is no minor lair but rather the planet's main address. . . . Broad writes with great enthusiasm and excitement. . . . Reading *The Universe Below* is more than a wade in the water. It is an immersion in a new view of our planet."—Harvey Webster, *The Cleveland Plain Dealer*

"Combining superb natural history . . . with military and economic facts, *The Universe Below* is fascinating reading."—*The Anniston Star*

"A richly rewarding overview of the recent flurry of deep-ocean exploration . . . *The Universe Below* is not only a well-written and even-handed book but also an engrossing one. It's probably the best available one-volume introduction to the deep-water world that some are calling the frontier of the 21st century."—Craig Ryan, *The Portland Oregonian*

"A chronicle of the dramatic changes sweeping one field of science . . . Broad has managed a front-row seat . . . takes the reader along . . . is above all a good storyteller. This is not textbook oceanography. It's science with a sense of the wonder that really drives us to explore."—Mike Vogel, *The Buffalo News*

"Intensively researched and crisply told, this is an illuminating, stimulating portrait of one of earth's last frontiers."—*Publishers Weekly*

"A epic history of exploration in the mold of John Noble Wilford's *The Mapmakers,* William Broad's *The Universe Below* weaves thorough research into a gripping narrative of the first human descents to the bottom of the ocean. From scientific expeditions in search of the unearthly creatures that inhabit volcanic calderas at 'unfathomable' depths, to cold-war missions in search of lost submarines, hydrogen bombs, and secret Soviet cables, to commercial dreams of mining the oceans' gold nodules, every chapter might serve as the plot for a Tom Clancy novel."—Walter A. McDougall, author of *The Heavens and the Earth*

"*The Universe Below* asks provocative questions about environmental stewardship and the possible demise of a part of the world that we are just beginning to understand."—*Outside* magazine

"The cold war might as well have been fought on a different planet for all we knew of its incredible secrets. This fascinating, compelling book reveals for the first time a technology of deep-ocean spying and confrontation even more amazing than the spy-satellite technology that loomed overhead."—Richard Rhodes, author of *Dark Sun* and *The Making of the Atomic Bomb*

"William Broad weaves a riveting tale that provides a startling perspective on human existence on our ocean planet. *The Universe Below* is destined to become an immediate classic about the marine environment and its importance to human history and well-being."—Roger E. McManus, President, Center for Marine Conservation

"*The Universe Below* is authoritative, devoted, manifold, and a book that anybody absorbed by the sea might wish to own."—Edward Hoagland, author of *Balancing Acts* and *Heart's Desire*

"Fascinating . . . a worthwhile gaze into the abyss: a beautiful, strange and biologically important world."—Osha Gray Davidson, *Chicago Tribune*

"Just as astronomers avow that at least 90 percent of the cosmos of galaxies remains beyond their powers of observation, oceanographers concede that 71 percent of the Earth lies unexplored beneath a world's worth of water. William J. Broad's informed, excited, authoritative voice makes the perfect vehicle for entering this alternate universe."—Dava Sobel, author of *Longitude*

"With an artful blend of sea history, personal experience, and the stories of today's oceanographic adventurers, William J. Broad's *The Universe Below* takes us into the fantastic realm of the deep. A fascinating account of a surprising frontier."—William Dietrich, author of *Northwest Passage*

"In his new book . . . [Broad] takes the reader to the depths of the cold, dark ocean, literally another world even though it is on the planet we call our own . . . [a] detailed, exhaustive examination of deep-sea exploration"—Ben Bova, *Hartford Courant*

"The universe below, William Broad shows us, is as saturated with adventure, beauty, scientific unknowns, technological challenge, and vulnerability to human foibles as the universe of space above and the more familiar land about us. Broad brings us face to face with this hidden world."—Charles M. Vest, President, Massachusetts Institute of Technology

"*The Universe Below* provides the reader with unique insight into the role the cold war played in the exploration of the deep. If you are interested in undersea exploration, it's a must read!"—Dr. Robert D. Ballard, President, Institute for Exploration

"Broad's descriptive writing paints vivid pictures which stir the imagination while imparting factual information."—Janice Shumake, Charleston (South Carolina) *Post & Courier*

"Once again William Broad has ferreted a gem of a story from the murky recesses of the cold war. Technology developed to play underwater cat-and-mouse with the Soviet Union and to retrieve our own lost 'assets' is now in the hands of scientists and explorers. Their discoveries are just as exciting and portentous as those made possible in outer space by the conversion of ballistic missiles into launch vehicles."—Alex Roland, Professor and Chair, department of history, Duke University

"William Broad's sea stories are elegant, engaging, amazing, sometimes preposterous but always true, and laced throughout with the spice of first-hand experience."—Sylvia A. Earle, Chairman, Deep Ocean Exploration and Research, Inc.

"Offers an excellent personal overview of current explorations . . . a readable introduction to deep-water oceanographic research and recovery techniques."—Jean E. Crampon, *Library Journal*

"The importance to science of this oceanic revolution cannot be overestimated. Broad tells absorbing stories of the investigators who unravel secrets of the deep ocean, and of the entrepreneurs who seek to extract its animal and mineral riches."—Laurence A. Marschall, *The Sciences*

"Broad's sympathetic portrayal of the new breed of underwater scientists and explorers, and of their state-of-the-art equipment, make this a fascinating account of the newest frontier."—*Kirkus Reviews*

“ ‘We fell for an hour through a sea of luminous creatures.’ begins one of the most fascinating works of science writing that our decade is likely to see. In an age of moon walks and space probes, the world below the sea has, in all its mystery, been largely ignored. Yet for sheer fascination, the present exploration and the wonders it reveals rival all known findings since the exploration of our universe began.”—Elizabeth Marshall Thomas, author of *Certain Poor Shepherds* and *The Hidden Life of Dogs*

“A fascinating, educational look at the bizarre creatures and strange natural phenomena that coexist with us, virtually unnoticed, off our shores.”—*The Staten Island* (NY) *Advance*

“Reading like a first-rate adventure story, William Broad’s *The Universe Below* describes the unbelievable potential of Earth’s last frontier, the mysterious and essentially unexplored vast deep sea, which makes up almost 80 percent of the world’s habitable ‘space.’ ”—Alfred S. McLaren, President, the Explorers Club

ALSO BY WILLIAM J. BROAD

TELLER'S WAR:
The Top-Secret Story Behind the Star Wars Deception

STAR WARRIORS:
A Penetrating Look into the Lives of
the Young Scientists Behind Our Space Age Weaponry

BETRAYERS OF THE TRUTH:
Fraud and Deceit in the Halls of Science
(with Nicholas Wade)

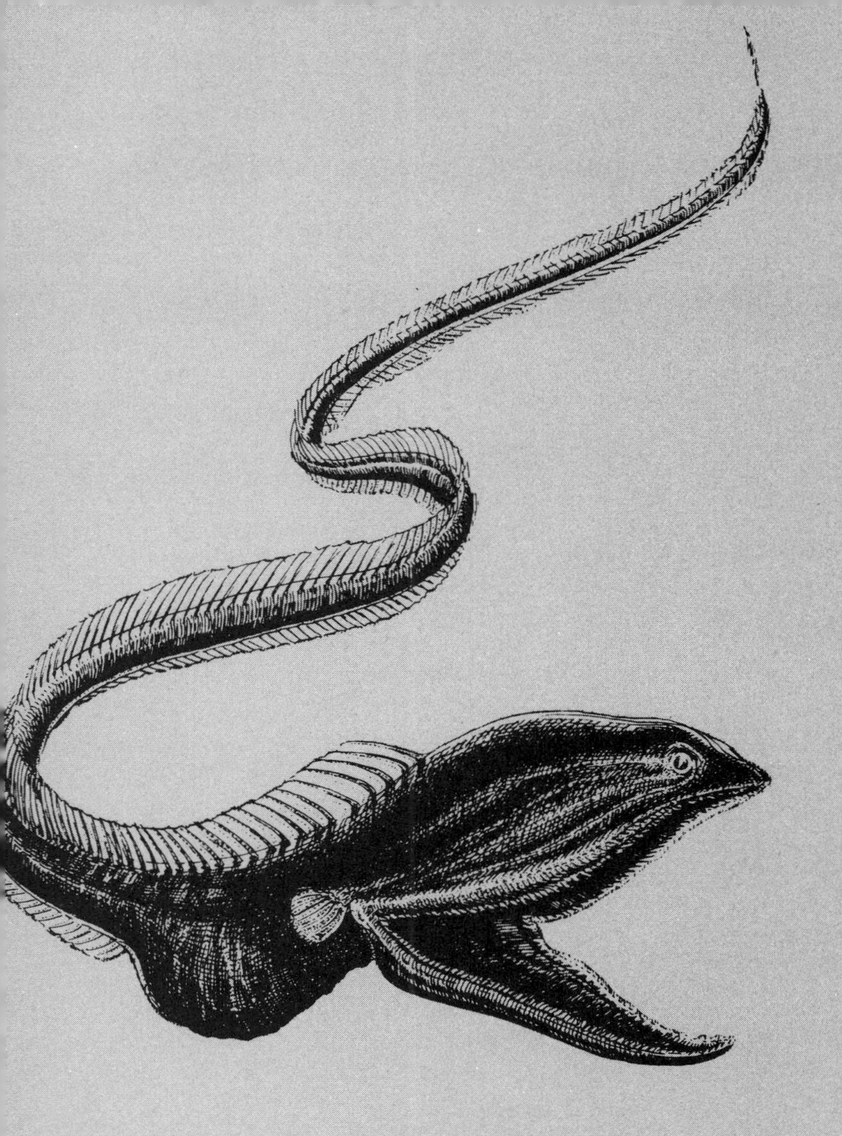

The

Discovering the Secrets of the Deep Sea

A TOUCHSTONE BOOK
Published by Simon & Schuster

William J. Broad

Universe Below

ILLUSTRATIONS BY

DIMITRY SCHIDLOVSKY

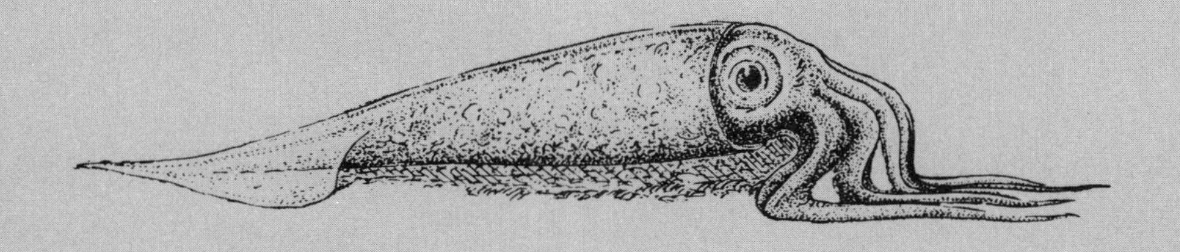

For Bug, Izzie, and Nana

TOUCHSTONE
Rockefeller Center
1230 Avenue of the Americas
New York, NY 10020

First Touchstone Edition 1998

TOUCHSTONE and colophon are
registered trademarks
of Simon & Schuster Inc.

Designed by Edith Fowler

Manufactured in the United States of America

10 9 8 7 6 5 4 3 2 1

The Library of Congress has cataloged the
Simon & Schuster edition as follows:
Broad, William J.
The universe below : discovering the secrets of the deep sea / William J. Broad ; illustrations by Dimitry Schidlovsky.
p. cm.
Includes bibliographical references (p. –)
and index.
1. Oceanography. I. Title.
GC11.2.B76 1997
551.46—dc21 96-50337 CIP
ISBN 0-684-81108-1
ISBN 0-684-83852-4 (Pbk)

The illustration on page 270 is based on illustrations by William H. Bond, © National Geographic Society, and Pierre Mion, © National Geographic Society.

And God said, Let there be a firmament
in the midst of the waters.

—GENESIS

Contents

List of Illustrations

Prologue

WE FELL FOR AN HOUR through a sea of luminous creatures, some a few inches in length, others a foot or two long or longer. Of the thousands we passed, many shimmered and pulsated with light, especially when startled by the passage of our tiny submersible. Out my observation port I sometimes saw a bright flash as we sped downward or watched a lengthy blur of radiance curl into a smaller shape. Every so often one of the living lights, caught in an eddy, would pirouette just outside my window, a swirl of luminescence in a sea of darkness.

"The ones in the distance are lighting up on their own," said John Delaney, a scientist from the University of Washington. "Some have recently been discovered that are twenty to thirty meters long."

We were searching for something far stranger than animals that glow in the dark. The three of us—a pilot, a scientist, and I—were looking for an alien world powered by a volcanic gash at the bottom of the sea, a kind of deep oasis lush with outlandish forms of life. Until a few years ago, such oases were beyond the dreams of science. Our hunt had taken us to the eastern Pacific and inky waters a mile and a half deep—far beyond the range of scuba divers and submarines and all but the most advanced exploratory craft. It was a place no human had laid eyes on before.

The bottom finally came into view. Barely illuminated by our lights were fields of gnarled lava. Once red hot and violently astir, the liquid rock had frozen fast. Now the shapes were rounded and bulbous and swollen with energy, bespeaking a kind of uneasy surrender as the eruption had come to a standstill amid crushing pressures and icy temperatures. Flat, smooth flows of the kind found on land were nowhere to be seen. Our expedition had searched similar terrain without luck for more than two weeks. Now it was the last day, the last dive. We had to hit pay dirt or all the meetings and hunts and sleepless nights would seem to have been for nothing.

After an hour of exploring the dim expanse, our pilot, Bob Grieve, reported a jarring find, given the gravity of the situation.

"A tennis shoe," he said, his voice tense, his eyes glued to the front window.

"No."

"Yes."

"No."

"A tennis shoe, a Reebok tennis shoe. God, I can't believe it!"

The shoe lay atop gray sediment near mounds of dark lava, sitting upright, in perfect shape. Bob raised one of the sub's twin robotic arms to retrieve the lost object.

Slowly our cabin grew colder. Carrying the bare essentials of life support, the submersible could supply three people with plenty of oxygen but had no heat to combat the chill from waters that were nearly freezing. It was like working in a refrigerator.

The breakthrough came after a working lunch of peanut butter and jelly on whole wheat, after hours of searching.

"I have a chimney," Bob sputtered as he eyed the gloom. "It's hot. It's got tube worms all over it."

Looming ahead of us was a rocky monolith three or four stories high. Its surface was a tangle of tube worms and bristly creatures and strange growths that swayed in the currents, making the whole thing seem alive. Seething hot water rippled up its sides and shimmered in our lights like waves of heat over desert sand. In spots we measured the ripples as more than twice as hot as boiling water. But eerily, no bubbles formed.

Adjacent to the monolith in the clefts of neighboring rocks were orangish-brown spider crabs the size of dinner plates that fidgeted and sulked and tried to hide in the shadows—clearly predators ready to resume the feast we had so rudely interrupted. The rocks were topped by white sea anemones and pinkish colonial growths, the arms of both arcing toward the monolith as if the creatures were at worship.

Nearby in the gloom we discovered other chimneys enveloped by odd forms of life. In all, we examined five of the towering structures, one discharging water hot enough to melt tin.

"The real question we have to address is whether or not these things are new," John, the scientist, said shortly before we had to surface. "We don't know how fast they evolve."

Suddenly, Bob noticed that the temperature of the submersible's skin was starting to rise. By accident, we had positioned ourselves over a hot vent, a dangerous thing to do because of the possibility of melting the sub's plastic windows. Quickly, we began our ascent, happy and exhausted.

THIS BOOK is about the largest unexplored part of our planet, the deep sea, and how we are illuminating its dark recesses in a rush of discovery that is shattering old myths, rescuing lost treasures, and laying bare secrets of nature hidden since the beginning of geologic time. It might seem that such a watershed of human history would be widely recognized and discussed. Surprisingly, that is not the case. The accelerating movement into the oceanic depths is poorly known and understood, at times even by specialists in the field. I stumbled on the speedup quite by accident as a science reporter for *The New York Times.*

Like many people, I knew little of the deep except that it is mysterious. My midwestern childhood and education left me essentially in the dark about its known wonders—its mountain ranges longer than the Andes, its fissures deeper than the Grand Canyon, its beasts more extravagant than the Komodo dragon. I knew nothing of the triplewart seadevil and how its dancing lights lure victims into needlelike teeth. I knew nothing of the deep's living fossils, of creatures long thought to have vanished from the face of the Earth. Like most people, I had only vague impressions of the sea's great size and knew nothing of its exact dimensions—that it covers 71 percent of the Earth, averages more than two miles deep, and plunges down in places to a depth of nearly seven miles, deeper than the tallest mountain is high. I had snorkeled in coral reefs and sailed the Atlantic off New England. I considered myself a water person. But I had no interest in what lay at the bottom of the sea, and perhaps some apprehensions about it.

My first professional contact with deep-sea inquiry came in 1985 as I reported for the newspaper on the discovery of the *Titanic,* the luxury liner that sank on its inaugural voyage in Atlantic waters two and a half miles deep, killing more than fifteen hundred people and burying its gargantuan hulk in darkness for nearly three-quarters of a century. Of course, the discovery of the most famous of all shipwrecks made headlines around the world. But the news reports tended to overlook the military's role in the discovery. As I found and wrote for the *Times,* the feat was done with military gear and military funding and was an adjunct to a top-secret mission. History also highlighted the military tie. An American entrepreneur had failed three times in the early

1980s to find the lost ship with conventional tools, his frustration driving home the difficulty of such things for civilians.

At first the discovery of the *Titanic* was an anomaly, a pleasant diversion from whatever else the military did in the depths. But then, in the late 1980s and early 1990s, it was joined by an inconspicuous wave of similar feats around the globe achieved by scientists, businessmen, and sometimes private individuals. Among the deep discoveries were faded swastikas and geological prizes, tons of gold and bizarre new forms of life. Experts began to say that no part of the seabed was unreachable, that no object was unrecoverable—a fundamental shift that hinted at the opening of a new frontier.

I was hooked. What was going on? And why had things previously been so backward? At first, I reasoned that the force behind the rush was robots. Influencing me were their designers, who declared these sleek devices with camera eyes and monsterlike claws to be revolutionary. Their unique feature was long tethers, which linked them to the surface and allowed them to send up crisp undersea images and draw virtually unlimited power and guidance from operators up top. New parts and materials were letting the robots dive deeper, grab more, and see further into the darkness. Robots fired into space had succeeded in probing the distant planets. Perhaps their brethren were making similar advances miles beneath the waves.

But as my understanding grew, and as the pace of discovery picked up, I saw that the shift was more fundamental than that. No single technology, however advanced, could prevail over anything as unyielding as the deep.

For all the sea's romance and poetry, for all the books and songs over the centuries, seawater itself turns out to be pretty nasty stuff. It is corrosive and viscous enough to pose a major challenge for naval architects and engineers. An even bigger difficulty is the sea's opacity, which dims sunlight to nothingness in a few hundred feet and accounts for the deep's icy temperature and inky darkness. Bright lights can push back the gloom for no more than tens of feet and, even then, blizzards of debris often rain down from above to scatter light and hamper viewing. Virtually the only way that experts glimpse deep terrain is with sound, which speeds easily through seawater. The echoes, when picked up by undersea microphones and processed into images, can reveal canyons and peaks but, by nature, miss most smaller objects and features. In short, explorers of the deep are blind to much of what they want to see.

The other big problem—a really big one—turns out to be the sheer heaviness of the sea. Water is almost one thousand times denser than air. By weight, every ten cubic feet of seawater equals roughly one cubic foot of lead. A consequence of water's density is that pressures mount rapidly during descent. At the resting place of the *Titanic,* the pressures on any object are equal to a tower of solid lead overhead that rises to the height of the Empire State Building, the one hundred stories trying to crush any void. Sea creatures are made primarily of water, which is virtually incompressible, and thus are largely immune to destruction. So are inanimate solids such as trawl nets, bottom scoops, and long drills that tear blindly into the seabed, which is why scientists rely on such tools so extensively.

But I found out that anything possessing a hollow or a cavity (including humans and robots and most deep-sea gear) is at risk of collapse, with the danger increasing as you go down. Thus, people with air tanks on their backs safely descend on most dives no more than one hundred feet or so. Military submarines go down a few thousand feet but are largely blind to their surroundings. Small research submersibles with lights and tiny portholes and cramped passenger spheres made of thick metal descend to some parts of the deep seabed but are few in number and extremely limited in their travels.

To me, the physical challenge was surprisingly great compared to that in the exploration of outer space, which I knew about from years of reporting. The void around our solar system is relatively simple to peer through, to pass through, and to beam electronic messages through. It has no great opacities and pressures. Most space probes are made of relatively thin metal and can transmit over vast distances signals that are remarkably weak. The hard part is getting into outer space, which is why rockets and winged spaceships are so big and costly.

As I learned about inner space and the field of deep exploration, I eventually came to understand that the surge of discovery was driven not by a single factor but by thousands of them.

THIS BOOK is structured more or less on chronological lines, looking at the past, the present, and the future. At its heart, it is a chronicle of my journalistic explorations over more than a decade, a venture involving much reading, travel, and many hundreds of interviews.

WE HAVE probed the physical world from the nucleus of the atom to the spiral arms of the distant galaxies. We have landed men on the Moon and sent spacecraft to all the major planets of our solar system and beyond, so that the receding Sun appears to be one of the numberless stars. These are great feats of the human spirit. Even so, it has always been the nature of science to do the easier things first.

The sea may cover 71 percent of the planet. But what dominates, and does so in utter anonymity and invisibility, is the deep, which lies beyond the shallows that border the continents and in total accounts for about 65 percent of the Earth's surface. This domain is so wide and so deep that by some estimates it comprises more than 97 percent of the space inhabited by living things on the globe, dwarfing the thin veneer of life on land.

Human eyes have glimpsed perhaps one-millionth of this dark realm. Perhaps a thousandth or a billionth. No one knows, and in any case the precise number is immaterial. The truth is that our planet has managed to remain largely unexplored, until now.

ONE

Lair

THE SURFACE of the sea, with its play of light and changing moods, its sudden storms and fiery sunsets, was seen from the beginning as a thing of beauty and terror, a giver and taker of life. People in small boats learned slowly and often painfully of its boundlessness and bounty and how to take advantage of both. They fished, migrated, and transported goods. They braved storms and thundering waves and learned to love the sea despite its toll. Its mercurial nature made it almost human. The sea, friend and foe, was praised and damned in speech and song.

But the world far beneath the waves was different. It had just one face. And though inscrutable, that countenance was judged to be dark with menace. People recoiled from a place that swallowed men, ships, and whole fleets without effort. Its position alone was ominous. Up was associated with God and goodness, down with evil and the devil. The opacity of seawater made the deep impossible to see and understand, adding to the apprehension. In comparison, the sky was a familiar friend. Man gave names to the stars and took comfort in the regular motions of the Moon and planets, marking the seasons with their passage. In time, the stars became central to navigation and sailing the sea.

But the deep was unknown and unknowable, unconnected to anything remotely human. No part of nature seemed more foreign. One word for it fairly resonated with sinister energy—*abyss,* from the Greek *a*, without, and *byssos,* bottom—a synonym for dark infinities and primal chaos. Magellan in 1521 was said, perhaps apocryphally, to have attempted to sound the central Pacific as he sought to become the first man to circumnavigate the globe. After splicing together six lines and lowering a cannonball more than four hundred fathoms (or, at six feet to a fathom, 2,400 feet) without hitting bottom, he declared that the Pacific was immeasurably deep. The sea was literally unfathomable.[1]

Despite the uncertainty about this bottomless world, people felt they understood a few of its attributes. It was cold. It was dark. Worst of all, it was well known to harbor the most loathsome aspects of creation, some of which from time to time crept upward from the unilluminated depths to bedevil man.

The ancient world was one in believing that huge creatures inhabited the deep. The Greeks populated it with a strange assortment of monstrosities. The Old Testament made frequent references to deep-dwelling ogres and alluded to an aboriginal combat between God and a colossal adversary of the deep named Leviathan or Rahab. "When he raiseth up himself," the Book of Job said of the beast, "the mighty are afraid." Medieval Arabs spoke of the sea of darkness and the menace of its monsters. Western Europeans during the golden age of exploration sailed far and wide in tall ships but did little to banish the demons, and often ended up reinforcing their reality. Sea serpents were reportedly big enough to smash brigs and hungry enough to gobble up hapless sailors. Mapmakers populated the waters beyond *terra incognita* with freaks and beasts whose wide mouths bore teeth that were extraordinarily long and sharp.

Modern psychologists would say the deep was simply a blank slate for the expression of man's subconscious fears and insecurities, a kind of Rorschach test that revealed more about the viewer than the viewed. Maybe so. But enough scary things actually arose from the depths to give the old legends some credence—fearsome beasts that defied comprehension, devilfish and worse. That was the thing. Whatever managed to live in such a wet, cold, dark place surely had to be worse than almost anything man could imagine. Tennyson in "The Kraken" had no need to describe the hideous creature that lay for ages "Below the thunders of the upper deep," feeding upon "huge seaworms in his sleep." The beast came alive by dark implication as Tennyson sketched an eerie habitat where "faintest sunlights flee" and "sponges of millennial growth" thrive alongside "enormous polypi." So too Edmund Spenser. In the final lines of *The Faerie Queene,* he conjured up that queasy sense of uncertainty that is the foundation of all fear:

For all that here on earth we dreadfull hold,
Be but as bugs to fearen babes withall,
Compared to the creatures in the seas entrall.

The layers of myth and superstition were slowly stripped away over the centuries by scientists and naturalists, their apprehensions subdued by curiosity. Struggling to fill the void of ignorance with a modicum of fact, dismissing warnings of danger and death, they sailed the globe to examine the deep sea with a skeptical eye, to drop lines and nets and scoops into supposedly bottomless depths, to examine writhing specimens on the decks of heaving ships. Progress was erratic. For every myth that died, a new one often arose to replace it and confuse the emerging picture. And the process of exploration was painfully slow, perhaps partly owing to lingering fears over what lurked below. The North and South poles were familiar places by the time an explorer descended into the sea's darkness to observe what was down there.

Despite such backwardness, advances were made. Scientists found that many of the old legends were built on exaggerated descriptions of such familiar animals as squids and whales, which, though often mighty by man's standards, were hardly evil or supernatural. In addition, the depths were found to harbor many unknown creatures that looked far from frightening when studied in broad daylight, even though, truly, they were often demonic in appearance, with enormous mouths and needlelike fangs. Some scientists, seeing a glimmer of truth in old legends, believed that undiscovered beasts still roamed the depths and found evidence to back their suspicions. So strong did such beliefs run that many expeditions were mounted to search out the monsters, even into the twentieth century. In some places the hunt goes on today.

DURING THE EIGHTEENTH and nineteenth centuries, increasing curiosity about the natural world and a run of biological luck brought observers into contact with a small zoo of odd creatures that washed up on the world's beaches, giving tantalizing hints of what else lurked below. Morton Brunnich, a Danish naturalist, found one of these enigmas near a coastal farm in Norway. It was improbably long and thin, with a fiery red dorsal fin that ran the length of its body and exploded into a feathery crest atop its head. In 1771, Brunnich described it in the scientific literature, calling it a bizarre type of fish. Today the creature goes by the rather pedestrian name of *oarfish.*

The eerie, serpentine beast can grow to lengths of up to fifty-five feet and weigh more than six hundred pounds. It is believed to live at depths of up to one hundred fathoms, or six hundred feet. Its lifespan is unknown, as is much else about the animal. Its common name derives from the way its brilliant red pectoral fins rotate like oars as it swims. In ghoulish fashion, oarfish feed by moving their upper jaws far forward, enlarging their mouths some forty times. The beasts are apparently rare. Over the centuries sightings of oarfish have been reported only about two dozen times, though the animal undoubtedly accounts for some sea-serpent tales, both ancient and modern.

In 1963, Carole Richards of Malibu, California, was walking her poodle on the beach around midnight when she came upon the lifeless body of an oarfish. She screamed in terror. Houses all over Malibu lit up as news of the monster spread. Today, little more is known of the oarfish than when Brunnich first stumbled on this marine phantasm more than two centuries ago. For all we know, it could be on the verge of extinction. Or it could be flourishing, moving silently through the deep waters off Malibu and Santa Monica.[2]

Other beasts cast up from the deep were even stranger. In the 1830s, Lieutenant Commander C. Holboell of the Royal Danish Navy was in Greenland when he noticed a grotesque fish that had washed ashore. It had an enormous gaping mouth and razor-sharp teeth. Gingerly, he gathered it up. In subsequent years he found two more, different but clearly related. What these creatures had in common, aside from their fierce visages, were ugly threads and projections around their mouths. The animals were fish that fished. The tips of these fleshy projections glowed to lure victims. Today the creatures are known as *anglers.* The living lights, dangled near gaping jaws, tempt unsuspecting prey into teeth that are extraordinarily long and sharp. The bizarre fish live down to depths of nearly two miles and can grow in length to three or four feet, though many varieties are smaller. Their bodies are jet black or dark brown, and their heads are often huge. Their repulsiveness, at least to some humans, is suggested by a few of the common names bestowed upon them over the years—*blackdevil, blacktail netdevil, triplewart seadevil.* The body of the latter fish is covered with warts and spines and furrows, and its large mouth is turned down in a perpetual frown. Today anglers are virtually icons of the deep.

Yet it took a long time to discover the most exotic of their secrets—why a little parasite was often found attached to an angler's body. The freeloader was actually a male angler, and the host invariably a female. The two sexes sought out one another in the deep and then merged for life in the most extreme kind of monogamy, apparently to increase the chances of successful reproduction in a world so vast and dark. The male, never big to begin with, upon sinking his teeth into the female largely fuses with her and loses many of his normal organs, including his mouth and digestive tract. What develop are his testes. Though his meals are free, his bloodstream having merged with hers, his new life is that of a parasite. He exists only in the service of procreation, having diminished in stature to little more than a reproductive organ.[3]

The greatest of the serendipitous finds suggested that the depths har-

ANGLER FISHES. Waving rods tipped with glowing lures to attract prey, angler fish are accomplished hunters with daggerlike teeth. Like all creatures of the deep sea, they live and die in a world of perpetual darkness.

GIANT SQUID. A legendary monster, the giant squid is now thought to be the Earth's largest invertebrate, a colossus that long eluded efforts to uncover its deep lair.

bored a creature that was slimy and huge and utterly unfishlike, a living nightmare with a number of exceedingly long tentacles that squirmed like a nest of snakes. It was the giant squid. In 1853, parts of one washed ashore in Denmark and eventually came into the hands of Johan Japetus Seenstrup, a Danish zoologist. He published a description of the parts and gave the creature its genus name, *Architeuthis,* Greek for *chief squid.* Two decades later, cod fishermen off Newfoundland stumbled on one of the giants at the surface, and, during what was said to be a life-and-death battle, succeeded in hacking off two of its arms, the longer one measuring nineteen feet. Soon after, other fishermen caught a complete animal. When laid out, the creature was found to have tentacles that were twenty-four feet long.

An accumulation of such evidence slowly turned the mythological beast into a scientific star—the largest invertebrate known to man, a colossus that can grow up to lengths of at least sixty feet. Today little is known of its deep habits and habitat. But accidental captures by fishermen have revealed much about its anatomy. Like small squids, the giant has eight stout arms and two

thinner but very long tentacles, whose ends bear sucker pads reinforced by toothlike rims. The arms and tentacles carry prey to its central mouth and sharp beak. The appendages can also battle some of the sea's mightiest creatures, judging from sucker marks found on sperm whales. Giant squids have relatively large brains, implying intelligence. Even more striking, their unblinking eyes are as complicated as man's and exceptionally large, typically the size of a human head.[4]

The tide of biological luck revealed startling things to early investigators, but, overall, the revelations were few. In time, the chance encounters with the deep's inhabitants would be eclipsed by intentional ones as man shed his old inhibitions and began to actively probe the sea's sunless regions.

The initial work focused on measuring the sea's depth and often was driven by the sheer curiosity that marked the age of discovery. A long line with a cannonball or lead weight on its end would be lowered, as Magellan had done, to try to find the bottom. Such exploratory probings went much deeper than the routine soundings near the coasts, which usually penetrated no more than one hundred fathoms. By the eighteenth and nineteenth centuries, diverse groups of mariners and scientists were often dropping sounding lines around the continents and, on long voyages, into the deep. Such undertakings turned out to be anything but easy. False readings were legion as the subtle twitch of a rope that signified touchdown on the seabed was often overlooked and coils of line simply kept paying out and piling up on the bottom. Depths of up to eleven miles were reported, exaggerating the truth nearly twofold. The best of the sounding work tended to be British, aided by that nation's nautical might and scientific acumen. Piano wire replaced hemp.

Little by little, accuracy increased and the idea of bottomlessness was thoroughly discredited, though the inability to find the seafloor in some spots kept the question of ultimate depth alive well into the twentieth century.[5]

A simple extension of such work was to put some kind of gathering device on the line's end, so bottom ooze and life could be brought to the surface for examination. Sir John Ross, commander of the H.M.S. *Isabella,* in 1818 lowered a whaling rope for a distance of about one thousand fathoms (or six thousand feet, slightly more than a mile), down to the bottom of Baffin Bay, the run of sea between Greenland and Canada. On the rope's end was what Sir John called a "deep-sea clamm." Its metallic jaws automatically closed tight as the sampling gear hit bottom. The clamm, hauled back on board, disgorged greenish mud and three worms. The prize, however, was found entangled on the sounding line—a basket star, *Caput medusae,* a beautiful relative of the common sea star. Its Latin name was inspired by its writhing arms, which were subdivided so many times that they resembled the head of the mythical Medusa. Sir John concluded that his work proved that animals lived on the ocean bed "notwithstanding the darkness, stillness, silence and immense pressure produced by more than a mile of superincumbent water."[6]

Such gems of fieldwork were undermined by poor communication of the findings around the world as well as by misunderstandings about the deep that confused the situation for decades, as historian Susan Schlee has shown. Remarkably, many biologists of the 1830s and '40s came to hold that the depths of the sea were utterly barren, despite accidental and purposeful finds to the contrary. Wrongly, they assumed the deep had no currents and no temperature shifts, which appeared to rule out nutrient and oxygen exchange and the physical prerequisites for life. The bottom was assumed to be sterile and stagnant, a view known as the *azoic theory.*

It is true that the depths see very little change compared to land. But the pioneers took the idea too far. Today we know that the deep has currents, eddies, and even storms that mix waters of different temperatures and oxygenations and keep its physical constituents astir, supporting a diverse fauna. But the scientific leaders of the day were keen to challenge the myths and fears of the past. The most influential champion of the azoic view was Edward Forbes, an energetic young naturalist at the University of Edinburgh. As samplings went deeper and deeper, Forbes wrote, the inhabitants became "fewer and fewer, indicating our approach to an abyss where life is either extinguished, or exhibits but a few sparks to mark its lingering presence."[7]

THE AZOIC THEORY and its confusions might have lingered longer than they did but for a burst of exploratory activity in the middle of the nineteenth century that eventually laid them to rest. The excitement was driven by an unusual confluence of forces—a hunt for living fossils, a growing demand for the charting of navigable waterways, and a rush to find the best routes for undersea cables.

The telegraph, made practical by Samuel F. B. Morse in 1837, rather quickly showed its talent for knitting together peoples and nations. The growing network was a boon to railroads, newspapers, weather forecasts, and commerce in general. Following success on land, telegraph wires were strung beneath the waves on short runs between England and France, Corsica and Italy, New Brunswick and Prince Edward Island. Bigger things were clearly possible, and perhaps profitable.

In 1853, Cyrus W. Field, an American industrialist, wrote the Navy to ask if the seafloor between Newfoundland and Ireland was fit for a telegraph line. The depths of the North Atlantic, replied Lieutenant Matthew F. Maury, head of the Navy's Depot of Charts and Instruments, concealed a plateau "which seems to have been placed there for the purpose of holding the wires of a sub-marine telegraph, and keeping them out of harm's way." Maury's portrayal of the two-thousand-mile-long route was based on about thirty soundings—a rather bold interpolation.

The truth, discovered slowly, was that "Telegraph Plateau," as Maury called it, was largely a fiction. Eager to be the first to tie continents together, Field made two voyages himself to sound the Atlantic's depths before he sailed to London in 1856 to raise capital and found the Atlantic Telegraph Company. In 1857, the British Admiralty sent the H.M.S. *Cyclops* to survey the bottom in greater detail. Soon a cable was laid. After many trials and tribulations, breaks and false starts, the line finally came to life in August 1858, carrying salutations between President James Buchanan and Queen Victoria.

Though the cable soon failed, the success was enough to inspire more efforts—always preceded by new soundings and samplings of the bottom. At Field's request, the British Admiralty made a new survey. In 1860, the H.M.S. *Bulldog* took over one hundred soundings of the north Atlantic, one time bringing up thirteen sea stars from a depth of nearly a mile and a half. Also in 1860, a Mediterranean cable between Sardinia and Africa, submerged for three years in more than one thousand fathoms of water, was pulled to the surface for repairs. It was found to be encrusted with all kinds of marine life, including corals, snails, bivalves, and bryozoans. In short, more by accident than design, in both the Atlantic and Mediterranean, the depths of the sea were surrendering no little evidence that life flourished in the icy darkness.[8]

As the laying of the first transatlantic cables got underway, Darwin's theory of evolution came along to give undersea exploration an additional surge of momentum by prompting an audacious hunt for living fossils. *The*

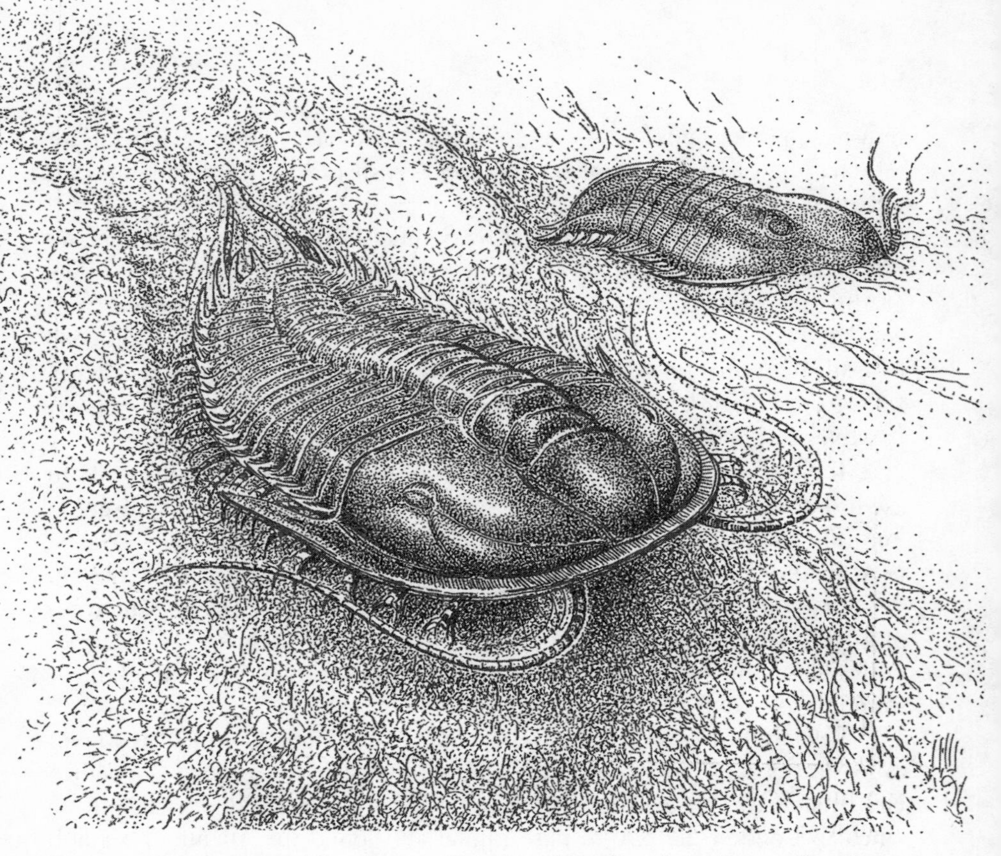

PAIR OF TRILOBITES. After its unveiling in 1859, Darwin's theory of evolution inspired a hunt for living fossils in the deep sea. Among the targets were trilobites, vanished animals up to two feet long that ruled the primordial seas.

Origin of Species, published in 1859 to wide praise and denunciation, was an intellectual watershed that contradicted the biblical view of creation and linked man to the monkeys and a long succession of lower creatures into the dim past. The whole tree of life, it said, had evolved from a single seed. The suspicions about the preservation of ancient sea life arose because the theory suggested that a main evolutionary force was environmental change, such as climate shifts and the rise and fall over geologic time of physical barriers on land and sea, including mountains and islands. Darwin held that these changing barriers not only created new physical challenges for all living things but rearranged the kinds of predators and competitors that came in contact with one another in particular regions and thus, over time, fundamentally and repeatedly altered the rules of survival. It was like a board game in continual upheaval. Plants and animals, drawing on the inherent variability of offspring over generations, either adapted to the new circumstances or died off, as the dinosaurs and countless other creatures had done since the beginning of time. The winners thrived, passing on their successful traits to descendants.

Victorian debates raged over such ideas. But it was the inconspicuous corollary, little noted publicly at first, that gave deep exploration its new fervor. Perhaps, just perhaps, the thinking went, the relatively stable conditions of the deep sea had nurtured types of primitive life that underwent little or no biological change over the eons. The fossil record showed that trilobites, sea scorpions, and orthocones—with thirty-foot shells and tentacles even longer—once ruled the seas. Evolutionary theory suggested that perhaps their descendants were still out there, concealed in the darkness.

Species, Darwin wrote, can be preserved for a long time if they inhabit "some distant and isolated station." He noted for example the ancient ganoid fishes, which include paddlefish, garpikes, and sturgeons, the latter having elongated snouts and barbels for locating food in bottom mud, a rudimentary means of feeding. Darwin also pointed to brachiopods, or lampshells, a primitive kind of bivalve "but slightly modified from an extremely remote geological epoch," with some species having lived more or less unchanged for more than half a billion years. Because of land's constant barrage of upsets and rearrangements, Darwin wrote, "the productions of the land seem to have changed at a quicker rate than those of the sea." And as the scientists of the nineteenth century were learning, nowhere was the sea more constant than in its unilluminated depths.[9]

States of stability, previously seen as responsible for deep wastelands, now quite suddenly were seen as favoring the opposite. So it was that in the 1860s a hunt began for living fossils, evolutionary throwbacks, and missing links, often with barely a nod to the fact that, by the azoic theory, such creatures of the deep were impossible. The idea of ancient monsters, however unorthodox by the day's standards, was just too tantalizing to ignore.

Michael Sars, a Norwegian naturalist teaching at the University of Christiania (now Oslo), was one of the first scientists to demonstrate that the conjectures about the preservation of ancient life actually had considerable merit. In the early 1860s, he cast dragnets and scoops into cold Norwegian fjords and dredged the bottom down to depths of several hundred fathoms. Among his most extraordinary finds were sea lilies, a stalked type of crinoid. Ancient and forgotten, living in darkness, remnants of a lost race, the sea lilies were the descendants of creatures that had thrived more than one hundred million years ago, during the Cretaceous period, a time when the Earth was ruled by the dinosaurs. Strangely, these relics of another age looked very much like plants, growing on long stems with an upper profusion that appeared to be leaves or petals. In fact, the petals were arms that captured particles from the water and passed them down ciliated grooves to a central mouth. Sea

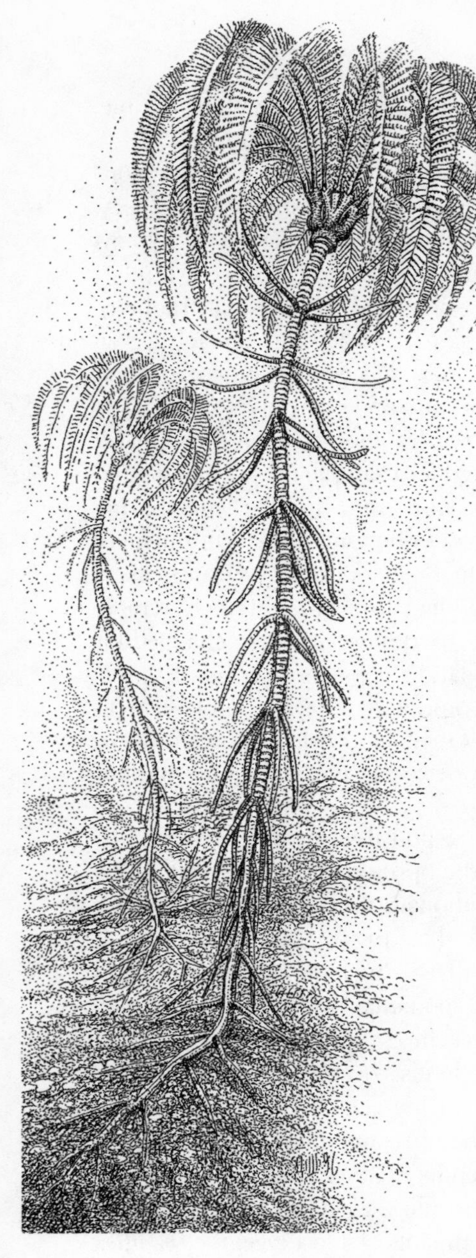

SEA LILY. An animal whose ancestors thrived in the days of the dinosaurs, the sea lily was one of the first living fossils discovered by deep explorers of the nineteenth century.

lilies of the type hauled up by Sars had never before been recognized while alive, only in fossils.[10]

Another ancient organism that came to light was a strange and delicate type of sponge that anchored itself to the seafloor with glassy filaments. In the exploratory wave of the 1860s, these so-called glass sponges were found in deep waters off Portugal and Japan, including one reportedly from a depth of five hundred fathoms. As naturalists hauled up more primitive sponges, sea urchins, and sea lilies, the belief grew that, in time, and with sufficient effort, bigger and better things from the prehistoric past would be hauled thrashing and writhing into the light of day. It seemed that man was about to meet the dinosaurs of the deep.

Interest became so great that the field's leadership passed from individual naturalists and fishermen to the Royal Society of London, a century-old group that aided the British government in science matters. Eager to move beyond coastal areas, the society asked the Admiralty for the loan of a ship so its investigators could conduct a summertime round of deep-sea exploration. In August 1868, the H.M.S. *Lightning,* a small paddle-wheel steamer originally built for hauling mail, sailed from the Outer Hebrides and headed northward for the Faroe banks, an area of moderate depth around the Faroe Islands whose flanks were within easy reach of the day's collecting apparatus. The ship had a derrick over the stern for deploying trawls and dredges, as well as numerous glass jars and bottles of preserving spirits. Up from the depths came a large new type of sea star, crimson and fiery orange. Also raised were sea urchins, shellfish, brittle stars, glassy sponges, and delicate sea pens, whose feathery arms made them look like the writing quills of the day. Odd crustaceans also were pulled up, including one with large eyes the color of burnished copper.

The deepest haul was made in waters of 650 fathoms (or seven-tenths of a mile), and the overall results showed that animal life in the depths was varied and abundant. That finding was important, given the old azoic beliefs, even if the voyage made no major progress in regard to living fossils. Significantly, the trek of the *Lightning* dealt a sharp blow to the notion that deep seawater was uniformly cold—one of the mistaken assumptions of the azoic theory. Careful measurements showed that nearly contiguous areas had slightly different temperatures, implying that great masses of water were circulating at the bottom of the sea.[11]

Happy with the expedition's results, the Admiralty freed up two larger ships for expeditions during the summers of 1869, 1870, and 1871. The H.M.S. *Porcupine* and the H.M.S. *Shearwater* dredged down to 2,435 fathoms, or nearly three miles. New animals were discovered by the bucketful. The scientists also marveled at how some organisms glowed in the dark, at times brightly enough to illuminate a pocket watch at night. One haul brought up masses of stalked animals that blazed with a pale, lilac-colored light, suggesting "a wonderful state of things beneath," C. Wyville Thomson, an organizer of the expeditions, wrote in his popular book *The Depths of the Sea.* In a reverie, Thomson envisioned the creatures in their dark abode as "scintillating and sparkling on the slightest touch, and now and again breaking into long avenues of vivid light indicating the paths of fishes or other wandering denizens."

The expeditions, while finding no trilobites or sea scorpions, repeatedly confirmed the general idea of antiquity. A bizarre sea urchin of scarlet hue was dredged up that panted, like a dog, its shell articulated and quite unlike the typical rigid ones. Nothing like it had ever been seen before—except as fossils dating to the Cretaceous period one hundred million years ago, the dinosaur era. Every haul of the dredge, Thomson wrote, "brings to light new and unfamiliar forms—forms which link themselves strangely with the inhabitants of past periods." The hunt, he added, had barely begun. "Notwithstanding all our strength and will, the area of the bottom of the deep sea which has been fairly dredged may still be reckoned by the square yard."[12]

The effort to explore the deep was exceptionally modest by the standards of the day. On land the crush of civilization was carving up things at a remarkable pace. Rushes for gold shook Australia and New Zealand. Livingstone and Stanley opened the interior of Africa, presaging a scramble for European colonies. The French completed the Suez Canal, its waterways linking the Mediterranean and Red Sea. Few mountains were so lofty, few islands so remote, as to remain untouched by man. Ideas, goods, and people were moving with increasing dispatch, thanks largely to the railroad, steamship, and telegraph. If, in contrast, the world of undersea exploration seemed to be progressing slowly, it was nonetheless undergoing changes that would eventually prove major. The azoic theory was under siege, and evolutionary theory was on the rise, even if no vanished monsters had so far come to light. For deep inquiry, the season was one of transition, if not turmoil.

IT WAS AGAINST this backdrop that a tale of deep intrigue captured the world's imagination. Jules Verne's *Twenty Thousand Leagues Under the Sea,* published in Paris in 1871, had everything—a malevolent hero in the form of Captain Nemo, a futuristic diving machine known as the *Nautilus,* and enough monsters to rivet the most jaded reader. Despite the book's richness and visionary glimpse of future machinery, a close reading shows that Verne bent over backward to embrace the main misconceptions of his day about deep life, even while avoiding some of the newer and more accurate ideas.[13]

Verne presented his fiction through the eyes of Professor Aronnax of the Paris Museum of Natural History, author of *The Mysteries of the Great Submarine Depths.* The naturalist, taken captive by Captain Nemo for a submarine tour of the world's oceans, finds himself in the enviable position of being the first scientist to actually see alive the things he has been studying about for so long as abstractions or withered specimens.

During a headlong dive to the bottom of the Atlantic, Aronnax tells his companions of the latest discoveries of deep life, accurately reporting the *Bulldog*'s raising of sea stars from depths of more than a mile. Verne peppered the book with such real-life references to give his fiction a sense of verisimilitude. But rather than citing the *Bulldog*'s work as a blow to the azoic theory, the book simply moves the border of the lifeless region downward, as its proponents of the day tended to do.

The passengers of the *Nautilus,* at a depth of more than eight miles, finally glimpse the bottom. Black peaks tower above forbidding valleys. In this dark wilderness, Aronnax spies some seashells and sea stars. "But soon these last representatives of animal life vanished," as the submarine dove ever downward. "The *Nautilus* passed the limits of submarine life just as a balloon rises into heights where no one can breathe." Aronnax offers no explanation for why such depths should be creatureless, but simply states it as a fact. Finally, the craft comes to rest at a depth of nearly ten miles (three miles deeper than today's deepest-known depth), prompting a professorial outburst. "What a situation to be in! Sailing such depths where humanity has never been! Look, captain, look at those magnificent rocks, those grottoes with no life in them, these final global repositories where no life is possible."[14]

AS VERNE'S FICTION was being admired and emulated, British scientists in search of hard facts were quietly embarking on the most ambitious voyage ever to probe the mysteries of the depths, including such questions as how far down deep life could go. No submarine was available for the job, that art being in its infancy. But a big ship was. The voyage of H.M.S. *Challenger* was no summertime jaunt, as Britain's earlier expeditions had been. It was a three-and-a-half-year assault on the riddles of the deep, through howling winds and raging seas, employing the best equipment that the Royal Society and British Admiralty could muster. Nothing like it had ever been attempted before. The ship's charter was to investigate, top to bottom, the waters that bounded the vast majority of the Earth, especially the dark regions far beneath the waves. The expedition was backed by the British Admiralty, Treasury, Museum, and Royal Society.

As might be expected, its aims were both abstract and practical. One was to advance the laying of telegraph cables in the deep—to learn of the seabed's terrain, to find the kinds of temperatures that cables might encounter, and to discover any deep animals that might attack cable covers. A more esoteric goal was to settle the question of whether the deep was home to lost races of living fossils, as Darwin's theory suggested. The discoveries in relatively shallow waters had fed hopes that fauna of even greater size and antiquity might be found in deeper regions. The voyage's lead scientist was Thomson, the British naturalist who had helped organize the cruises of the *Lightning* and the *Porcupine* and who had written that deep life seemed to be strangely related to forms that were thought to have vanished long ago from the face of the earth.[15]

The H.M.S. *Challenger* sailed from Portsmouth in December 1872 against a strong wind, all canvas and rope. She carried 240 sailors and scientists. The warship, a naval corvette, was 226 feet long. On close inspection, this vision of British might was revealed to have been transformed into a floating scientific laboratory. All but two of her guns were gone. The main deck was crisscrossed with bridges and platforms for handling dredges and trawls. The aft section had been subdivided to create extra space for cabins and shops. The ship in general was laden with microscopes, spirit jars, sample bottles, and hundreds of miles of hemp line for probing the deep. The main deck was fitted with an eighteen-horsepower steam winch. Aided by sailors, it was to be the primary means by which trawls and dredges were hauled from the depths of the sea. The netlike trawls were to sample inhabitants of the midwaters and the bottom while the heavier metal dredges were to dig into the bottom a bit to scoop up material there as well as any organisms living in the ooze or buried slightly beneath its surface.[16]

Between 1872 and 1876, *Challenger* circled the globe. The wooden ship plied the North and South Atlantic, sailed eastward through the Indian and Antarctic oceans, crossed to the Pacific and visited some of its numerous isles, sailed to South America, rounded its tip through the Strait of Magellan, and then headed for home. All the way it probed the bottom.

Among its discoveries were hints of a long, sinuous mountain range that snaked down the middle of the Atlantic, cutting Maury's Telegraph Plateau in two. Scientists were startled by its size. Newspapers declared that it was the lost continent of Atlantis. As for deep life, *Challenger*'s scientists found some areas poor and others rich. From a depth of more than half a mile off the coast of Argentina, a wide-mouth dredge hauled up sea anemones, brittle stars, sea cucumbers, sea urchins, a feather star, a jellyfish, snails, sea slugs, seven types of worms, two kinds of barnacles, and ten different types of deep-sea corals. In all, this single bit of the ocean floor surrendered 127 different species of marine life, 103 of them new to science.

In contrast, a haul from a deep basin near the Canary Islands brought up nothing but one hundred pounds of muck. Nonetheless, such barrens yielded much of interest over time, including potato-shaped rocks in vast numbers. These nodules were found to be mainly oxides of manganese, often accompanied by varying amounts of iron, nickel, tin, copper, cobalt, and other metals. The center of larger nodules usually held a bit of volcanic glass, an ear bone of a whale, or the tooth of an extinct shark. *Challenger* scientists were baffled by the plentiful rocks. Their origin is still mysterious, though most scientists believe the nodules form as metals slowly precipitate on small objects in layers over the ages, growing much as do pearls.[17]

The expedition was no lark. One seaman was killed when a line got snagged on the bottom and tore loose a block and tackle, hurling it across the deck to crush and maim. Two men drowned. One was accidentally poisoned. One scientist died of infection. Two men went insane. One committed suicide. The biggest problem of all was sheer boredom caused by lowering and raising the various dredges and trawls in a monotonous, mind-numbing routine, day in and day out. Sixty-one sailors deserted ship, and many of those who were persuaded to stay aboard were chronic complainers. Henry Moseley, a British naturalist on the voyage, described the change from excitement to apathy. "At first, when the dredge came up, every man and boy in the ship who could possibly slip away, crowded round it, to see what had been fished up," he later wrote. "Gradually, as the novelty of the thing wore off, the crowd became smaller and smaller, until at last only the scientific staff, and perhaps one or two other officers besides the one on duty, awaited the arrival of the net on the dredging bridge."[18]

Despite the troubles, the expedition was an extraordinary success. Its influence was such that the scientific study of the sea was thereafter seen as a legitimate discipline and was given the name *oceanography.* The renown of the ship was such that, in the twentieth century, its name was equated with exploration and was bestowed upon one of the American shuttles that probed outer space. The first and foremost of *Challenger*'s accomplishments was giving humanity its first large-scale understanding of the sea's physical structure, based on 492 soundings. Most land was found to be surrounded by shallow areas that grew slowly deeper with seaward travel, until, at some point, these continental shelves dropped away rather precipitously to deep water. Most of the ocean lying beyond these shelves was found to be an enormous plain a little more than two miles deep. The gentle hills and valleys of this seabed were sometimes riven by sharp mountain chains that rose to within about a mile or so of the surface and, over time, were eventually found to meander all over the world's oceanic deeps. At the other extreme, *Challenger* discovered ocean trenches that descended to depths greater than five miles, leaving the crew in awe and careful to redo the measurements so as to rule out mistakes.[19]

Most interesting of all, the expedition found that the pitch darkness of this unfamiliar world teemed with life. The crew lowered dredges down to the bottom 133 times and trawls to various depths 151 times, hauling into the light of day literally tens of thousands of animals, some writhing and squirming on deck. All told, the *Challenger* scientists identified 4,717 new species—giant worms and slugs, spindly crabs and prawns, delicate sponges and sea lilies. Many of the creatures were sessile, meaning fixed to the bottom, or had legs or other means for moving over and burrowing into bottom ooze. Some were surprisingly large, including deep-sea shrimp nearly the size of lobsters, their antennae and body parts huge.

A serious disappointment, given all the prevoyage excitement, was that the expedition failed to discover any major evolutionary throwbacks or living fossils. One exception was a small type of squid known as *Spirula,* a lifeless specimen of which was hauled up off the Banda Islands in the Pacific from a depth of nearly half a mile. Unlike contemporary squids, which have no shell, *Spirula* had a delicate coiled skeleton that was partly inside and partly outside its body. It was clearly a missing link between the creatures of the remote past and contemporary squids. Of all the scientists, Thomson, the team's leader, was most especially let down at their failure to catch trilobites or dinosaurlike beasts. Yet he still believed that the ancient ones lay out there in the icy darkness and apparently thought that *Challenger*'s gear had simply proved too feeble to mount an effective hunt.[20]

As the sheer abundance of deep life became clear, so did the issue of how the creatures fed. The main clue was sediment. *Challenger* brought up and analyzed hundreds of samples of bottom muck. The ooze—even from the deepest sampling sites more than five miles down—was found to be riddled with organic matter and waste, especially the shells and skeletons of planktonic flora and fauna. Plankton are the drifters of the sea. Their name is derived from the same Greek root as *planet,* with both words denoting wanderers. Plankton are often small or microscopic. They live mainly in sunlit waters and represent the first link of the sea's gargantuan food chain, especially the phytoplankton, which are tiny plants that tap the sun's limitless energy. *Challenger* found deep layers of sediment composed of planktonic waste. These surface creatures upon dying were streaming down as a gentle rain into the depths, providing a source of continuous nourishment to the denizens below. The rainfall was eventually found to include molts, plant debris, fecal pellets, and dead organisms, the detritus falling by the ton over every square mile of ocean floor. In short, the feast above was providing a steady diet of crumbs and wastes for waiting mouths below, mainly those of scavengers and filter feeders, who in turn were eaten by predators.[21]

But the greatest discovery of the voyage was the finding that life, once thought wed to land and sunlit waters, was in fact waiting to be found wherever scientists looked, no matter how deep, no matter how cold or dark. It flourished even in the most remote of recesses, contrary to the theorization of the preceding decades. "The distribution of living beings has no depth-limit," Thomson wrote in an expedition report, emphasizing that creatures of the deep were found to "exist over the whole floor of the ocean." After decades of confusion, the azoic theory was finally dead, even if many questions remained about the nature of life below.[22]

WILLIAM BEEBE was a lean, balding scientist at the New York Zoological Society in the early decades of the twentieth century. He had grown increasingly frustrated with remote sampling of the sea's middle region, that enormous mass of dark water between seabed and surface, assuming, like other scientists, that his nets and trawls were missing much of the action. So he innovated, coming up with an uncommonly bold technology that he used with great personal courage.

Rather than exploring the global sea, Beebe focused his energies on a particular patch of ocean—a circular area eight miles in diameter off Bermuda. The cold Atlantic waters there were more than a mile deep. Year after year, Beebe and his colleagues cast nets into the depths, laboring to do a thorough assessment of the area's midwater fauna. Between 1929 and 1937, they caught more than 115,000 animals representing 220 species of deep life, many of them new to science. At the time, his work represented some of the most comprehensive and methodical sampling of one part of the ocean that had ever been done.[23]

Amid these conventional labors, Beebe in the 1930s pioneered the use of a vehicle he called a *bathysphere,* taking *bathy* from the Greek word for *deep.* The device, though crude in appearance, represented a turning point in the history of deep exploration. It was basically a metal ball just big enough to hold two men. And it had two thick windows so the men could peer out. Lowered into the depths by cable, the unpretentious sphere quickly shattered all records for manned descent. Divers of the day, wearing heavy metal helmets supplied with air from above, could go down a few hundred feet before crushing pressures ruled out further descent. The same was true of submarines. They could travel down to four hundred feet or so before the pressures threatened to crush them. But the bathysphere that Beebe pioneered with Otis Barton, a creative engineer who devised many of the technical advances, succeeded in taking the two men down more than a half mile. Never before had people consciously entered this realm of unending night.

The bathysphere, as its name implied, was perfectly round—the strongest of all possible shapes for withstanding the strain of large, evenly applied pressures. It was small in order to maximize the ratio of wall to hollow. All other considerations being equal, the smaller the sphere, the greater the strength of its walls. A hollowed-out baseball would be stronger than a basketball. The bathysphere was made of steel. Its walls were an inch and a half thick, and its interior space was four and a half feet across. On the outside, with its legs and single large cable fitting, the sphere was about as tall as a man. The entrance hole, a body-scraping fourteen inches narrow, was capped by a thick steel door that weighed four hundred pounds. Ten large bolts sealed it tight. The bathysphere's windows were six inches wide, crystal clear, and made of fused quartz some three inches thick. A single spotlight outside could be switched on or off.

Oxygen for breathing was supplied by pressurized tanks, and carbon dioxide was removed from the air by an open tray of chemicals. In the bathysphere's original setup, the passengers waved palm-leaf fans over the chemicals to speed the cleansing reaction. Later, that job was taken over by an electric fan. The ride was sometimes rough. Tethered to the surface, the ball would rise and fall as waves caused the support ship to pitch and heave. At times, the sphere was shaken so hard that the chemicals would fly out of their trays and the two men had to cling onto the bottom to keep from being thrown about and bruised by the hard metal walls.[24]

NEW YORK
ZOOLOGICAL
SOCIETY

The bravery needed to enter this contraption and descend into the depths was driven home one day when the bathysphere was lowered more than a half mile on a test run, unmanned. It came up heavy and, as it turned out, full of water under extremely high pressure. Deck hands began to unscrew a large brass bolt at the center of the door. Jets of water screamed out. Suddenly, with no warning, the bolt tore loose and shot across the deck like a shell from a gun, hitting a nearby steel winch and gouging out a half inch of metal. Meanwhile, the sphere spewed a cannonade of roaring water.[25]

Manned dives began in June 1930, and eventfully produced a torrent of words as Beebe struggled to communicate the beauty of the twilight world man was seeing for the first time. Translucent blue, dark blue, then even a blacker shade of blue met his eyes as he descended ever deeper. "On earth at night in moonlight I can always imagine the yellow of sunshine, the scarlet of invisible blossoms," he wrote in his engaging book *Half Mile Down.* "But here, when the searchlight was off, yellow and orange and red were unthinkable. The blue which filled all space admitted no thought of other colors."

Deeper down, in utter darkness, Beebe was surprised and amazed to see a bright and continuous display of flickering lights from fish, jellies, and animals he was unable to identify. After a while, peering into the pitch blackness, he began to be troubled by the sheer vitality of the show, which suggested that much of the scientific wisdom of the day was wrong and that the inhabitants of the deep had abilities far different from those usually imagined. "No wonder," he wrote, "that but a meager haul results from our slow-drawn, silken nets when almost all the organisms which came within my range of vision showed ability to dart and twist and turn, their lights passing, crossing and recrossing in bewildering mazes." Many of the creatures that he sketched and later described in his book turned out to be new to science.[26]

Once, four-tenths of a mile down, Beebe spotted a fish bigger than the sphere. Its general shape was that of a large surface fish, but with much bigger eyes and shorter jaw. Its mouth was held wide open the whole time Beebe viewed the creature, and its numerous fangs were lit either by luminous mucus or indirect internal lights. Dots of pale blue fire lit its body, and two long tentacles hung down, one tipped in reddish light, the other blue. These glowing lures, Beebe noted, twitched and jerked. He saw enough distinguish-

DRAGON FISHES. William Beebe, the first scientist to view the deep sea, discovered new kinds of dragon fish, with long barbels dangling from their chins, and viper fish, with long rays extending from their dorsal fins. These barbels and rays have luminous tips that lure prey into needlelike fangs.

ing characteristics to place the unknown creature in the general category of sea dragons, or stomiatoids.[27]

Beebe also encountered strange, indistinct things that were much harder to describe and categorize. More than a third of a mile down, he came upon something he could only describe as a "pyrotechnic network" about two by three feet in size. "I could trace mesh after mesh in the darkness," he wrote, guessing it was a type of life "so delicate and evanescent that its abyssal form is quite lost if ever we take it in our nets."[28]

During his record dive of 1934, Beebe came upon a true giant of the abyss as he neared the depth of a half mile. The animal's outline, flicking in and out of the spotlight, was dim but not indistinct. Finally, its whole shadow-like contour passed through the far end of the beam. At the very least, the creature seemed to be twenty feet long and very wide. Its skin had no color or texture. Beebe could make out no eye or fin. It was just a wall of alien flesh moving silently through the dark water. "Whatever it was, it appeared and vanished so unexpectedly and showed so dimly that it was quite unidentifiable except as a large, living creature," Beebe wrote. In contemplating this encounter, which left him slightly shaken, he recalled the old reports of sea serpents. Despite his cool eye and scientific calm, Beebe was quick to see the legends in a new light, observing that his own experience was "a pretty good beginning" in the history of encounters between man and beast.[29]

THE DEPTHS OF the sea, as well as its middle regions, continued to yield a number of surprises over the decades of the twentieth century.

Puzzled fishermen off South Africa in 1938 pulled up a creature some five feet long, fishlike but odd. It was steel blue in color and had stubby, limblike fins that were tasseled on their ends, apparently for creeping through bottom muck. After excited study on land, the oddity was pronounced to be a type of coelacanth, a member of a large class of fishes thought to have gone extinct some seventy million years ago after a long and successful reign in the seas. The discovery made Thomson of the *Challenger* voyage seem at least partially right. There *were* other large relics of evolution out there in the darkness, living in icy solitude.[30]

From 1950 to 1952, the Danish ship *Galathea* probed the deep with stout steel lines up to seven miles long and large dredges with mouths up to twenty feet wide. From a depth of more than two miles, the ship pulled up an archaic type of mollusk previously thought to have gone extinct at the end

COELACANTH. Steel blue and five feet long, the coelacanth, once thought to have accompanied the dinosaurs to extinction, stunned the scientific world in 1938 by turning out to be very much alive.

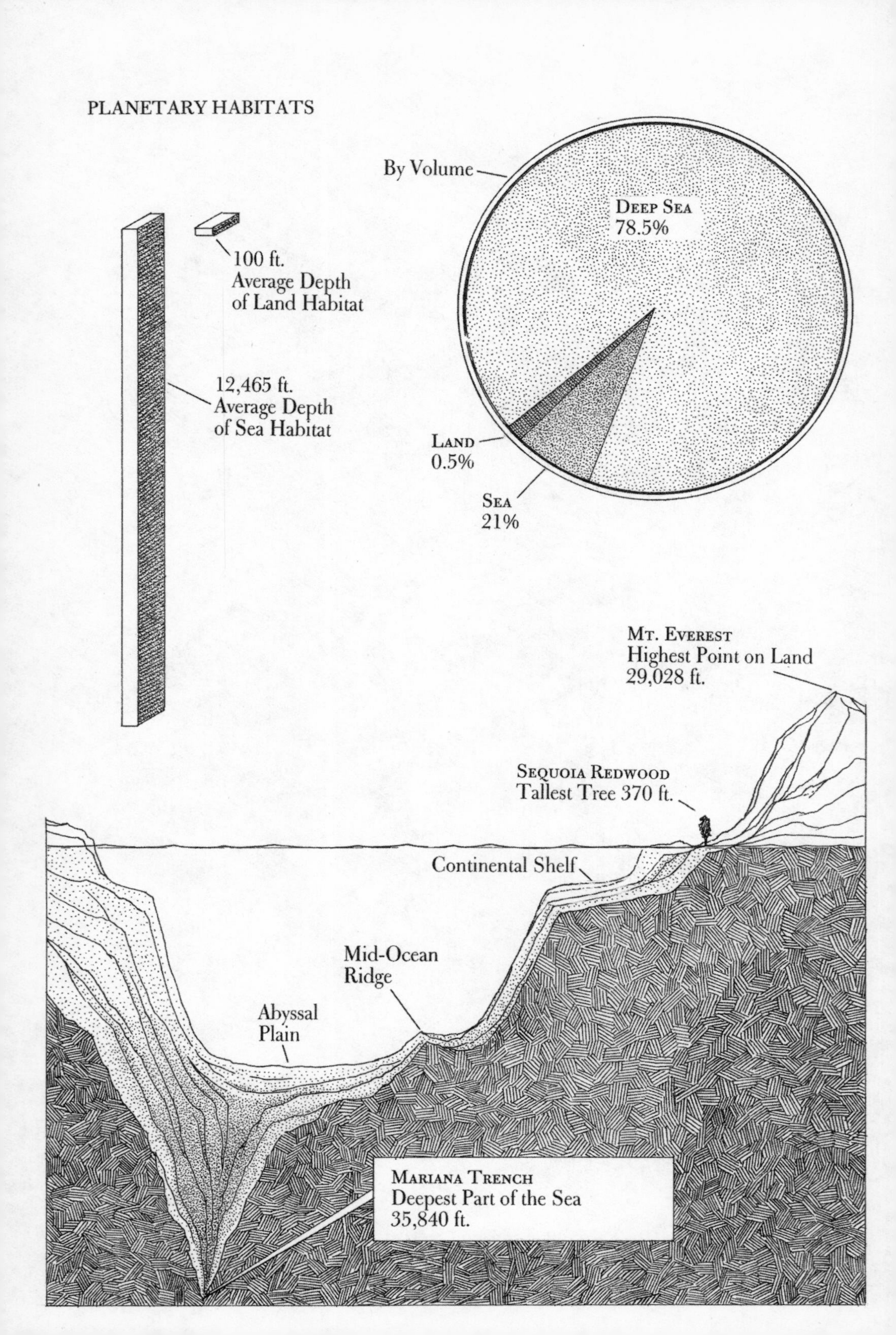
PLANETARY HABITATS
By Volume
DEEP SEA
78.5%
100 ft.
Average Depth
of Land Habitat
12,465 ft.
Average Depth
of Sea Habitat
LAND
0.5%
SEA
21%
MT. EVEREST
Highest Point on Land
29,028 ft.
SEQUOIA REDWOOD
Tallest Tree 370 ft.
Continental Shelf
Mid-Ocean
Ridge
Abyssal
Plain
MARIANA TRENCH
Deepest Part of the Sea
35,840 ft.

of the Devonian period 350 million years ago, a time when amphibians were just invading the land. Most notably, *Galathea* probed the icy blackness of the sea's deepest known fissures and discovered in these recesses not just hints of life, as *Challenger* had done, but heaps of it—thousands of sea anemones, sea cucumbers, sea lilies, bivalves, tanaids, amphipods, bristle worms, shrimps, and squat lobsters. Most of the species were new to science. Eventually, experts came to estimate that such creatures were part of a global mass of deep diversity whose ranks held as many as ten million species—far more than the million or so varieties of life so far identified on land.[31]

THE ADVANCES of the centuries, in which regular probings began at twenty fathoms and eventually extended down to the sea's deepest fissures, produced no discovery more important than the fact that the sea is a continuous habitat from top to bottom. To be sure, there are fluctuations and gradations in the density of living things. Some areas of the deep hold swarms of life and others appear to be populated by loners and strays. Some parts of the dark may conceal monsters. But there are no azoic zones, no biological deserts, no voids of lifelessness. Rather, the sea is filled with life all the way down into the deep trenches, extending into a region of unimaginable cold, pressure, and darkness. Even the bleakest of the Earth's environments turns out to have the ability to seethe with life.

This realization worked a slow revolution in scientific thinking about the nature of habitats. That the living sea covers nearly three-quarters of the globe was eventually judged, as a measure of a biological richness, to be so understated as to be misleading. What was seen as more revealing was the sea's volume, every square inch of which turned out to be susceptible to the phenomenon of life.

The merit of the volumetric approach was usually illustrated by a comparison to life on land. Fertile dirt was inhabited down to a depth of only a few meters and forests rose no more than a hundred meters or so, according to such analyses. Air had no significant life of its own since winged creatures depended on the ground for food. Land, in essence, was seen as a two-dimensional habitat in which life was restricted to a relatively narrow band. But the sea, by contrast, appeared to teem in three dimensions over its entire breadth and depth, down through the dark middle waters, down to the bottom, down to the recesses of the canyons seven miles deep. It was animate space. The sea in its living bounty was unique.[32]

LAND AND SEA AS HABITATS. Because of its innate properties and enormous volume, the deep sea is the largest area of the planet that supports complex life. It is estimated to make up somewhere between 78½ percent and 97 percent of the global biosphere.

As this truth hit home, scientists worked to get an intellectual grip on the issue, which, in typical fashion, meant they tried to quantify it. The results were extraordinary. By volume, land was calculated to make up a meager half of one percent of the Earth's biosphere. In other words, all of Africa, all of Siberia, all of the continents and archipelagoes and mountains that humans had slowly discovered over the centuries, were estimated to have a volume that made up one two-hundredths of the planet's total habitable area. In contrast, relatively shallow seas and continental shelves were judged to make up 21 percent of the whole. But the overwhelming leader was the deep sea, which was found to account for a phenomenal 78½ percent of the volume. While perhaps seeming to be an overstatement, this figure was in fact extremely cautious because it assumed that the deep sea began at a depth of one kilometer—a very conservative view. Other estimates put the leading edge of the abyss beneath the photic zone, the one hundred or so meters of the upper sea that get enough sunlight to support the bulk of the sea's plants and photosynthesis. If that definition is adopted, then the deep-sea fraction soars to become more than 97 percent of the global biosphere.[33]

Put simply, the deep is the Earth's largest habitat. And its residents are undoubtedly the most typical forms of life on the planet, despite what seem to be their odd looks and unfamiliar ways, despite their often being denigrated as devils and demons, as freaks and monstrosities of nature. Over the ages man has shown no shyness about displays of vanity. In the nineteenth and twentieth centuries some parts of the scientific community found their work beginning to undermine such self-confidence, to reveal the real position of humanity and all terrestrial life in the biological order of things. The deep is no lair. It turns out to be the planet's main address.

WHAT OTHER SECRETS do the depths conceal? No doubt there are more, perhaps many more, given the extreme superficiality of our investigations to date. Thomson, the deep pioneer who led the *Challenger* scientists, observed in 1874 that the explored part of the seabed could be "reckoned by the square yard." More than a hundred years later, things are better, but not that much better if the strides of the past century are carefully measured against what remains to be done.

In 1991, a group of leading oceanographers from around the world did just that, addressing the question of what part of the seabed had been explored to date. The figure they settled on was something between one thousandth and one ten-thousandth of its total area. That is obviously a rough estimate. The actual figure might be a hundredth or a millionth. No one knows.

And whatever the truth vis-à-vis the seabed, the situation gets murkier in corresponding estimates for the investigation of the sea's middle waters. The question is almost meaningless, since the fauna and physical constituents of the middle realm are constantly in flux. Pulling a net through one area may help identify some of its residents but in no way gives anything approaching an exhaustive census, since many creatures simply slip out of range, as Beebe concluded. In any case, the intermediate zone is far larger and less studied than the bottom, which deep expeditions have tended to focus on. It is conceivable that, in total, taking both the bottom and intermediate zones into account, humans have scrutinized perhaps a millionth or a billionth of the sea's darkness. Maybe less. Maybe much less.[34]

The main impediment to better understanding has been a shortage of instruments that can withstand the deep's crushing pressures and illuminate its inky darkness while advancing the exploratory job. A cannonball endures but tells little. The entire history of deep exploration bristles with frustration, with jibes about having to blindly grope the sea's boundlessness "by the square yard." Like astronomers before the invention of the telescope, oceanographers often found the available tools inadequate for addressing the great questions at hand. Much was learned from lucky finds and bottom scoops, from baited baskets and whale bellies, from bathyspheres and long trawl lines. But the findings in general produced only disparate hints as to the nature of the whole, arousing interest and curiosity.

Yet, institutionally at least, there was no real urge to speed things up, no rush to sink money into scientific endeavors that tended to be extraordinarily costly. The work over the centuries was funded by a patchwork of groups and organizations both public and private that had differing agendas and philosophies, none of them high priority. The main exception was the *Challenger* expedition, a pioneering venture inspired by an unusually potent mix of state, commercial, and theoretic interests. But once the newness and romance of deep exploration faded, so did governmental funding. To a degree that now seems remarkable, the bill in the late nineteenth and early twentieth centuries was picked up by unencumbered royalty and rich philanthropists, whose interest and patronage often developed while they sailed the globe in yachts and schooners.

Prince Albert the First of Monaco devoted himself to deep studies, starting his explorations in the 1890s. Beebe had many wealthy benefactors, including bankers and industrialists. The Scripps newspaper dynasty put up the money to found and finance what eventually became known as the Scripps Institution of Oceanography, based in La Jolla, California. Rockefeller money was responsible for the founding of the Woods Hole Oceanographic Institution, nestled in a sleepy Cape Cod harbor near the summer homes of New York City financiers. Both of these organizations, which grew to become global leaders of deep inquiry, did much of their early work from the decks of sailboats. Even the *Galathea* expedition of the early 1950s, hobbled throughout by lack of funds, was heavily beholden to affluent citizens who set up the Danish Expeditions Fund to channel gifts and grants to the endeavor. To a surprising extent, deep exploration in the first half of the twentieth century was supported by rich individuals. And that tended to define its character and limit its scope in light of the great cost of penetrating the deep with capable instrumentation.[35]

As the age of rocketry took off and man got ready to go to the Moon and probe the planets, most research around the world that delved into the mysteries of inner space was still done as it had been a century earlier, by lowering various kinds of nets and trawls on long lines.

All that changed, however, and with relative haste. The exploratory gear began to advance dramatically as the mightiest nation on earth discovered that the deep was no mere diversion but a realm whose otherworldly characteristics offered intriguing new possibilities for the conduct of war.

TWO

Battle Zone

THE FIRST GLIMMERINGS began in a most unlikely spot—the cluttered workshop of an elderly Swiss physicist who wore wire-rim glasses, a slight mustache, and long gray hair curled down his neck. Auguste Piccard was a world-class inventor. A genius at turning ideas into things, he distinguished himself early on by helping Einstein make instruments for physics experiments. He did his own research, too. Curious about what happened when cosmic rays from outer space hit the upper atmosphere, Piccard invented a high-altitude balloon with an aluminum gondola, in which he and other scientists did experiments while breathing pure oxygen from a pressurized bottle. In 1931, he rose to a height of nine miles, nearly twice as high as man had ever gone before.

His restless mind soon turned to another frontier, the deep. In 1933, he met Beebe at the Chicago World's Fair and learned of the wonders below and the apparatus that had laid them bare. By the 1940s and '50s, with typical acumen and daring, Piccard had invented and piloted his own manned vehicle for descent into the oceanic depths. It took Beebe's idea a giant step further and in so doing opened new worlds of exploration. Piccard's craft was similar to its predecessor in that it had a steel ball. But its key innovation was that it had no tether, making the ride more comfortable than Beebe's and in some respects safer. Gone were the up-and-down jostlings as the support ship pitched and heaved in the waves. Most important, the elimination of the tether meant Piccard's vehicle was free to land on the bottom.[1]

Beebe opened man's eyes to the mysteries of the middle waters. Piccard did the same for the deep ocean floor, starting humans on their exploration of the planet's last frontier.

Piccard called his invention the *bathyscaph,* from the Greek for deep boat, *bathos* and *scaphos.* But in truth it was more like a balloon, rising and falling at will. The secret of its independence was an enormous upper tank filled with gasoline, which was lighter than seawater and counteracted the heaviness of the seven-foot steel personnel sphere. The crew cabin was not only larger than Beebe's but its walls were thicker, so the vehicle could go deeper. Descent began when two chambers of the upper tank were flooded with seawater, making the overall craft negatively buoyant and allowing it to sink. Ascent from the bottom began when tiny pellets of iron ballast were dropped in mass, making the craft positively buoyant and bringing it floating back to the surface of the sea. Another of Piccard's innovations that proved key to the craft's success was his adapting the tough new plastic known as Plexiglas to make thick windows, which turned out to be quite sturdy and ideal for viewing the abyss.

The vehicle's big limitation was that it went up and down but not much else. Once on the bottom, the behemoth could be nudged forward only reluctantly by the craft's giant propellers. And even limited travel over the seabed came only after years of experimentation and false starts. The first dives were all straight up and down, often violently so. At times the action tended to be like that of a runaway elevator.

The construction of the first bathyscaph was funded by Belgium, where Piccard worked at the Free University of Brussels. The Belgian king over the years had become interested in Piccard and his exploits and Belgian institutions had helped fund them. The prototype device was tested for the first time in 1948. It went down deep but carried no passengers, and the dive revealed the need for a number of costly improvements, which the Belgians declined to underwrite. The French Navy was interested in the work and acquired the first craft.

Eager to press ahead with an improved model but having no money to do so, Piccard and his son Jacques slowly won financial support from a mix of Swiss patrons (cantons, schools, institutes, industrialists, citizens), as well as from the Italian city of Trieste, which wanted to strengthen its industries by building the innovative craft. *Trieste,* as the new model was named, was completed in 1953. Its giant gasoline tank was fifty feet in length, making the craft more than twice as long as the original prototype and far more seaworthy.

And it was judged safe for people, or at least for adventurers bold enough to enter its dim confines.

The deep debut of *Trieste* came in 1953 off Naples as the sixty-nine-year-old Piccard and his thirty-one-year-old son dove straight down almost two miles, into a realm of pitch darkness that humans had never experienced before. They landed with a bump. The underside of the sphere was mired in ooze and the front porthole was covered with mud. But the two men, peering through the aft window, aided by the ship's external lights, could discern the seabed. It was a disappointment. There were no frenzies of life such as Beebe had witnessed. There were no dramatic seascapes as Verne had imagined. In fact, there was precious little at all. Whether because of the craft's hard landing or the bright lights or the region itself, the bottom was lifeless, or apparently so. And it was featureless. All that was visible to the explorers was a light-colored plain of ooze that faded in the distance to the deep's usual state of unremitting darkness.[2]

Trieste languished for lack of operating funds, its backers having lost enthusiasm for the endeavor. It performed only shallow dives in 1954 and no dives at all in 1955. Desperate to keep up the momentum, Piccard turned to the Italian Olympic Committee, which was persuaded to contribute a modest sum to the project on the theory that exploration is the sport of scientists.[3]

Then a new patron appeared out of nowhere, one whose pockets were very deep. The United States Navy or, more specifically, some of its scientists, were curious about the abyss and had become fascinated with the Piccards and their pioneering craft. After a slow courtship, the Office of Naval Research early in 1957 signed a contract for fifteen dives in the deep Mediterranean off Naples. Saved from insolvency, *Trieste* was a blur of activity that summer, repeatedly ferrying Navy scientists down to the seabed and everywhere in between for scores of observations and experiments. Young Piccard piloted all the dives.

Things began with a jolt off Capri, where the undersea slope near the island is quite steep. A soft bump at a depth of a quarter mile signaled that *Trieste* had settled down. But a glance out the window showed that the craft was poised precariously on a narrow ledge that dropped away into darkness. Suddenly, the ledge crumbled. *Trieste* began sliding down as the crew cabin filled with the sickening sound of metal scraping against rock. Another glance out the window showed that a deep avalanche was in progress. Mud and sand poured over the descending craft, with clouds of muck billowing up to obscure vision. Eager to abort the dive, Piccard flipped the switch to drop the forward iron ballast. No reaction. The mechanism was jammed with sand and mud. He quickly flipped the aft ballast switch. This time, to the relief of

Piccard and his passenger, tiny pellets of iron ballast began to pour out. Slowly, very slowly, with half the positive buoyancy usual for an ascent, *Trieste* lifted out of the tumult and rose toward sunlit waters.[4]

Other dives proved to be less eventful and more enlightening, especially about the prevalence of deep life. During one downward plunge, Robert S. Dietz of the Navy Electronics Lab in San Diego, who later won geological fame as a father of plate-tectonics theory, was fascinated by flashes of bioluminescent life on the way down, including a strange bluish-green flare and a long row of mysterious lights that went swimming by. On the bottom, at a depth of nearly a mile, Dietz was excited to see darting shrimp and a sinuous foot-long fish with a bulbous head thrash its way through the muck looking for a meal.

Even though the Mediterranean is a comparatively barren sea, the bottom that summer was found to harbor all sorts of life—rattail fish alive with undulations, tripod fish with outlandishly long fins, and a writhing eel about six feet in length. During many dives, the lights of the submersible attracted swarms of white isopods, tiny, insectlike creatures that were one source of sustenance for the larger fish and animals of the region. Other animals went unseen, but their presence in the ooze was marked by countless hummocks and perforations and holes, some an inch or more in diameter. Apparently the thump of the landing bathyscaph was enough to send the nameless bottom dwellers deep into their burrows.[5]

Not all was biological wonder that summer. A fundamental part of the agenda was the investigation of new weapons for the conduct of undersea war, including the nuclear kind. Eight of the Navy's dives were devoted to experiments on sound propagation through the sea. Layers of water whose temperatures and pressures are unlike, the Navy had discovered, produce a kind of wall at their interface that can bounce sounds back and forth with such regularity and strength that the noise can reverberate over hundreds and even thousands of miles and still be strong enough to be picked up by sensitive microphones beneath the waves.

Such startling discoveries were starting to be exploited by the Navy, which was building a global network of deep microphones moored on the seabed that could monitor Soviet ships and submarines over long distances, eventually known as SOSUS, for SOund SUrveillance System. The Navy was also studying sound transmission to improve its hunter-killer submarines, also known as attack subs, which conducted their hunts in the dark by

SPY NETWORK. Built for $16 billion over four decades, the United States Navy's global network of deep microphones spied on Soviet warships and submarines by monitoring their undersea sounds over distances of hundreds and thousands of miles.

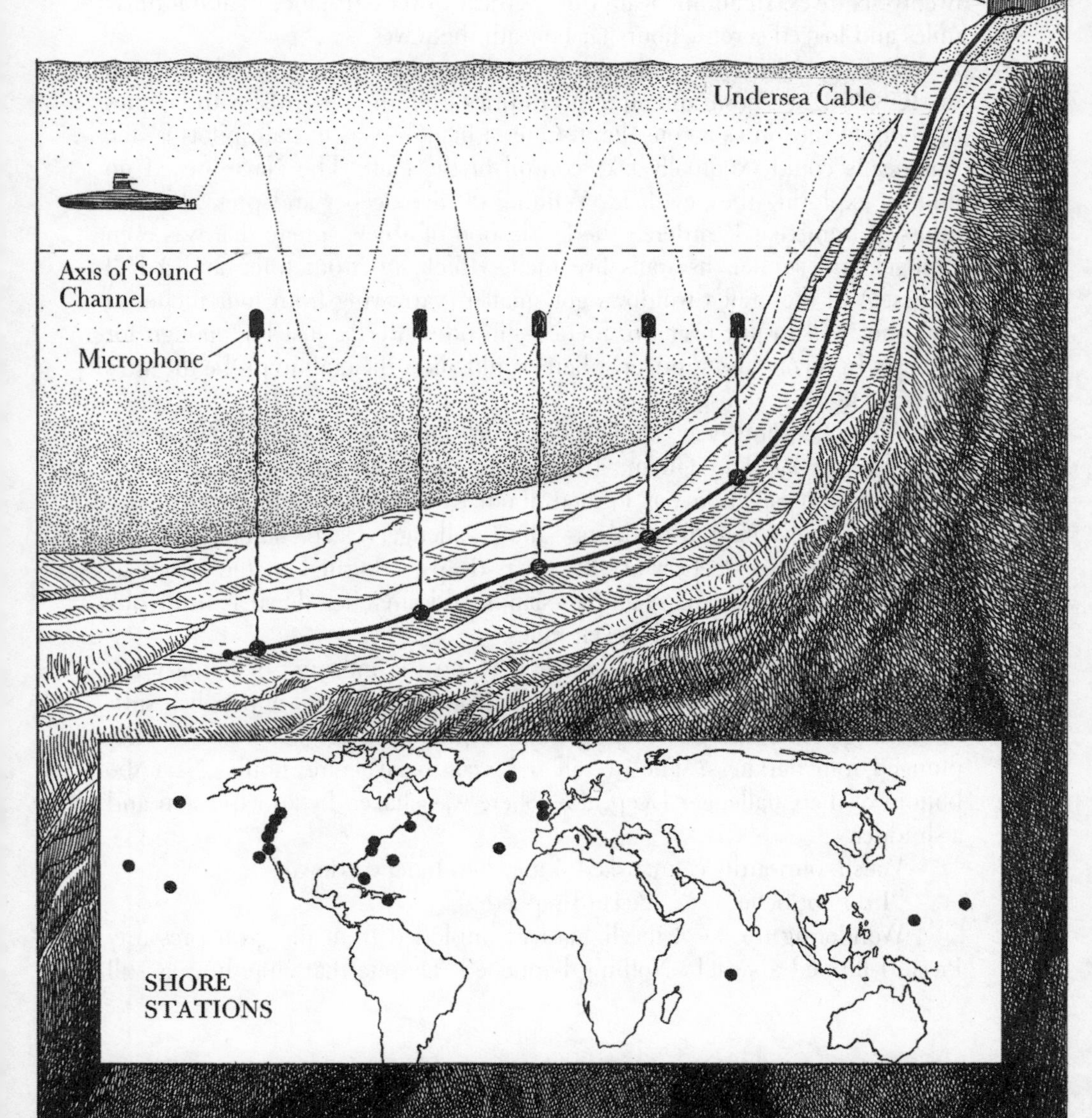

Shore Station
Undersea Cable
Axis of Sound Channel
Microphone
SHORE STATIONS

homing in on faint enemy noises. Their main job was grim. At the start of a nuclear war, American attack subs were to destroy the fleets of Soviet submarines that carried ballistic missiles, thus preventing them from firing their nuclear armaments at American cities. Working in *Trieste* that summer, Navy scientists helped refine such armaments by moving up and down through the Mediterranean with great ease, pinging and ponging and making all kinds of noises that were carefully monitored. They also listened carefully. Among the discoveries was an unexpectedly deep sound channel lying at a depth of nine-tenths of a mile.[6]

For both patron and pioneer, the sense of discovery that summer was so great that the agreed-upon fifteen submersions were increased to a total of twenty-six dives. In all, the team dove vertically over a distance of about thirty miles and logged seventy hours far beneath the waves.[7]

Delighted with the work and the findings, the Office of Naval Research in 1958 bought *Trieste* and all its accessories from the Piccards, with Jacques agreeing to stay on as a consultant. One of his jobs was to train pilots so the Americans could eventually take control of the craft. The Navy wasted no time in exploring the new field. Wanting to dive deeper and press the art's limits, it immediately ordered the fabrication of a new sphere that was even stronger and stouter, its walls five inches thick, up from three and a half inches. And the craft's windows got smaller, narrowing from four inches at the inner wall to just over two inches—in other words, just big enough for one eye. In theory, the stronger sphere would allow the craft to probe the sea's deepest known regions.

After checkouts near San Diego, the new, improved model of *Trieste* was taken in late 1959 to Guam, a sprawling American base in the Western Pacific. The installation was located about two hundred miles from one of the world's great natural wonders—the Challenger Deep, named after the famous ship. This deepest of all known deeps is located in the Marianas Trench near the Mariana chain of islands and stretches down beneath the waves for a distance of nearly seven miles.[8]

Seas were high and the sky was dark on the morning of January 23, 1960, as Jacques Piccard and Lieutenant Don Walsh, a spit-and-polish submariner, got into the shuddering bathyscaph. Soon all was calm as they plunged into darkness. The overall trek was to take nine hours. Near the bottom of the Challenger Deep, the sphere was shaken by an explosion and a shudder.

Walsh, outwardly calm, asked if they had touched down.

"I do not believe so," Piccard replied.

Wondering if one of the lights had imploded from the great pressure, Piccard flipped a switch. Nothing happened. Despite that clue, he was still uneasy. Finally, the seafloor came into view. It was a wilderness of light-colored ooze. More important, the bottom of the world turned out to shelter life, and not just swarms of tiny invertebrates. A foot-long flatfish that resembled a sole lay in the muck. Amazingly, given the dark habitat, it had eyes. Slowly, extremely slowly, the fish rose and swam out of the spotlight back into perpetual night. A red shrimp flitted by. Examining the ooze through his tiny window, Piccard noticed a number of undulations on the bottom suggestive of animal tracks. Unfortunately, during the dive he was unable to photograph any animals, allowing skeptics to later dismiss them as deep hallucinations. After twenty minutes of observations, the allotted time was up. The two men began a long ascent. Upon breaking the surface, they climbed out quickly, eager for fresh air. Two Navy jets screamed overhead out of nowhere, dipping their wings in salute, startling the explorers.[9]

The fighting Navy, in spite of the advances and insights, was ambivalent about exploring the deep. Such work was seen as appropriate for limited experimentation and analysis but was no candidate for real money. Instead, the Navy during this period devoted itself to amassing jets, warships, and submarines, to procuring more powerful nuclear bombs for the arms race with Moscow. Deep achievements with no clear military aim were seen as a potential distraction from the real work at hand. So it was that *Trieste*'s thicker sphere was retired in 1961. After the fact, Navy analysts decided that it was incapable of safely withstanding the pressures at the bottom of the trench. No replacement was built.

The growing corps of Navy deep explorers was furious. The pilots and scientists of the bathyscaph lobbied for more *Trieste* dives and for a totally new vehicle that was smaller and more maneuverable and that had bigger windows. They were smitten with the new world. They wanted more of it, not less. The head of the Deep Submergence Program at the bathyscaph's San Diego base, Andreas B. Rechnitzer, a witness to the deep avalanche off Capri, began talks with a potential builder of a new submersible, but he was quickly shot down. "The Navy said it was not in our mission to go deep," he recalled. His ire was directed particularly at Vice Admiral Hyman G. Rickover, the father of America's nuclear Navy. "Rickover said anything deeper than 2,500 feet was a waste of money. I was fed up." Rechnitzer quit the Navy and went into private industry, sure that one way or another a new age was about to dawn.[10]

The fight within the Navy slowly intensified as officials split for and against the new field. One expert in Washington who was convinced that the military had a deep calling was Charles B. "Swede" Momsen, Jr., Chief of Undersea Warfare at the Office of Naval Research. Momsen argued that a deep submersible, if nothing else, could help the Navy install its global arrays of underwater microphones, noting that new ones were about to be strung around Bermuda in waters more than a mile deep. Most importantly, he took action.

During 1961 and 1962, Momsen conducted a one-man war that bent rules, twisted arms, ignored directives, and spent money. Plans for a small vehicle took shape. It would hold three people. It would have at least three windows, each big enough for a pair of human eyes. It would be small and comparatively sleek, able to move over relatively long distances on the bottom. Its streamlining was achieved by eliminating the huge gasoline float of the bathyscaphs, instead surrounding the metallic crew sphere with a new kind of miracle material known as syntactic foam. Hard as brick but extremely buoyant, the foam was filled with millions of tiny, hollow glass beads. Plain old air would thus give the needed lift.

An important ally in Momsen's fight was Allyn C. Vine, a scientist at the Woods Hole Oceanographic Institution on Cape Cod, who had long argued, contrary to most oceanographers, that people, not probes, were the best way to fathom the deep. Woods Hole would be in charge of the new Navy sub, which in Al Vine's honor was to be named *Alvin.* Woods Hole was chosen as its home because no other marine lab in the country wanted it. There was just one problem with all this planning—how to get the sub built. It was too new, too radical in design, too antagonistic to the status quo. No big company like General Dynamics, which made submarines for the Navy, wanted to take on a project disparaged by both the Bureau of Ships and Admiral Rickover, the gods of naval patronage. Finally, a willing company was found. It had no experience in building submarines and knew nothing about the deep but had an equipment division that was eager to learn.[11]

In 1963, the Minneapolis company General Mills, maker of such breakfast cereals as Cheerios and Wheaties, began to assemble the one-of-a-kind submersible in the same room where it built the machines used to grind out breakfast flakes by the billions.

THE AMERICAN NAVY'S love-hate relationship with deep exploration lasted half a decade, from the day it bought the bathyscaph. Then a crisis changed everything. And, as is often the case, the great tragedy turned out to hold the seeds of great opportunity.

The *Thresher* was the nation's most advanced submarine, the best war machine that American money could buy and that Yankee ingenuity could build. Nonetheless, on April 10, 1963, while conducting test dives off Nova Scotia, the sub inexplicably sank, its 129 men lost to icy waters more than a mile and a half deep. It was peacetime's worst submarine disaster. President Kennedy and Queen Elizabeth sent messages of condolence to the stunned families of crew members. Newspapers demanded to know why there had been no means of rescue. Cardinal Spellman held a special mass at St. Patrick's Cathedral in New York City.

Beyond the human tragedy, the incident was a political blow to the United States. The loss of America's most advanced sub was seen as a major setback in the cold war against Communism, in the global effort to project strength and promote alliances opposed to the socialist threat.

Named after a tough species of shark, *Thresher* had been the ultimate attack submarine, the first of a new breed that would target Moscow's growing fleet of ships and submarines, including those based in Cuba. The Navy wanted twenty-four attack subs just like *Thresher.* Her special skill, which in time became the general goal of all American subs, was stealth, not speed. Powered by a single nuclear reactor, she had an array of ingenious silencing methodologies that dramatically cut the usual din of turbines, pumps, and gears that made other submarines sound like locomotives. She was a sound laboratory. As important, her thick hull meant she could dive deeper than any previous American submarine, greatly increasing her ability to hide and fight back.

The exact operating depths of submarines are always secret, but in hindsight we know the historical trend. Subs in World War Two routinely went down to about four hundred feet. After the war they went perhaps to eight hundred feet. *Thresher* was said to be rated for thirteen hundred feet, a very deep figure for the day. An important job of attack subs such as *Thresher* was to protect America's growing fleet of submarines that carried Polaris missiles. These had fundamentally altered the geopolitical calculus by allowing the basing of long-range missiles to be moved from land to sea, making them nearly invisible to the enemy yet close to his shores, expanding the number of targets. Polaris warheads were fast becoming the cornerstone of Western efforts to frighten the Communists and deter aggression. *Thresher*'s powers of quietude and deep submersion were slated to be adopted not only by the nation's attack submarines but by the missile-carrying ones as well. In short, *Thresher* was meant to inaugurate a new level of deadliness for the nation's arsenal.[12]

Like a rock thrown into a tranquil pond, the sinking of the nation's premier submarine sent out ever-widening shock waves. The Navy was alarmed as years of planning were thrown into disarray and the rationale for its submarine buildup was eroded. Morale went into free fall. Resignations from the fleet rose and the recruiting of new submariners became difficult if not impossible, as fears swirled of accidental death in the deep. Exploiting the propaganda opening, the Soviets charged that the sunken reactor was sure to poison the North Atlantic. Some allies grew distant. Worst of all, no one was sure why *Thresher* had gone down. Her last messages to an escort ship had been garbled, something about attempting to come up from test depths. The last thing heard over the hydrophones of the escort ship was the thud of collapsing compartments. An investigation found no evidence of sabotage or enemy action, but did come up with possible causes of the accident. Ultimately, whatever sank *Thresher* was a mystery.[13]

The Navy's humiliation was compounded by months of frustration in trying to locate the sub's wreckage, which in theory could solve the riddle of its demise. The hunt over a ten-mile square of ocean was carried out by two submarines, three dozen ships, and thousands of men. In late May 1963, a month after the sinking, the Navy announced that a camera lowered from a ship on a long line had photographed the ruptured hull of *Thresher.* The next day the Navy reversed itself in a fit of embarrassment. In fact, the camera had taken pictures of its own anchor, which looked sufficiently blurry in the photos to have initially fooled the experts.

A different source of embarrassment was the difficulty of knowing the location of cameras that were dangled on long lines in the deep. It was like lowering a fishing line from the top of a tall building at night in a squall and hoping to know the location of the hook dangling somewhere down below in the darkness. Ocean currents pushed the cameras and their long tethers every which way. In fact, the cameras were usually judged to be anywhere but right below a ship. Finally, in June, bits of some kind of wreckage were photographed on the bottom, though the main body of the sub still eluded searchers. The Navy did the logical thing. It sent for the decade-old *Trieste,* the only manned vessel it owned that could plumb such depths. The bathyscaph, now equipped with a robot arm, was far from ideal for the hunt. At best, its three small propellers could nudge the behemoth across the bottom at a molasseslike crawl, and its tiny windows were a major frustration to observers.

Indeed, five *Trieste* dives in June found nothing of significance a mile and a half down. Floodlights swept the dark bottom to reveal only endless muck and occasional sea stars, sea anemones, and worm holes. In late August, *Trieste* returned to the site for another round of searching. The Navy's degree of desperation was highlighted by the fact that preparations for the new series of dives included the dropping of 1,441 numbered plastic signs at regular intervals across the seabed to try to help the bathyscaph keep track of its position.

On the third dive, nearly five months after the disaster, *Trieste* found the remnants of *Thresher* on the ocean floor. It was a mess. The intense pressure of the deep sea had shattered the 278-foot-long submarine in a violent implosion that left it shredded over the seafloor. Enormous hunks of twisted metal were interspersed with jumbles of smaller debris, including battery plates, lead ballast, and tattered cables. Already living amid the torn hulks of the sub's superstructure were sea urchins and sinuous fish several feet long. *Trieste* roamed the wreckage. On one of the many dives to the site, Lieutenant Commander Donald L. Keach of the Navy took control of the bathyscaph's robot arm. It resembled a human arm—with a hand, wrist, elbow, and shoulder joint, each controlled by a separate electric motor. After much trial and error, including repeated freezing of the arm in arthritic immobility, Keach succeeded in grabbing a four-foot section of twisted pipe and returning with it to the surface. The disaster site, he later told reporters, "is like an automobile junkyard."[14]

Within days of the *Thresher*'s loss, even before the embarrassment of the long hunt, the Navy began a round of soul-searching about its virtual inability to do serious work in deep waters and its general ignorance about the oceanic depths. Somehow, the orders went, this folly had to be corrected so such disasters would be less devastating in the future. *Thresher* was the Pearl Harbor of the deep and, as with the destruction of the Pacific fleet in World War Two, the Navy vowed to do everything in its power to avoid a similar crisis in the future.

Two weeks after the sinking, the Navy Secretary, Fred H. Korth, appointed a panel of fifty-eight people to review all the Navy's deep-sea capabilities and suggest improvements. Known as the Deep Submergence Systems Review Group, it was led by Rear Admiral Edward C. Stephan, the Navy's Oceanographer and a veteran submariner. Among its members was Al Vine of Woods Hole.

The Stephan panel was slow and meticulous. Its deliberations took almost a year, but its top-secret report, issued in March 1964, was a virtual encyclopedia of what the Navy had and lacked by way of deep equipment and philosophies. The panel called for a spectrum of improvements, including a major effort to improve the Navy's ability to recover personnel from sunken subs, to investigate the ocean floor, to recover small objects, and to salvage submarines. Its advice was read and heeded, despite the tumult caused by the assassination of President Kennedy a few months earlier. The new Navy Secretary, Paul H. Nitze, hailed the Stephan report in a speech and predicted that the *Thresher* tragedy would become a turning point "in our quest for knowledge and mastery of the undersea world."[15]

In June 1964, the tiny submersible *Alvin,* about which the Navy had originally been so ambivalent, was christened at Woods Hole on Cape Cod before an audience packed with attentive Navy personnel in their best dress uniforms. The tiny white sub, about the size of a bread truck, looked more like a toy than a key military vehicle. It nonetheless was festooned with red, white, and blue bunting, an outcast no longer. "Everybody was there," recalled John Cooper, a Woods Hole scientist. "It was as if they were commissioning the battleship *Missouri.*" Reflecting the new importance acceded to matters of deep exploration, the main speech was given by James H. Wakelin, Jr., Assistant Secretary of the Navy for Research and Development. Also present was Swede Momsen, the naval expert who almost single-handedly brought *Alvin* to life. Before climbing the rostrum, Momsen turned to Wakelin and asked, with a grin, whether the audience should be told that the Assistant Secretary had once tried to kill the project.[16]

That same month, Wakelin took to the pages of *National Geographic* to reassure the American public that the Navy was learning from the tragedy and was committed to mastering the deep. "*Thresher* left behind her the realization that we are ignorant, a condition once described by Benjamin Disraeli as a great step to knowledge," Wakelin wrote. "We are trying to build on this realization to become more knowledgeable."[17]

What followed during the next decade was a rush of deep-sea planning, research, and construction, initially driven by nothing more substantial than the Navy's new attitude. The nation making conspicuous plans to send men to the Moon in response to President Kennedy's 1961 call was now, in 1963 and 1964, under the new administration of President Johnson, gearing up to go as boldly as possible in the opposite direction as well, probing the hidden recesses of inner space, often with eerie, crablike machines easily mistaken for monsters. The total cost of the endeavor eventually ran to untold billions of dollars. Moreover, the Navy's investments were amplified as private industry poured money into the perfection of undersea gear in anticipation of big Federal contracts related to the conquest of inner space.

No longer was the cutting edge of deep exploration under the control of royalty and philanthropists and quirky committees. The richest and most powerful nation on Earth had decided to make the field its own. Its attitude was very different from forerunners' in that it had little or no interest in the traditional goals of deep research, in exploring for its own sake, in laying undersea cables, in searching for orthocones and trilobites, in admiring angler fish and other wonders of the sunless realm. Its aims, by contrast, were military. It was fighting a war it intended to win. In short, the new patron was not only wealthy: it was driven.

Old equipment such as the bathyscaph *Trieste* was overhauled, improving its powers and range. A rugged new version of the bathyscaph, *Trieste II,* was built. Bearing little resemblance to its homely predecessor, *Trieste II* looked like a powerful, hefty submarine, its aft section tapering down to giant fins and propellers that forced the behemoth forward. Its body bristled with lights, television cameras, and robotic claws, and its crew compartment was jammed with electronics. Big pads kept its crew cabin from sinking into the mud. Joining the two *Triestes* and *Alvin* was a small fleet of new underwater vehicles. The deadly seriousness of the enterprise ensured that the main contractors were no longer breakfast-cereal makers but military giants, such as Lockheed and General Dynamics.

One of the most advanced vehicles was the *NR-1,* a nuclear-powered submarine that could view the seafloor with sonars, video cameras, still cameras, and human eyes peering through any of three ports. For picking up objects, it had not only a big claw but a dexterous robotic arm that could reach over distances of nearly nine feet. *NR-1* dove deep, carried seven people, and stayed down for weeks. Strangest of all, it could roll along the bottom on wheels.

Also developed by the Navy were a pair of cigar-shaped Deep Submergence Rescue Vehicles, known as *Mystic* and *Avalon,* built to be easily transportable around the globe by road, boat, plane, or submarine and able to rescue up to twenty-four people at a time from sunken submarines. A pair of *Alvin* look-alikes named *Turtle* and *Sea Cliff* were constructed that were heavier and more advanced than their forerunner, moving slightly faster. New materials were also tried for deep subs, although long devotion to steel and strong metals limited the spread of such innovations. *Deep View* was a tiny two-man submersible whose front end was capped by a thick glass hemisphere, eliminating portholes and giving greatly enhanced viewing of the deep. The crew capsule of *Makakai* was made entirely of transparent plastic, giving its sole occupant a wondrous panoramic view.

To support the new undersea fleet, a small armada of surface ships was also created, including a floating dry dock, twin catamarans known as *Asr-21* and *Asr-22,* and vessels such as the *Laney Chouest,* which could conduct oceanographic research as well as launch submersibles. An altogether different kind of surface ship was the *Mizar,* which was converted from a cargo ship into a unique vessel for probing the depths. Through the well-like central opening in its deck, crew members could lower miles of cables, whose ends were heavy with lights and various types of cameras. Film was processed up top in *Mizar*'s suites, which were crammed with advanced photographic gear.[18]

In addition to the new ships and submersibles, dozens of new instruments were invented for deep exploration, in many cases usable by any member of the new armada. One of the first robots was known as *CURV,* for Cable-controlled Underwater Recovery Vehicle. It saw with television cameras, moved with propellers, and grappled objects with a robot arm. Its many offspring were increasingly advanced and compact. The Navy also pioneered such exotic technologies as underwater lasers, eager to have the thin, concentrated beams of light flash in innovative ways to illuminate large parts of the dark seafloor.

Ingeniously, the Navy and its contractors also came up with a deep navigation technique far better than the plastic signposts of the *Thresher* search. It used sound. A series of electronic buoys were moored near the bottom to form a network. When a robot or search craft sent out a "ping," each buoy would answer with a "pong." The elapsed time between ping and pong allowed the calculation of a vehicle's exact position, despite the inky darkness. Such devices were known as transponders or pingers. They allowed deep vehicles to move progressively across large areas of the seabed without retracing their steps or becoming lost.

Another advance in the use of sound, often from the surface, was a new type of sonar that better revealed underwater terrain. Old sonars bounced a sound signal off the bottom as a ship moved forward, the echo giving a crude picture of bottom topography. The new sonars used multiple thin beams, eventually dozens of them, to increase the sharpness of sound-produced pictures. The beams spread out in a broad swath like the ribs of an oriental fan. The result was that, for the first time, large parts of the ocean floor were mapped to reveal hidden worlds of hills and valleys, mountains and fissures. Such maps quickly became vital to any detailed search of the abyss. By the late 1960s, the Navy was mapping the ocean floor and probing its mysteries with a fleet of twenty-seven oceanographic ships, including the *Knorr, Melville, Dutton, Bowditch, Michaelson,* and *Compass Island.*

Finally, the Navy in its underwater push made great use of space-age technology in the form of navigational satellites, which were just beginning to orbit the earth. The radio beacons of these high-flying craft gave new precision to navigation on the high seas, allowing ships and submarines to calculate their positions down to an accuracy of feet and meters instead of miles and kilometers. The Transit class of navigational satellites, which became operational in 1964, had originally been built to improve the accuracy of Polaris missiles fired from submarines. But the satellites were quickly exploited by all aspects of the undersea rush. Ships surveying the blackness of the deep now were able to know their positions much better and thus produce seafloor maps of much greater accuracy. As a result, deep searches became more systematic and efficient, with less aimless wandering.[19]

The deep push was motivated by much more than the operational needs of the Navy's submariners. Indeed, they were more interested in escape systems on board their subs than in bizarre-looking craft that would swim toward them through the dark during a crisis. So ambivalent were the submariners that their support for the broad agenda at times turned to active opposition when it threatened to come at the expense of submarine budgets. Instead, among the main patrons of the buildup were the Navy's scientists and oceanographers, the people who all along had been fascinated by the deep but frustrated at the lack of strong official support. For them, *Thresher* and its crisis atmosphere were a great opportunity. The Stephan panel had been packed with science types who worked hard to influence developments. After its report came out, one panel member, Edwin A. Link, a noted inventor and underwater pioneer, argued that the *Thresher* incident would lead to a new era of deep exploration for the government and general public alike. "Riches await us under the sea," Link wrote in *National Geographic,* his article thick with page after page of lavish illustrations.[20]

The science community, with its raw enthusiasm, was not the only source of strong support. Nor would its advocacy have been sufficient, since scientists lacked the political clout needed to drive through many of the buildup's costly elements. A quieter and more influential source of patronage was the American intelligence community. Among other things, it eventually saw the advantages of being able to examine on the seabed the military and commercial hardware of other nations. The nation's spies played an important role in defining and funding the new wave of exploratory gear and made sure that many of the items had powers that far exceeded the needs of the ostensible users, Navy submariners. So it was that rescue vehicles with their powerful lights and robot arms were designed to operate much deeper than the collapse depth of any submarine.[21]

A main architect of the overall endeavor was the first director of the Navy's Deep Submergence Systems Project, which lay at the heart of the undersea push. John P. Craven was fond of poetry and intrigue. Born in Brooklyn of an old Navy family, a Ph.D. mechanical engineer and lawyer, Craven had risen through the Navy ranks to become the chief scientist of the Special Projects office, which had won much respect in Navy circles for its development of the Polaris missile. He knew the Navy's bureaucratic maze, the people, and how to make them work together as a team in the cause of technical innovation. Forty years old as he took up his new duties, he was extremely energetic and had a clear idea of where things should go. From long association with the Polaris project, and from meditations about what might come afterward, he had personally decided that the future of the Navy —and the nation—lay in the wilderness of the deep sea.

The impact of the new submersibles and other undersea gear, he argued in a 1966 article entitled "Sea Power and the Sea Bed," would be no less fundamental for the Navy than the advances produced by the introduction of steam engines and atomic power. The makings of a major step, he wrote, "are present and at the outset are favorable to the United States." He declared that America had a responsibility to lead the advances "so that when the inevitable and Gordian problems of sovereignty on the ocean floor are raised at the international conference table, the ability to resolve them on terms favorable to international peace and stability is matched by the capability for enforcement." Put less delicately, it was to be a new phase of Manifest Destiny. Only this time the expansion would be across the continental shelf and the deep seabed instead of the wilds of North America.[22]

The design philosophy for much of the deepwater gear envisioned by Craven and his teams was known as "two and twenty." At a minimum, equipment was to function to a depth of two thousand feet. At the other extreme, it was to go to twenty thousand feet, or 3.8 miles, if that depth was "technically and economically feasible," according to Craven. The latter figure fell short of the ocean's deepest trenches, but was sufficient to cover about 97 percent of the seabed, enough to deal with the vast majority of the exigencies that might arise. Still, some inkling of the vastness of the sea is conveyed by the fact that the 3 percent left out of the Navy's operational equation was equal to 4.2 million square miles, an area bigger than Europe.[23]

Much of what was done with the new armada was top secret, despite the very public trauma surrounding its origins. No detailed maps of the seafloor were released publicly, no intelligence coups hinted at. So, too, much of the new gear was virtually invisible, with its existence or exact powers never mentioned in public. The heart of much action was Submarine Development Group One in San Diego, which had little or nothing to do with developing submarines and everything to do with covert operations.

Other aspects of the Navy's work were widely advertised. After all, the rescue capabilities were meant at least in part to reassure prospective submariners that the service was taking every precaution to ensure their personal safety. In 1964, even as the Stephan panel finished its advisory work, Navy artists were drawing up graphic depictions of new rescue vehicles at work in the deep, saving submariners from icy death. Such renderings were made available to the public and press, even though it would be years before the rescue vehicles actually made the transition from blueprints to working machines able to plumb the sea's depths.[24]

For the Navy, the steady movement into the deep represented more than hard work and a new focus. It was also fun, despite the dangers. The scientists and experts who ran and piloted the vehicles became fascinated with how the abyss could crush everyday objects. A ritual developed in which cups and other objects made of plastic foam would be sent down on the outside of robots and piloted submersibles, to be uniformly squeezed down in the deep to a tiny fraction of their former size. Any drawing or writing on the objects would shrink as well, creating an opportunity for imaginative play that was cherished during long, monotonous voyages. On the deep-survey ship *Mizar*, crew members marveled at how dour scientists would erupt into fits of laughter as foam heads meant to hold wigs came back looking like objects out of a voodoo ceremony.[25]

THE FIRST MAJOR TEST of the Navy's new determination arose on January 17, 1966, while much of the envisioned gear was still in the process of being invented and built. A hydrogen bomb, the deadliest armament of the American arsenal, was lost at sea. The incident stirred international anxiety, frayed alliances, and handed the Communists yet another propaganda lift. Moreover, one of the imponderables was whether the repercussions might include the thermonuclear incineration of part of Europe.

The loss occurred as an Air Force B-52 bomber on a mock bombing run was being refueled by an aerial tanker over the Mediterranean. The two planes collided and burst into flames. Amid a rain of smoking debris, three of the B-52's hydrogen bombs fell to earth near Palomares, Spain, a dirt-poor coastal town. The last of the hydrogen bombs fell into the Mediterranean. No one knew exactly where. As with the other bombs, its power was equal to more than one million tons of high explosive, making it about seventy times as deadly as the bomb that leveled Hiroshima. It simply had to be found. President Johnson ordered that the search succeed no matter what its scope, duration, and cost. Pressures mounted as tourists fled Spain and Soviet trawlers shadowed the hunt. At a disarmament conference in Geneva, the Soviet Ambassador rose and solemnly declared that "a densely populated Mediterranean area is now in grave danger." A Paris newspaper ran a map whose concentric rings of potential ruin radiated out from Palomares to cover much of northern Africa and all of Spain, Portugal, and a large part of southern France. Even if unarmed, the lost weapon was an international time bomb.[26]

In its hunt, the Navy worked from a number of clues to narrow the initial search area from a sprawling 125 square miles down to a more manageable zone of 27 square miles. Still, the job was like trying to find a needle in a haystack, and initial sweeps of shallower areas by divers were disappointing. Finally the new gear began to arrive. One of the first pieces was *Mizar*, the deep-search ship, whose upgrades in the wake of the *Thresher* disaster had been cut short so it could sail to the Mediterranean. In February, *Alvin* arrived, ready to serve the men originally so opposed to its birth.

The sub and other gear repeatedly dove into the deep and found nothing of significance. The dark waters teemed with life—red and white lobsters, shrimps that swarmed like insects, luminescent pencil fish, chunky hatchet fish. At one point cannonballs and a wrecked galleon were found. But the underwater terrain was craggy and ideal for concealing a bomb. Undulating plains were punctuated by sharp slopes and deep canyons, some more than four thousand feet deep. Visibility was bad. On a good day, observers in *Alvin* could see perhaps twenty feet. The muddy bottom was easily stirred into blinding storms. Sometimes *Alvin* sank into the muck, covering some of its lights. The acoustic networks for deep navigation had yet to be invented, so the sub usually had no idea of its exact location. Nor were its labors always appreciated. The fleet commander was feverish with pneumonia, barking orders, pounding tables, demanding action no matter how bad the weather. The *Alvin* men worked up to fifteen hours at a stretch. One man was found asleep on deck, standing upright, his arms wrapped around a clipboard.[27]

The bad luck broke on *Alvin*'s nineteenth dive as the sub's three-man crew found a deep furrow in the mud that led off into darkness, perhaps made by the bomb. The team tracked it down a ravine. The date was March 15, 1966, almost two months after the accident. *Alvin* was down nearly half a mile. The pilot, Marvin "Mac" McCamis, a grizzled ex-submariner, followed the furrow down a slope so steep that the sub's big aft propeller kept hitting the bottom and stirring up blizzards of muck. So McCamis turned the sub around, driving backward to avoid stirring up the blinding clouds. It was a risky maneuver. He was driving blind, possibly backing into a hydrogen bomb or, more likely, its parachute, which could foul the sub's propellers.

Art Bartlett, an *Alvin* electrician, was looking through the sub's bottom porthole. Suddenly, he saw the ghostly, billowing shape of a parachute, partly covered by a squid.

"That's it!" Bartlett shouted. "I've seen a lot of parachutes and this is a big son of a bitch."

The men burst into nervous laughter.

They were straddled atop the world's deadliest weapon, unsure what state it was in.

McCamis turned the sub around and edged closer, mindful that the current could tangle the sub in the parachute. The bomb lay in the steep gully, its straps and lines undulating like cobras. The men talked of the threat of entanglement. If caught, McCamis said, they could dump the sub's heavy batteries and hope *Alvin* would be buoyant enough to haul them and the bomb back to the surface. The men kept the sub from making any contact with the weapon, parts of which could be seen beneath the parachute. Slowly and very gingerly they clipped an acoustic pinger onto the parachute. Like a beacon in the dark, its beeping would let them relocate the target later.[28]

New dives over the next few days were difficult, the weather bad. Finally a long line was attached to the parachute. The fleet commander ordered *Mizar*'s winches into action, but the line snapped after a few minutes. A new dive to the area found that the parachute and its deadly cargo had vanished. And the acoustic beacon had come loose. The whole nightmarish operation was right back to square one. After days of new searching, *Alvin* relocated the parachute at the bottom of a steep bank, 360 feet from its original position. The crew celebrated with a lunch of soggy sandwiches. On a subsequent dive *Alvin* tried to move the parachute off to one side, but things went poorly. Each action of the robot arm triggered awkward reactions in the sub. In frustration, eager for stability, the pilot set *Alvin* down on top of the bomb. ("You what?!" a colleague later asked in disbelief.) No matter what the team tried, it seemed impossible to disentangle the bomb from its chute.[29]

The fleet commander sent for the *CURV* robot. It had been rushed in from California after a quick upgrade for dives beyond its rated depth of two thousand feet, which was shy of the bomb by about eight hundred feet. The robot was nothing if not ugly, bearing no resemblance to the sleek, anthropomorphic slaves of science fiction. It looked warlike. Weighing more than a ton, it was six feet high and fifteen feet long. Atop its welded metal frame were four dark flotation tanks that resembled torpedoes. Its front end was a tangle of wires, lights, cameras, and a massive claw with foot-long curved jaws. Its rear end boasted three propellers, two horizontal and one vertical, all powered by big electric motors that allowed the robot to speed through the water almost twice as fast as *Alvin.* Best of all, it carried no human passengers, to the relief of the fleet commander, who was eager to avoid compounding the bomb crisis with a death.

Finally, all was ready. The robot was hooked up to the tether that supplied it with electricity and control signals. Everything, even the big claw, could be operated from up top. The robot was gently lowered from the support ship *Petrel* into the depths, while a team of controllers tracked its progress on the television monitor. On the bottom, it located the parachute and after some labor dug in a hook that was attached to a long line. The controllers up top decided to try to pull the parachute to one side so the robot's television camera could see the bomb—a move considered vital to finding out whether the weapon was fractured or whole, and thus whether a simple recovery was even possible. But a yank on the line succeeded only in stirring up a cloud of sediment. *Alvin,* sitting nearby to monitor the work, was engulfed in the muddy cloud and nearly buried in billowing cloth. Later, the robot successfully attached a second line.

Then came a new snag. To their dismay, the crew in *Alvin* saw that the bomb was slipping further down the muddy slope toward the edge of an underwater cliff. Time was running out for a third line, which had been deemed vital to lessen the risk of another snapped line. Lieutenant Commander J. Bradford Mooney, Jr., ordered surface controllers to drive the robot into the parachute as it billowed wide and high. In a flurry of motion, it hurled itself into the hovering mass of cloth, entangling itself deeply, its tether providing a last-ditch connection to the bomb. After heated debate, Mooney persuaded the fleet commander to go ahead with the improvisation. Up top, the pale sun was rising as the winches began to turn. On the bottom, the ooze clung to the bomb and then finally let go. The thermonuclear weapon began to rise. In less than two hours the drama was over. Robot, parachute, and bomb sat dripping wet on deck.[30]

It turned out that the weapon had only a few minor dents and scrapes after a fall of five miles through the sky and a plunge of a half mile through the sea. The search had taken eighty days and cost millions. But it was successful. Spain and neighboring countries breathed a collective sigh of relief, as did the White House and the Pentagon. The nightmare was over.

The Navy and its growing ranks of undersea experts gleaned a number of lessons from the crisis. Some were minor, such as the need for better means of navigation in the eternal midnight of the deep. Others were major, such as the revelation of the rich new potential of robots for aiding jobs of deep-sea recovery. Without the emergency, the experimental *CURV* robot might have remained a laboratory curiosity and minor tool. Initially its giant claw had been used to recover torpedoes on firing ranges. But during an international crisis with potentially deadly repercussions, the robot had done remarkably well, performing deep work deemed too dangerous for men. The Navy, taking the lesson to heart, quickly began to develop all kinds of advanced robots. Directed through long tethers, peering with camera eyes and manipulating heavy objects with mechanical claws, they would eventually roam the deep for miles and stay under for days and weeks as their operators enjoyed the usual shipboard comforts, safe and dry, watching the undersea action on television monitors.

Perhaps the most important lesson was geopolitical. Military hardware that was lost at sea, be it nuclear or conventional, backward or advanced, respected no national claims or boundaries. It was simply there on the wilderness of the seafloor, ready to be inspected or seized by whoever had the wherewithal. The presence of the Soviet trawlers during the operation had driven home this simple truth. In the aftermath of the bomb recovery, Navy planners, newly sensitized to the geopolitics of the deep sea, gave increasing thought to how such facts could most readily be turned to the West's advantage.

ON THE ARID plains of Kazakhstan, just east of the Aral Sea, lay a sprawling center for rocket launchings from which Moscow routinely fired its astronauts into space and its intercontinental ballistic missiles on test flights. It was called Tyuratam. All its operations were conducted in great secrecy and were of immense interest to the West. Tyuratam had set all the records, launching the first warhead into space, the first satellite, the first astronaut, and a whole string of other firsts. Moscow may have been backward in many aspects of modern technology. But in rocketry it was a gifted and deadly competitor.

The long-range missiles that flew with alarming regularity out of Tyuratam carried payloads known as reentry vehicles or, in military jargon, RVs. The acronym was a euphemism for envoy of death. Generally cone-shaped but often softened with subtle curves, the RVs were carefully designed to hold and protect nuclear warheads as they sped through space and fell back to earth during the jarring descent through the atmosphere. RV designers had many aims—avoiding destruction of the device during the fiery reentry, minimizing the effects of wind, rain, and aerodynamic drag, and keeping overall speed and accuracy as high as possible. On some test flights, the core of an RV was instrumented with electronic gear for taking and transmitting back to earth a range of measurements. On others, the RV was packed with dense metals meant to mimic the weight and center of balance of a real nuclear warhead as closely as possible, so engineers could be reasonably sure the speeding RV would fly in war as it did in peacetime.

Belching fire and smoke, the rising missile would fly upward and eastward over the Asian landmass. Engine stages would fall away one by one until all that was left was the RV speeding through space. After traveling thousands of miles, the projectile would zero in on its target—often on or near the Kamchatka Peninsula, a wild region that juts out from Siberia, surrounded by the Sea of Okhotsk on one side and the Pacific Ocean on the other. A military base on Kamchatka tracked the incoming RVs with powerful radars. Sometimes they fell into impact zones in the Sea of Okhotsk, or flew far beyond Kamchatka into the deep waters of the blue Pacific. Especially in the 1960s, Moscow kept up a steady beat of long-range shots that landed further and further out to sea. Like its annual parade of missiles on Red Square, these shots were explicitly meant to show that the weapons were being continuously modernized and improved, that they were increasingly powerful and deadly.[31]

In theory, RVs that plunged into the depths were lost forever, adding to the wasteland of junk already scattered over the seabed. But in the West these military discards were considered priceless bits of potential intelligence. Their shape and composition would give insights into their aerodynamic properties and ultimately into the accuracy with which they could be fired over long distances. Their interior would reveal the space available for a bomb or bombs, from which an analyst could judge explosive yield. Their instrumentation would disclose the state of Soviet electronics. Even RV fragments would be valuable, since their composition and state of wear and tear could denote the degree of fiery ablation during descent, which also bore on the issue of accuracy. In short, photographs of lost RVs or, better yet, physical recovery of the mock warheads, would substantially aid the West's struggle to understand and lessen its nuclear vulnerability. The question was how to get these vital pieces of the intelligence puzzle.

The Soviets claimed control over the waters around the Kamchatka Peninsula and all of the Sea of Okhotsk, which was up to two miles deep. Even American operations on the high seas were a problem if the Soviets were to be kept in the dark about activities that, at the very least, would prompt technical countermeasures if not deadly reprisals. Obviously, the delicate jobs of espionage and retrieval were unsuitable for any vehicle that worked on or from the ocean surface, such as *Mizar, Alvin,* and *Trieste.* Their movements and those of their support ships would easily be spotted by Soviet border patrols or spy satellites.[32]

Thus began one of the stealthiest operations in the annals of cold-war espionage, one whose feats are still under wraps today. Its code name was Winterwind. To carry it out, the Navy decided to deploy the most furtive of its underwater assets, one unknown to Moscow and other foreign powers because it was designed and made in extreme secrecy. It was a unique submarine. The nuclear-powered sub traveled submerged over long distances, giving no indication of when it entered or left an area of operations. More important, it had remarkable powers of deep inquiry.

As illustrated by the *Thresher* episode, all regular submarines had to work in the upper reaches of the ocean or risk being crushed by the enormous weight of accumulated water. But the spy submarine had a special way—an obvious way, once you thought about it—of getting around that problem. While holding a steady position, it simply lowered a long cable downward into the darkness. The cable's front end bore lights, cameras, and other gear, giving its controllers the ability to probe the deep's secrets. As was the case with all submerged submarines, it was completely invisible to reconnaissance satellites high overhead.[33]

The spy sub was the *Halibut,* named after the large type of flatfish that lives on the seabed. The vessel, longer than a football field, had been commissioned in 1960 as a launching pad for a kind of low-flying cruise missile known as Regulus. Oddly shaped, *Halibut* had a cavernous internal hangar for Regulus missiles that gave the submarine's bow a sizable bulge. It looked a bit like a snake digesting a big meal. As it turned out, *Halibut* lost its missile-launching job not long after commissioning, when the task of Regulus was usurped by Polaris, the long-range missile. But the ungainly sub lived on.

As momentum gathered in the Navy's push to master the deep, Craven and other naval experts decided that *Halibut* and its big internal bay were ideal for a new job. Beginning in 1965, the sub underwent a major overhaul at a dry dock in Hawaii. The floor of the cavernous hangar was cut open to form a commodious well that penetrated the sub's hull. Retractable doors were added so that the seawater below could be sealed out when so desired. Also added to the bow compartment were overhead winches, big spools carrying miles of cables, and racks of reconnaissance gear that could be lowered through the well into the sea below. The camera-laden spy systems were referred to as Fish, a common nickname for oceanographic probes. Up top, thrusters were added to the hull to help the big sub maneuver.

The plan was to have *Halibut* lower gear-laden cables into the depths for reconnaissance, recovery, and manipulation, just as *Mizar* and other ships were now doing from the surface. The revolutionary change was that its work would be totally surreptitious. And its labors promised to be easier in some respects since the sub would be untroubled by waves. The search for the *Thresher* and the hydrogen bomb had been frustrated by storms that stopped all work. Even in mild weather the up and down of waves often translated into jostlings of long cables that could make deep operations either difficult or impossible, as Beebe had discovered in his midwater explorations. But a submarine had to descend only a hundred or two hundred feet to find waters that were perfectly calm.

Not that everything on *Halibut* was easy. Experience at sea showed that line lowerings and raisings could be a headache. "That damn cable would get tangled every once in a while or it would start to fray, and we wound up pulling every bit of it a foot at a time inside the boat," one crew member recalled. Fixing such problems, he added, could take up to several days. But when it worked, the reconnaissance system worked very well. Some of the photographs that *Halibut* took of the seabed were in vivid color, revealing the nature of the deep and its debris in extraordinary detail.[34]

So secret was the *Halibut*'s work that it was undisclosed to even such a high-ranking naval official as Admiral Rickover, the nuclear czar. On one occasion Rickover made Craven fly nearly halfway around the world, from Hawaii to New York, for a heated and unsuccessful cross-examination meant to elicit the nature of the sub's activities. So, too, the Navy bureaucracy went out of its way to hide the vessel's real mission with false stories. At one point, *Halibut* was advertised as the first mother submarine for the Navy's Deep Submergence Rescue Vehicles, and photographs of the sub carrying a rescue-vehicle simulator on its back were widely distributed to the press.[35]

At sea, *Halibut* hit pay dirt as it carried out operation Winterwind, its cameras and other gear finding lost Soviet RVs. For Craven, the success of the top-secret project was a great prize that he quietly advertised around the

Pentagon to build support for his larger agenda. In late 1966 and early 1967, he briefed a number of high military officials on its feats, showing them, among other things, sharp photographs that *Halibut* had taken of RVs in the abyss. Among the officials privy to the briefings were Robert A. Frosch, Assistant Secretary of the Navy for Research and Development; Paul H. Nitze, the Secretary of the Navy; John S. Foster, the Pentagon's Director of Defense Research and Engineering; and finally, at the top, Robert S. McNamara, the Secretary of Defense. These briefings were a turning point. After them, "black" funds from the nation's intelligence budget were increasingly directed into the deep program, eventually becoming an important and sometimes dominant force in the military's probing of the new frontier.[36]

THE AMERICAN NAVY during this period suffered one of its worst setbacks of the cold war. The *Pueblo,* a spy ship whose advanced antennas were designed for espionage, had been sailing in the Sea of Japan close to the shores of North Korea, monitoring coastal activity and trying to eavesdrop on transmissions out of Wŏnsan, a port recently transformed into a major Soviet sub base. The United States at the time was increasingly mired in the Vietnam War and attracting all kinds of hostile reactions from Communist states and clients. The North Koreans objected to the ship's presence, violently. On January 23, 1968, they seized the *Pueblo* in a rain of gunfire. One wounded sailor died in captivity, while the rest of the eighty-three crew members were tortured and abused.

Politically, the incident was a heavy blow. The North Koreans, and presumably their Soviet patrons, now had in their possession a Navy espionage ship bristling with the latest spy gear and code books. No intelligence loss like it had ever happened before. Washington was furious and fired off a series of protests, but nothing could be done.[37]

Then, out of nowhere, came a tantalizing prospect for retribution on a grand scale. A Soviet submarine on routine patrol in the North Pacific sank just a few weeks after the *Pueblo* incident, in early March 1968. The stagnant air and cramped quarters of the sub were home to about one hundred men ready for the possibility of war with the United States, ready to target Hawaii, the heart of American military strength in the Pacific. If the Kremlin had sent a coded message, the sub would have fired its missiles skyward to precipitate a rain of thermonuclear ruin.[38]

But something went wrong on that winter patrol. The time of the sub's regular radio transmission came and went, with headquarters at Vladivostok hearing nothing but static on the assigned frequency. After twenty-four hours of silence, the Soviet Pacific fleet declared an emergency and began to hunt for its missing sub. The first forces to arrive in the search area were battered by gale winds and dense snowfall. The sea was rough, with waves running up to forty-five feet high. Four submarines crisscrossed the area, back and forth, back and forth. They found nothing. After a month, the search was called off. Whatever its cause, the accident was a major blow to Moscow. Never before had it lost a submarine loaded with men, missiles, warheads, code books, and all the other paraphernalia designed for the waging of nuclear war.[39]

For Washington's spies, and for the White House, the incident was a potential windfall, an undreamed-of bonanza. The code books and encryption gear could aid in deciphering all kinds of intercepted messages gathering dust. The thermonuclear arms, not just RVs, might reveal design secrets that could make possible all kinds of countermeasures, including ways to try to knock out the warheads in time of war. Perhaps most significant of all, the sub's missiles could be analyzed to reveal their accuracy. This factor, mainly involving devices known as inertial guidance units, was a daunting mystery for America. It determined whether Moscow's missiles were crude blunderbusses or precision arms that could dig out underground missile silos, military command posts, and presidential bunkers, vastly increasing the odds of ruin. It was *the* military question of the day. With accuracy, the Soviets might forge a war-fighting plan meant to prevail quickly in times of tension by striking first. American analysts of the 1960s thought Soviet guidance of missiles was too primitive for that kind of first-strike precision. But it was improving. If it ever got really good, the vulnerability of American leaders and forces on land would soar.[40]

The Navy had monitored the sub's sinking from its intelligence post at Pearl Harbor, which eavesdropped on the wide Pacific with a network of underwater microphones. In subsequent weeks it became clear that Moscow had no idea as to the grave's location. But Washington did, and intended to do something about it.[41]

Fresh from its RV accomplishments, *Halibut* was assigned to the hunt, which turned out to be no picnic. One man went overboard. Another went crazy. The secrecy was so great that only the captain and his executive officer knew what the daunting operation was all about. For the rest, the blind tedium was agonizing. Worms in the breakfast cereal did nothing to ease the pain. For months the hunt went on, unproductively searching the dark seabed more than three miles down. One reconnaissance Fish had to be cut loose when a snag ruined its cable. The sub's reactor ran dangerously low on fuel, threatening to automatically shut down. Overall, the main problem was the size of the search area. The Navy's eavesdropping system in the Pacific, with the nearest microphones hundreds of miles from the disaster site, had supplied only the roughest of coordinates. Up close, the search region seemed like a boundless wilderness. To build morale and enthusiasm for the hunt, a crewman passed around photographs of the bottom showing sea slugs and bizarre fish with ghoulish faces. The seabed seemed to be alive with oddities. But everyone knew that the target had eluded them.

After months of frustration and repeated failure, *Halibut* during the summer of 1968 began yet another of the interminable hunts. From its internal bay, dubbed the "bat cave" by the crew, the sub lowered a thick cable downward for miles into the icy darkness. This time, instead of endless muck, the lights and cameras illuminated a grim scene of death and catastrophe. The ensuing survey and photomontage showed that the lost Soviet sub had broken into pieces, with some of its guts spilled out. While the dead sub was in better shape than the *Thresher,* a junkyard of debris, it was nowhere near intact. Its rear engineering section was detached from the central and forward compartments, which held the all-important missiles and were about two hundred feet long. The debris field around the broken submarine was rich in wreckage and lost artifacts.[42]

Exactly what transpired during the *Halibut*'s mission, and the existence of the operation itself, remain high American secrets to this day. No one has publicly detailed what intelligence was gathered. Cameras were unquestionably brought into play, and probably claws as well, if not on the initial mission then on subsequent ones. What is known with certainty is that the *Halibut*'s actions were deemed so significant that the Johnson White House bestowed upon the spy sub a much-coveted Presidential Unit Citation for "exceptionally meritorious service" in a secret operation of "great military value to the Government of the United States." Craven, a key architect of the sub and the whole operation, was awarded the Defense Department's Distinguished Civilian Service Award, the highest honor of its kind. While giving no details of what had been accomplished, Craven years later described the operation as an intelligence coup. "*Halibut* was able to locate, examine, and evaluate the accident and to obtain significant intelligence information concerning the submarine, its mission, and its equipments," he told the Senate. "It was the opinion of many in the Navy and the Defense Intelligence Agency that optimum recovery of intelligence information from this accident was achieved." The phrase "optimum recovery" is heavy with significance and suggests a collection of some consequence. Also, Craven in his testimony said *Halibut* used "investigative equipments," a turn of phrase that suggests much more than cameras. Tiptoeing around security strictures in an interview, Craven emphasized the operation's thoroughness. "We milked the submarine dry of really meaningful intelligence," he told me.[43]

The greatest accolade of all was that the spy sub, with back-to-back feats, became the prototype for a new class of American submarines designed for deep surveillance, recovery, and manipulation. Typically, aging attack subs, after retirement from the job of tracking foreign targets, underwent conversion for the less physically demanding role of deep spying. Advances in miniaturization, particularly of electronic gear and cables, eliminated the need for the kind of "bat cave" possessed by *Halibut* and allowed almost any attack sub to be converted for the surveillance job.

Over time, the spy submarines became the undersea counterparts to the nation's orbiting fleets of spy satellites, which often epitomized the cold war's technical glamour. Some analysts judged the invisible subs as more important. One reason was their stealthiness. Spy satellites, or any craft circling the Earth through space, have predictable orbits obvious to anyone who carefully examines the night sky. An enemy who knows one is passing overhead can try to fool it with camouflage, decoys, and misleading deployments of real weapons, or can skip sensitive operations altogether so as to avoid its gaze. During the cold war, Moscow turned satellite dodging and deception into high art. The other reason spy subs were so valued was their intimacy. Spy satellites peered at their targets over vast distances, sometimes many hundreds of miles. But surveillance Fish lowered into the abyss not only viewed targets at extremely close range, improving photo quality, but also in some cases picked up objects and manipulated them. Typical targets of deep espionage included all the paraphernalia of modern war—ships, planes, weapons, rockets, RVs, spacecraft, and nuclear gear that foreign powers lost to the deep —as well as equipment placed purposely on the seafloor, such as cables, microphones, and other sensors. In some respects, the United States during this period found the deep sea to be a gigantic and extraordinarily rich lost-and-found.[44]

During the presidential campaign of 1968, Richard Nixon charged that America had lost international prestige during the Johnson years and cited the *Pueblo* incident as evidence of this erosion. The few Democrats in the campaign who knew of *Halibut*'s swift retaliation could say nothing. Their lips were sealed. The developing world of deep spying was beyond politics. In the nuclear era, the new capability was a priceless secret in the calculus of national survival.[45]

NOT THAT AMERICA'S deep warriors focused exclusively on adversaries. A related issue was the mountains of equipment lost at sea by the armed forces of the United States. Be it paranoia or prudence, Washington decided that the most sensitive of these items had to be reconnoitered if at all possible in order to know their vulnerabilities to enemy action. Moscow appeared to be in the dark about what was going on at the bottom of the sea and to have none of Washington's deep capabilities. But, then again, perhaps it did have them. After all, such operations worked best when they were covert. Driven by this kind of logic, Pentagon officials put the deep armada to work on a variety of reconnaissance and cleanup jobs directed at its own forces. For instance, in Scottish waters a third of a mile deep, the *NR-1* investigated a wrecked F-14 jet fighter and used its giant claw to recover a Phoenix air-to-air missile, the workings of which were top secret. Addressing an even more sensitive issue, the Pentagon put into motion a secret operation known as Dragon Shield. At the oceanic impact zones of American RV tests, the deep flotilla searched for and at times recovered mock warheads that had previously been considered lost to the deep. Finally, the Pentagon gave its highest priority of all to the reconnaissance and retrieval of real nuclear arms, the deadly icon of the age. So it was that great care was taken to recover the hydrogen bomb off Spain. That crisis had been the responsibility of the Air Force. Just two years later, the Navy suffered a similar kind of loss.[46]

The *Scorpion* was an American nuclear attack sub carrying ninety-nine officers and men. It was an aging workhorse of the Skipjack class, not a thoroughbred like *Thresher,* meant to sire a new breed of subs. Nevertheless, it was armed and deadly. In May 1968, while coming home from exercises in the Mediterranean, *Scorpion* failed to report home by radio. An armada of ships, planes, and submarines scoured the Atlantic but failed to find the sub or any clues to its demise. Foul play seemed unlikely, given the lack of enemy forces in the area, but the disappearance was still a crisis because *Scorpion* carried two Astor torpedoes, which were designed to sink enemy submarines. Each torpedo was a shaft of metal nineteen inches wide and almost nineteen feet long. And each was tipped by a nuclear warhead.[47]

Craven by this time had his deep program in high gear and brought important skills to bear on the search. He gathered sound recordings from underwater listening posts off Newfoundland and the Canary Islands, found a suspicious series of explosions, and used them to pinpoint a likely spot some four hundred miles southwest of the island of San Miguel in the Azores. The area was more than two miles deep.[48]

The main search zone was a square twelve miles on a side. Surface ships were used since Washington wanted to fly the flag and send a clear signal to Moscow to stay away. This was an American issue. While ships such as *Compass Island* surveyed the target area's seafloor with the latest multibeam sonars, the deep-research ship *Mizar* photographed the inky depths, its teams of crewmen working in shifts around the clock, seven days a week. The reconnaissance ship towed a long cable and a Fish that bore lights, cameras, and a magnetometer to detect concentrations of metal. The cameras were loaded with enough black-and-white film to take pictures automatically every thirty seconds for as long as thirty hours.

The Fish was towed back and forth in the kind of progressive pattern used to mow a lawn. Amid endless gray swells, the *Mizar* kept careful track of its position with the aid of the Navy's navigational satellites high overhead. It also periodically interrogated deep networks of acoustic transponders that had been moored to the seafloor. The pings and pongs transmitted by these devices not only helped *Mizar* pinpoint its location in relation to the seafloor, but, as important, the location of the search sled at the cable's end, which also was fitted with one of the acoustic devices. Overall, the gear reduced the *Mizar*'s positioning errors to about three hundred feet, a remarkable feat of navigation for the wilds of the featureless Atlantic.[49]

The search began in June 1968 and continued the entire summer, fruitlessly. The endless scenes of bottom muck revealed little but creatures of the deep, some lost human artifacts, and a big magnetic rock—but no submarine. Finally, the efforts paid off on October 30. Photographs rushed out of the processing lab revealed the hulk of the *Scorpion* broken in two, buried in mud where her bow had slammed into the bottom.

The gruesome scene, as revealed in a photographic mosaic painstakingly pieced together, was full of surprises. For one thing, the sub was largely intact. It had obviously undergone no catastrophic implosion, as had the *Thresher.* An implosion could take place only when a sinking vessel's inner spaces were full of air and then rapidly collapsed as external pressures soared. In contrast, flooded spaces tended to keep their structural integrity, since pressures inside and out were the same. *Scorpion* on inspection was judged to have flooded long before passing the crush depth. A series of other clues, including a gaping hole in the side of the sub's operations compartment, led experts to propose that a self-inflicted tragedy had occurred on or near the surface. They believed a conventional torpedo had come alive accidentally in one of the sub's bow torpedo tubes, had been jettisoned from the vessel, and had swung around to hit its nearest target of opportunity, the *Scorpion.*[50]

The Navy and its advisers judged that a submersible was needed to advance the forensic work, to ascertain the state of the nuclear warheads, and to clean up any sensitive weapons or debris that could conceivably be picked up by enemy forces. The site was too deep for *Alvin,* and none of the newer vehicles in the works, such as *Sea Cliff,* were yet ready. Only *Trieste*—the new *Trieste,* built specifically for such missions—could withstand the pressures of that depth. After gentler weather replaced winter and its rough seas, *Trieste II* and her support ship, a converted floating dry dock, headed across the Atlantic to the site of the tragedy.

It was July 1969, a busy time for the United States. President Nixon was settling into office. Gas guzzlers ruled America's highways. A musical group known as the Fifth Dimension sang of the Age of Aquarius and an impending era of spiritual harmony. American forces in Vietnam were starting a slow pullout, after reaching a peak of a half million troops, even as bombings picked up and waves of antiwar violence swept the nation. American astronauts were preparing to land on the Moon.

At the same moment, with no fanfare or publicity, *Trieste II* descended into the darkness of inner space. The downward trip of more than two miles to the hulk of the *Scorpion* was no guessing game, as had occurred with *Thresher*. Now all was science. *Trieste*'s descent was guided by computers that calculated the momentum and trajectory and slowed the underwater craft to a gentle landing on the seabed. That job, in Craven's estimate, was more challenging than putting astronauts on the Moon, since *Trieste*'s motion set up currents and sympathetic motions never found in the vacuum of space.

After the watery descent, the sub's broken hulk was closely inspected and photographed, especially in the area where the *Scorpion*'s hull had split in two. But the interior spaces of the sub were generally inaccessible, even for the small robot on a tether with a television camera that *Trieste* had brought along. In effect, the nuclear torpedoes were buried deep inside the sub, safe from prying eyes or claws that might come along. They were judged to be beyond recovery for the foreseeable future. However, *Trieste*'s robot arm did scour the debris field near the operations compartment, which had been torn open by the original blast. One item taken to the surface was a sextant, an astronomical instrument that the *Scorpion*'s navigator had used to help judge the sub's position as he stood on deck, sighting the sun and the stars. It was an antique of the sea in an age of high technology. Eventually, the instrument was put on display in a Washington museum, a memorial to the calamity.[51]

THE RECORD of accomplishment that America's deep forces compiled in these years casts a revealing light on one of the greatest enigmas of cold-war espionage—the case of the *Glomar Explorer*. The episode, long a topic of public conjecture and debate, began in great secrecy when the Nixon Administration in its early days decided to retarget the Soviet submarine that had sunk in 1968 to the Pacific seabed. Even though it had been reconnoitered by the Johnson Administration, the Nixon White House ordered the Central Intelligence Agency to build a special salvage ship to raise the disintegrating hulk for detailed analysis back on land. Some experts in the Navy and the Defense Intelligence Agency objected to the plan, saying the conspicuous nature of the operation would invite discovery. Moreover, they argued, it was wasteful. As Craven later put it in his Senate testimony, *Halibut* had already achieved the "optimum recovery" of intelligence information. But the plan went ahead nonetheless, in the end costing more than a half billion dollars. Craven suspected that its main objective was not intelligence but financial reward to the California contractors who had supported Nixon's political rise over the years.[52]

Much has been written about the *Glomar Explorer.* The giant, 618-foot-long vessel was built in the guise of a deep-mining ship for the industrialist Howard R. Hughes and in 1974 conducted its secret mission in the Pacific depths, lowering a long pipe capped by a giant claw. After the cover story was blown, hundreds of news stories enveloped the episode in a fog of conflicting claims. Some accounts said the whole sub had been recovered, some that the *Explorer*'s huge claw had broken and dropped the majority of the sub back into the deep.[53]

The truth, according to a project architect, was that the sub was a fragmented jumble right from the start. The only section deemed interesting enough for retrieval was an intact section of the bow and center structure that measured some two hundred feet in length. The *Explorer* was designed to retrieve this section, and could have been built longer if the target had been longer. As it turned out, the huge central bay of the ship was 199 feet long—slightly longer than the section meant to be retrieved. According to the expert, the claw did crack during the lifting operation, dropping the majority of this section back into the deep.[54]

The CIA broke its long silence in 1992 after CIA director Robert M. Gates told Russian President Boris N. Yeltsin a few details of the secret episode, in particular that the remains of six sailors had been recovered from the wreckage and later buried at sea. Six out of one hundred or so men implies a relatively small recovery of materiel. Another disclosure came in 1993 when a high-level Russian report said the *Explorer* had successfully grabbed radioactive material equal to two nuclear warheads, apparently from torpedoes located in the sub's forward compartment.[55]

No Washington official, past or present, has ever publicly detailed what the *Explorer* got or publicly defended the mission against the kinds of charges leveled by Craven. But the newly revealed skill of America's undersea warriors, of *Halibut* and her kind, as well as reports of an internal debate in the intelligence community over the wisdom of the costly operation, at the very least raise questions about its merit. It might indeed have been an extravagant sideshow, largely political in origin. That thesis is consistent with other Nixon Administration acts. Faced with the winding down of the Vietnam War and the loss of political leverage that came with big military contracts, the Nixon White House is known to have worked hard before the 1972 election to shower Federal largess on key battleground states, especially California. Among other companies, the Golden State was home to Hughes, Lockheed, and Global Marine, whose corporate throngs built the *Glomar Explorer* for the CIA in the early 1970s and later were happy to oversee its probings of the wilderness below.[56]

ONE AIM OF AMERICA'S deep armada had nothing whatsoever to do with lost gear. Rather, its goal was to gather raw information, which under some circumstances was more valuable. It was orders that set men and arms into motion, commands that ignited wars and initiated killing. Washington's spies knew well that captured information could turn the tide of battle, as it had repeatedly done during World War Two, when the monitoring of German and Japanese communications played vital roles in the winning of Atlantic and Pacific battles. So too the American leaders of the cold-war conflict constantly sought out the best possible information on Moscow's military moves and ambitions.

The greatest source for such intimacies turned out to be cables that radiated out from Russian naval bases across the dark seabed. Unlike communications relayed through space by orbiting satellites, the deep messages were typically unencumbered by ciphers meant to foil eavesdroppers. The unencrypted communications included not only high-level orders but everything from personnel matters and performance reports to logistics rundowns. It was like working for the Russian fleet and knowing all its activities and troubles, its foibles and secrets. Such information added up to another potential intelligence bonanza for the West, just lying there for the taking. The trick was reeling it in.

The key weapons in the information war were a pair of stealthy vehicles that greatly enhanced the powers of the Navy's undersea arsenal. They had no ability to roam the sea's abyssal plains miles below the waves, but they were perfect for probing the relatively shallow seas and continental shelves around the Soviet coastline and Soviet naval bases, their pinchers ready to perform the most delicate of operations and manipulations.

The undersea spies were the deceptively benign, cigar-shaped craft known as the Deep Submergence Rescue Vehicles, or DSRVs, which the *Halibut* had been associated with at one point. Each was fifty feet long and eight feet wide—roughly the size of a city bus. Beneath each craft's outer shell were three pressure spheres made of superstrong steel and linked together to form an inner habitat. The center sphere had a skirt that mated to a sub's hatch. DSRV-1, known as *Mystic,* and DSRV-2, known as *Avalon,* in theory were meant to rescue up to twenty-four people at a time from a sunken submarine, but in fact were designed to operate down to 6,500 feet, considerably deeper than the collapse depth of any submarine. They could scan wide areas of the seafloor with powerful sensors. They bore elaborate manipulators, winches, and claws, which were advertised as perfect for clearing away debris from the hatches of crippled submarines but had many unadvertised uses as well. Most importantly, they were stealthy. Hitching a ride atop a submerged submarine, they could be transported anywhere in the world's oceans, could perform a deep mission, and could return to the mother sub—all surreptitiously, with no chance of discovery by hostile ships or satellites. This capacity for complete secrecy made the DSRVs similar to *Halibut* and very different from *Trieste, Alvin, Turtle, Sea Cliff,* and even *NR-1,* all of which needed support ships of one sort or another.[57]

The DSRVs were the crown jewels of the Deep Submergence Systems Project. They were started first and finished last, the delay being mainly due to their complexity. Craven labored to give them the latest gear. He had viewports augmented by closed-circuit televisions. He had navigation aided by multiple sonars and a miniaturized inertial guidance system. He had overall vehicle control aided by special computers and displays. And he had it all put together by Lockheed Missiles and Space Company, which was admired in government circles for its advanced spy satellites. As the envisioned powers of the DSRV grew, the date of delivery fell back and expenditures soared. Originally, the Stephan panel had foreseen the rescue vehicles as costing about $3 million. Craven's group initially estimated $20 million. In the end, the total bill for two vehicles came to $220 million, including such items as spare parts and pilot training. As the costs rose, the Navy put less and less money into the DSRV project. By its end, virtually the sole source of financial support was the shadowy parts of the United States Government that specialized in espionage.[58]

The two vehicles entered service in 1971 and 1972 and instantly brought important new capabilities to the fleet of American spy submarines probing the dim recesses of the sea. *Halibut* and her sister subs had a hard time doing delicate manipulations over miles of lowered cables, which fell prey to snags as well as currents and other forces that caused jarring motions in cameras and gear. Such operations were easiest when done close up by a human in a rock-steady vehicle. The DSRVs could control their movements to within one inch—a steadiness important for hooking up its hatch with that of a stricken sub, and for other matters requiring some delicacy.

One of the intelligence coups of the DSRVs was tapping a cable lying on the bottom of the Sea of Okhotsk. The cable, lying between the Kamchatka Peninsula and the Soviet mainland, was a prime target for American intelligence because it was thought to carry communications of the Soviet Pacific fleet as well as Soviet missile tests. Again, it was the kind of data judged vital to helping the United States understand its nuclear vulnerability.

During the 1970s, in the Sea of Okhotsk, a DSRV set up the eavesdropping system. The deep-diving vehicle put on the cable a large pod with a wrap-around attachment that picked up electronic emanations without directly tapping the internal wires. Inside the pod, an advanced system of miniaturized gear, batteries, and tapes recorded the signals. Periodically the tapes would be picked up for analysis.[59]

The system worked perfectly and supplied a treasure trove of intelligence, and its feats were repeated at other locations, such as along the northwestern Soviet coastline around Archangel and Severomorsk, the base of the Northern Fleet. While a boon, this kind of tap had shortcomings that revolved around its operational difficulty and tardiness. Tapes were processed many months after the transmission of the Soviet messages, and the nature of the work forced a steady stream of risky missions to retrieve the clandestine recordings.

By the early 1980s, the intelligence experts of the Reagan Administration came up with a plan to circumvent such limitations and install underwater taps whose usefulness would be dramatically increased. The plan was costly and audacious, and its implementation would require all of the undersea muscle the Navy had built up over the decades, including the DSRVs and lots of other gear as well. The target was the undersea cables of the Northern Fleet. Earlier taps showed that they were an extraordinarily rich source of data, revealing such things as the movements of Soviet warships and submarines. The plan was to tap this prize directly by building a long undersea cable that would carry Soviet messages almost instantly to American spy agencies. With that accomplished, there would be no more lengthy delays between retrieval missions. Instead, the chiefs of American intelligence would hear the active heartbeat of the Soviet military. Indeed, the ultimate rationale for the on-line system was that it would be an alarm bell for Armageddon. If the Soviets were ever going to attack the West, the Northern Fleet would be involved and would get advance notice. Thus, the West would also get early warning. To the Reagan Administration, with its apprehensions about nuclear war, the system was deemed essential for Western survival. So it was that work was inaugurated and quickly grew to become the most expensive item in the intelligence budget.

Moving under arctic ice packs, attack submarines and DSRVs scouted out a workable route. This was no easy task since the whole system by definition was meant to be invisible, buried in the seafloor, so that any chance observations of the area by Moscow would reveal nothing. Special taps were developed that would hold Russian undersea cables from underneath, capturing their electronic emanations. The taps would be tied to fiber-optic cables that would run under the ice packs of the Barents Sea all the way to Greenland, a distance of about twelve hundred miles. There the lines would surface and the signals would be relayed to Washington.

The furtive lines were so long that special devices had to be developed to boost the signals. Custom equipment was developed to dig into the seafloor and bury the cable once it had been laid. In short, it was a massive industrial undertaking on the seafloor, the likes of which had never before been attempted, much less in secrecy. The cost of just burying the cable was estimated at $1 million a mile, or well over $1 billion for its total length. In all, the project's costs were estimated at nearly $3 billion.[60]

The plan collapsed in 1986 after American officials discovered the treason of Ronald W. Pelton, a former communications expert for the National Security Agency, who confessed to selling intelligence data to the Soviet Union. Among his $35,500 in sales to Moscow was information on the interception operation in the Sea of Okhotsk, which had mysteriously failed in 1981. Now that Washington had a clear understanding that Moscow was alerted to such covert activities, it saw little or no chance for the successful completion of a project that was far more ambitious. The gamble was too great. So the plug was pulled on the undersea tap after the expenditure of nearly $1 billion.[61]

DESPITE SUCH SETBACKS, America's deep warriors had no lack of targets. Sunken submarines alone constituted a rich opportunity. For Moscow, its loss of 1968 was compounded by disasters in 1970, 1973, 1986, and 1989. These sinkings consigned dozens of Moscow's missiles and nuclear arms to the ocean floor, often in the kind of chaotic disarray ideal for perusal by spy subs. After the cold war, Russian officials pulled aside part of the veil surrounding the disaster of October 1986, in which a Yankee-class submarine went down in Atlantic waters three miles deep, and revealed publicly that its demise had scattered a vast armory across the wilderness of the seabed. Of all the sinkings, this one was said to involve the greatest amount of nuclear gear, including two nuclear reactors, two nuclear torpedoes, and sixteen long-range missiles topped by nuclear arms. It was a gold mine just waiting to be worked. And it materialized during the crusading days of the Reagan Administration, a time when American officials were eager for opportunities to engage Moscow and investigate its arsenal.

For the United States, no other theater of the cold war offered comparable riches. No Russian bombers ever flew to American bases with hijacked nuclear arms. No missiles suddenly fell into the hands of Washington analysts. No submarines ever defected, contrary to the wishful thinking of *The Hunt for Red October.* By contrast, the bottom of the sea was extravagantly endowed with all the paraphernalia of war. To deal with the steady flow of lost material, the Navy set up a special processing center in suburban Washington, where the nastiest parts of Moscow's war machine were picked apart and analyzed.[62]

The details may never be known. But as the veil began to lift in the 1990s, some analysts suggested that the shadowy feats in the deep sea were the West's greatest coup in four decades of spying on Moscow, eclipsing the more glamorous exploits of human agents and spy satellites.[63]

It is unclear when Moscow caught on, belatedly, it would seem. After it did, the deep warriors were happy to gloat. In one instance, a lost piece of Soviet gear was picked up off the seabed and carefully copied—but only in its external appearances—and the replica was then placed back gingerly on the seabed. If Soviet forces came upon the item, it would appear safe and undisturbed. But retrieval would show that it was actually a dummy with a note inside telling Moscow exactly where to go.[64]

"THERE'S A HELL of a lot of stuff that went on," Craven told me as we sat at his home above Honolulu, sipping iced tea, enjoying the Pacific breeze. He leaned back in a rocking chair, his face weathered. After all, he said, "the whole object of life is to adapt."[65]

THE END CAME swiftly. After four decades, the East-West struggle that had come to dominate so many aspects of world affairs and that had managed to fill superpower arsenals with nearly seventy thousand nuclear warheads suddenly lost its momentum. A treaty was signed in 1987 that for the first time reduced nuclear arms, foreshadowing a period of more cuts and political upheaval. In 1989, the Berlin Wall fell. The Soviet experiment ended in 1991 as red flags atop the Kremlin were lowered. After three-quarters of a century, the workers' utopia had abruptly ceased to exist.

The cold war had been fought at a cost of trillions of dollars, giving birth to history's largest and most carefully guarded storehouse of scientific advances, including many dedicated to deep exploration. For decades, only sovereign states in possession of big navies and the very best equipment had been able to probe the darkness below with vigor, doing so mainly in the interest of science and warfare.

Now, as the Soviet Union collapsed and global tensions began to ease, the secretive arsenals of the deep warriors were thrown open, even as navies around the globe restructured their fleets and forces to take on new kinds of jobs. Out of the shadows came old gear, methods, and exploratory craft that created undreamed-of opportunities for a number of civilian fields. Not everything was released. Far from it. But enough was made public to leave deep exploration changed forever. Significantly, the wave included thousands of undersea experts who, as budgets dropped after the cold war, left the military for jobs in academia, industry, and the civilian side of government, often eager to advance the art of deep inquiry.

In short, it was a major if quiet conversion of military swords into civilian plowshares, one that shook the foundations of civil oceanography. The newly empowered civilians tended to pursue things that the military had long ignored and thus rapidly accelerated the overall pace of discovery in the dangerous art of deep exploration.

A key factor behind the peace dividend was a historic shift within the United States Navy that changed the focus of its fighting forces and intelligence work from the high seas to shallow waters, which consequently reduced the need for deep expertise. Years in development, the shift was formalized in 1992, just as the cold war and the Soviet Union began to fade from memory. Known as ". . . From the Sea," the new plan signaled a move away from operations *on* the sea toward power projection *from* the sea to influence events in the world's littoral zones, or coastal areas, where naval forces could strike, often with impunity. Put simply, the aim was to influence events on land, especially during regional conflicts in the developing world. The new strategy made irrelevant much of the costly infrastructure that had been built up over the decades of the cold war to probe the deep and spy on the Soviets there.[66]

Today the peace dividend is sometimes direct, as when governmental gear is sold, leased, or loaned to individuals and companies and becomes an exploratory force unto itself. Other times its nature is catalytic, as when military spinoffs boost the effectiveness of existing deep resources, including those of industry, academia, and government. Such amplification is felt especially in the commercial sector, which is already strong because of two early phases of development. It first boomed in the 1960s as an adjunct to the governmental push into the deep and advanced again in the seventies when the skyrocketing price of crude oil after the Arab embargo drove large companies to sink billions of dollars into offshore exploration, speeding the evolution of some kinds of undersea equipment, including robots. Because industry's gear tends to go shallow, and the military's deep, their strengths are often complementary.[67]

The most spectacular example of what raw naval power can do in civilian hands came during the closing days of the cold war with the discovery of the *Titanic,* the world's most famous shipwreck. Shrouded in darkness on the seabed for nearly three-quarters of a century, the luxury liner was discovered by a French-American team led on the American side by Robert D. Ballard, a civilian oceanographer at Woods Hole, who designed, built, and ran deep exploratory gear for the Navy. In 1985, after completing a secret mission to map *Scorpion*'s debris, Ballard joined with French colleagues and lowered Navy robots through more than two miles of water to find the lost liner. In 1986 he explored it with *Alvin,* photographing the hulk and documenting it as a relic of Edwardian affluence.[68]

Enterprising and ambitious, Ballard, a former Navy officer, turned himself into a major architect of the peace dividend. In showing the world how civilians could use military tools for deep exploration, he influenced American policy and similar developments abroad. He also reaped some of the dividend's greatest rewards, giving him the opportunity to explore a whole series of remarkable worlds.

Ballard in 1989 lowered a Navy robot into the deep and succeeded in finding the Nazi battleship *Bismarck,* lying nearly three miles down on an undersea mountain in the eastern Atlantic, a mass of deteriorating guns and fading swastikas. In 1992 he sounded the South Seas with the Navy's *Scorpio* robot and *Sea Cliff* submersible, finding a total of fourteen American and Japanese ships lost in the battle of Guadalcanal. In 1993 he maneuvered the Navy's *Jason* robot across the Celtic Sea to probe the *Lusitania,* whose torpedoing in 1915 by Germany killed more than one hundred American passengers and helped bring the United States into World War One.[69]

Ballard's greatest coup was winning access to the Navy's *NR-1,* the world's smallest and deepest-diving nuclear submarine, the one that can roll across the seabed on wheels. For Ballard it was a dream come true. He showed me the craft in 1995 as he prepared to crawl across the belly of the Mediterranean to bare the secrets of ancient trade, hunting for Roman wrecks, amphoras, pottery, lamps, anchors, cooking pots, coins, and even whole wooden ships preserved in the seafloor's mud.[70]

Like something out of a spy novel, the *NR-1* lay moored in the Thames River of Connecticut, low in the water, ominously dark and wet, its menacing air fitting the gray day. Most of it was concealed under water, with only the superstructure visible. We ran down a gangplank through icy rain to clamber on board. It was a sign of the times that I was there, no national journalist having been allowed inside before.

The interior was warm and well lit, humming with machinery and gleaming with pipes and electronic gear. It took only a few steps up and down the narrow corridor to go past rows of bunks, a toilet, and an eating alcove. Off limits at the rear was the nuclear reactor. The most spacious part of the sub was its forward control area, where two padded chairs faced steering wheels and a colorful blur of switches and glowing monitors, including ones for viewing the seabed. A circular depth gauge, high on the console, had been covered up and was unreadable.

Like a false wall in a haunted house, the floor behind the control center lifted up to reveal a hidden well in which observers lying on their stomachs could peer out of three small portholes. I crawled down. Nothing was visible but muddy water. From a briefing, I knew the viewing area outside the ports was contiguous with the big arm, so observers could watch and control its movements.

On the main deck, Ballard wandered up and down the corridor, pointing out the sub's systems and secrets. "Isn't this unbelievable?" he said, beaming with pride. "It's perfect for what I want to do. This is an unbelievable opportunity."

BALLARD'S FINDING of the *Titanic* was the start. But he was not alone in exploring the new field's possibilities.

Much of the action centered on lost ships, especially those carrying gold. One of the first to be located was the *Central America,* a sidewheel steamer heavy with riches from the California gold rush, which had gone down in a hurricane off South Carolina. In 1988, more than a century after its loss, the shipwreck was found under a mile and a half of water by a team of entrepreneurs who had taken advantage of the developing synergy in civil and military techniques. The wreck was found with the aid of a mathematical search method pioneered by Craven for the Navy. And it was picked apart by a huge, bottom-sitting, six-ton robot based on a mix of military and industry gear. In the end, the team recovered more than a ton of gold, including hundreds of coins and bars.[71]

The initial feats of the peace dividend were largely ad hoc affairs, coming in dribs and drabs. But in the early 1990s, as East-West cooperation held firm despite any number of crises and flare-ups, the transfer was formalized and often encouraged at the highest levels of government. Russia, facing severe economic hardship, began selling huge amounts of gear and renting whole fleets. Moscow's crown jewel, the *Akademik Keldysh,* began to sail the world on numerous jobs for foreigners, its twin *Mir* submersibles able to dive down nearly four miles. Surrounded by utter darkness, the Russian craft filmed the *Titanic*'s hulk, studied mysterious deep life, and hunted for sunken Spanish gold. In the United States, Congress as well as the Bush and Clinton administrations made forceful declarations of policy to spur the conversion of Navy assets into civilian ones or, short of that, to allow their dual use by civilian and military experts. The aim was to help old naval bureaucracies find new identities and to promote the transfer of high technologies into the private sector as an aid to investment and job creation. In addition, the Navy often found that its own interests were served when contractors diversified into private work, allowing them to remain solvent despite an overall drop in naval orders.[72]

The organized sharing began in earnest in 1991. The Navy that year signed an accord with the National Oceanic and Atmospheric Administration, a Federal agency, which gave NOAA formal access to a small fleet of naval robots and manned craft, including the *NR-1.* The plan was an expansion of the kind of liberality the Navy had shown in sharing *Alvin* with the science community ever since its commissioning. The new program began in 1992 off the Hawaiian Islands in an expedition featuring the *Sea Cliff* submersible and the *Advanced Tethered Vehicle,* an undersea robot the size of a small truck that can dive nearly four miles deep. Probed were such things as the Loihi seamount, where deep volcanic fury is giving birth to a new Hawaiian island. In 1994, a voyage off Oregon featured the same robot and *Turtle,* the Navy's other *Alvin* look-alike. In deep waters thought to harbor no volcanic activity, team members were surprised to find fields of smoky vents that formed gnarled mounds of minerals. Formal sharing culminated with wide access to one of the Navy's preeminent robots, *Jason,* which Ballard designed and made during the 1980s buildup. In 1996, the seven-foot-long robot began a new life as part of the civilian governmental fleet available to university and Federal scientists. Its first mission, to the Mid-Atlantic Ridge, was to hunt for hidden volcanism.[73]

Even attack submarines are now shared. In 1993, civilian scientists aboard one dove into the arctic depths to investigate the region's ecology, water chemistry, gravity anomalies, ice composition, and deep geology. After that test cruise, the Navy in June 1994 signed an accord with Federal civilian agencies to start a cooperative program of annual expeditions under the arctic icecap, a wilderness long off limits because American and Soviet submarines during the cold war hunted one another beneath the ice.[74]

One of the biggest windfalls is not a vehicle but a small fortune in deep equipment—SOSUS, the Sound Surveillance System of the American Navy. Moored on the seafloor, this global network of deep microphones had spied on Soviet ships and submarines over distances of hundreds and thousands of miles during the cold war. Its total cost came to $16 billion. In 1993, the Navy began letting civilian scientists at NOAA listen in so they could track seaquakes and other geological action hidden in the deep. It was also used by university scientists to track whales and other marine mammals. In 1996, the Navy began privatizing part of the spy system to a foundation intent on building an acoustic observatory in the Atlantic to study such things as the sounds of seaquakes, fish, whales, and other marine creatures. Amos S. Eno, executive director of the National Fish and Wildlife Foundation, called the step "a breakthrough in terms of applying top technology developed by the military to solve day-to-day natural resource problems."[75]

In all, the peace dividend has freed up hundreds if not thousands of military devices and technologies, some of which I saw displayed in January

1993 at a conference and trade show in New Orleans sponsored by the Marine Technology Society, a Washington-based professional group with close Navy ties. At the New Orleans Hilton, some ninety exhibitors vied to sell a vast array of underwater gear, much originally developed by or for the Navy. It was a high-tech bazaar packed with advanced lights, sonars, robots, acoustic pingers, electronic cameras, mechanical arms, and lifting devices.

The star of the show was an innovative camera known as a laser line scanner that was making its public debut after years of Navy secrecy. Rival models were displayed by Westinghouse Underwater Laser Systems of San Diego, and Applied Remote Technologies, Inc., of San Diego. Both lasers worked by sweeping thin, concentrated beams of light back and forth through deep water, tracking the reflections off objects, and electronically assembling the information into a computer image.

The resulting picture was startlingly clear, judging from the examples on display. Even more impressive was the range of operation. The laser could see up to ten times farther through water, even cloudy water, than the regular mix of floodlights and cameras. Moreover, the usual glare of back-scattered light from organic debris was virtually eliminated because the laser beams were so narrow. In short, the laser cameras rendered sharp images of distant objects and terrain that previously had been lost in darkness. It was a development, courtesy of the Navy, that promised to open up large stretches of the ocean floor to visualization for the first time.[76]

The Navy itself at the conference unveiled an unusual robot that it had quietly inaugurated a decade earlier. Unlike most of the world's undersea machines, the robot had no tether. The seventeen-foot-long, torpedo-shaped craft instead had batteries for ten hours of power and advanced computers that directed wide searches of the seafloor. Amazingly, the robot was able to keep in close contact with surface controllers (even sending back seafloor pictures) with pulses of sound.

The Advanced Unmanned Search System, or *AUSS,* as it is called, is designed for depths of up to 20,000 feet, or 3.8 miles. Its great advantage is speed. *AUSS* (pronounced A-use) can observe up to three square miles of seafloor an hour. At that rate, it could have conducted the *Scorpion* search of 144 square miles in a day, instead of the months it actually took. And great speed, of course, means low cost. Despite its remarkable powers, the robot had been thrown into governmental limbo by budget cuts, according to a Navy brochure. The handout ended by encouraging interested parties to contact the Naval Ocean Surveillance Center in San Diego for further information. The exhibit and brochure had the melancholy air of a going-out-of-business sale.[77]

At sea, the spinoffs are working not only individually but sometimes in pairs and trios to produce unusual synergies and advances. *NR-1* dove to the depths of the Gulf of Mexico in 1994 with a laser scanner tied to its belly, allowing scientists from Texas A&M University to illuminate a large brine pool dense with chemosynthetic mussels and oil seeps thick with tube worms. Until recently, such realms of life were wholly unknown and unexpected.[78]

Commercial beneficiaries of the peace dividend include contractors once dependent on the American Navy, who have become increasingly free to work for anyone with the cash. A good example is Oceaneering Technologies, Inc., of Upper Marlboro, Maryland, a giant of deep recovery located on Chesapeake Bay. The company built its reputation on Navy and Federal service (for instance, in 1986 helping retrieve the wreckage of the space shuttle *Challenger* from waters off Florida) and sometimes moonlighted on civilian jobs. But in 1990, as the cold war ended, the company made a decision to aggressively court the open market, and its private work rapidly soared to become 70 percent of its revenues in some years. Its many deep-diving robots include *Magellan,* a tethered vehicle that can dive more than four miles deep, and the *AUSS* tetherless robot, which it acquired from the Navy.

Throughout the 1990s the company did a number of remarkable but little-publicized jobs, often working for foreigners. In one, it plied the Arabian Sea to help locate and salvage from a depth of nearly two miles the *John Barry,* an American ship loaded with tons of silver that a German submarine had sunk in 1944. Another job involved an Italian DC-9 that had mysteriously plunged into the Tyrrhenian Sea in 1980, killing all eighty-one people on board. The company found the shattered plane more than two miles down and brought up some five thousand bits of its remains, which Italian investigators pieced together to address the crash puzzle. A job of unusual delicacy involved the *Lucona,* a freighter that inexplicably had sunk in 1977 in the Indian Ocean, killing six crew members. The ship's Austrian owner claimed $18 million in lost gear. But doubts swirled around the affair. In the 1990s, the recovery firm was hired to do some detective work, finding the shattered hulk nearly three miles down. Grisly pictures brought to the surface by the company suggested that responsibility for the loss lay not with the hand of God but a massive bomb blast. An Austrian court agreed and convicted the owner of insurance fraud and murder.[79]

One of the peace dividends is so comprehensive that it is shedding light on the entirety of the dark seabed. In 1995, scientists from NOAA and the Scripps Institution of Oceanography in La Jolla, California, released a striking multicolored map twelve feet wide and eight feet high that gave the first good overview of the rocky seabed. Previously, such maps were generally made with the aid of surface ships that bounced sound waves off the bottom to glimpse the wilderness below. Over the decades, mapping ships had managed to acoustically image less than 5 percent of the world's ocean floors, with coverage gaps often the size of Kansas. The resulting renderings were far less credible than they appeared, since artistic license often took the place of hard information.

The new map is largely based on declassified Navy data. From an orbit five hundred miles high, a Navy satellite in the 1980s had made gravity measurements over the world's oceans as part of a quiet effort to increase the accuracy of long-range missiles fired at sea from submarines. With the Navy's cooperation, the scientists turned declassified data from the $80 million mission, as well as readings from a European satellite, into a global map of the seabed that revealed all kinds of plains, fissures, ridges, mountains, volcanoes, and riddles heretofore hidden in the sunless depths. Still, the new map can reveal no feature smaller than six miles in diameter or one kilometer high. Anything smaller is invisible. In short, the Moon, Mars, and even Venus with its thick veil of clouds are more familiar to scientists. Even so, the new map is far better than any predecessor and thus has excited much interest. Fishermen are searching it to find seamounts, which cause upwellings of deep, nutrient-rich waters that attract aquatic swarms. Oil companies are using it to hunt for the kinds of rocks that indicate the possibility of oil fields. And scientists are pondering it for clues to the seabed's origins and how newly identified mountain chains might affect currents in the deep ocean, and thus global climate.[80]

So, too, the overseers of the naval storehouse are slowly releasing other physical data about the deep gathered inconspicuously during the cold war. The riches include global readings on magnetics, bathymetry, sediments, salinity, temperature, and bioluminescence. Over the decades, the American Navy deployed hundreds of ships and submarines and satellites to amass such information. Among other things, its release will aid scientists studying the global environment by helping them judge change over time.[81]

In the end, the cold-war spinoffs have had some of their most dramatic impacts on venturesome individuals looking for new ways to explore the darkness below.

GRAHAM HAWKES in 1993 sat in a small shop overlooking San Francisco Bay, surrounded by tools, a drill press or two, and a small, sleek white vehicle that looked very much like a miniature jet fighter. It had stubby wings and a transparent nose. Inside it was just big enough for one person. Contrary to its jetlike appearance, it was meant to carry a passenger not up into the blue sky but rather down into the sea's blackness, to fly him there. Hawkes called it *Deep Flight.*[82]

He embarked on this inner spaceship as one of the most successful designers in the field of commercial submarines and robots. Hundreds of his small *Phantom* robots prowl the global deeps. British by birth, fascinated by airplanes as a boy in London, Hawkes had learned marine engineering in the 1970s amid the boom in offshore oil exploration in the North Sea. He developed a number of mini-submersibles and eventually branched out into robots. In the United States, his exploratory interests were strengthened by long collaboration with Sylvia A. Earle, a marine biologist and underwater explorer who eventually became NOAA's chief scientist. Together, the two founded Deep Ocean Engineering, based in San Leandro, California. In the 1990s, he became a main beneficiary of the Navy's spinoffs and a good example of their catalytic effect. He culled them carefully, traveling repeatedly to San Diego, the mecca of military undersea work. Though bespectacled and not appearing like the venturesome type, Hawkes intended to use them personally, taking a giant step that was unthinkable for civilians just a few years earlier. His goal was to travel to the Challenger Deep, nearly seven miles down, the ocean's most reclusive spot. He wanted to go there not as the Navy had done in 1960, to gird for war, but to probe its mysteries and luxuriate in its peacefulness.

"As you go down in a vehicle, the ocean goes from light blue, through dark blue, to indigo, to blackness," he said, his hand tracing the arc of descent. "It's a beautiful transition. If you're really lucky, you get into a blackness that is really black and then cut out all the lights and fall through a bioluminescent cloud of plankton. Movies like *The Abyss* and *Jaws* make people think the ocean is threatening. It's not. It's very tranquil. Afterward you get yanked out into blinding sprays, waves, and a heaving ship. But down there it's peaceful. You never want to come back. Ever since my first dive, I've gone back every chance I get."[83]

Deep Flight was a prototype of the vehicle Hawkes wanted to pilot on the Challenger dive. Its body made of reinforced fiberglass and its nose of acrylic, the unfinished prototype was to fly as deep as four thousand feet, which would allow Hawkes to set a new world record for a solo dive. (He holds one record already.) The secret of *Deep Flight II,* the intended successor and his ultimate diving machine, was alumina-ceramic casings that the Navy designed and quietly developed over the years but in the nineties began to advertise far and wide. The casings are as sturdy as steel or titanium but weigh far less. Hawkes and the Navy engineers calculated that they could easily withstand the rigors of the trench, where seven miles of water crush down with a pressure of eight tons per square inch.

The lightness of the casings made them neutrally buoyant, so his submersible could be very small. *Trieste* had a huge superstructure filled with gasoline that counteracted the great weight of its metallic crew cabin, and

Alvin, Turtle, Sea Cliff, Mystic, and *Avalon* did the same thing with thick layers of syntactic foam. But Hawkes needed none of that added buoyancy—or bulkiness. His design was a uniting of form and function, not a clash between the two. His sleek craft, based on a single large ceramic casing topped by a transparent hemisphere, is to be neutrally buoyant (neither rising nor sinking). It is to be small and fast, its descent powered. Unlike its many predecessors, all its structures are to be intimately tied to the job of exploring the deep ocean.[84]

"The pieces are in place, courtesy of the cold war," Hawkes told me. He said the whole field was accelerating because of the release of military gear and ideas that were aiding the creation of piloted submersibles as well as new kinds of deep robots.

At the beginning of the twentieth century, machine-shop daredevils flew into the sky on a shoestring and, almost as an afterthought, founded the field of aviation. Hawkes said something like that was now happening with undersea gear at the dawn of the twenty-first century. Thanks to the end of the cold war, a wave of advanced technology was sweeping through the civilian worlds of deep engineering, leading creative minds to produce all kinds of new machines. It might, he said, lead to a new age of exploration.

"In the early days of aviation, no one saw the future clearly," Hawkes said, sitting by his undersea jet, toying with a wrench. "Even the proponents of aircraft said, 'Maybe we'll carry mail.' That was it. They had no idea of where it would lead. Right now it's the same kind of thing. I don't think we have a realistic idea of what's ahead, except that it's going to be big, whatever it is."[85]

THREE

Garden of Eden

WE SAILED OUT the mouth of the Columbia River into a Pacific Ocean that at least in one region was living up to its good name. It was an October morning in 1993. The water had an agreeable chop and the breeze was fresh. The sun peeked through a hazy autumnal sky. Our ship, the *Atlantis II,* a little more than half a football field in length at 210 feet, was alive with last-minute preparations and the electricity that marks the start of an adventure.

Much of the action was located aft and centered on *Alvin,* the submersible that, over the next two weeks, was to repeatedly carry three people down a mile and a half into icy darkness. The sub, bright white except for a conning tower of red, was about twenty-five feet long and ten feet high. Though small, it worked like a powerful magnet to attract not only workers but groups of interested bystanders. It worked on me, too. The only journalist among fifty-two people, I was curious about the celebrated craft, which had achieved so many firsts over the decades. I had been promised a chance to join the ship's scientists on one of their daylong dives to the bottom of the sea, schedule and weather permitting. The conditions were not insignificant as storms could arise quickly in October and make the North Pacific a hash of big waves, stopping all work for days on end. The biggest and deepest of the planet's oceans, the Pacific could also be the roughest. No matter what prompted a delay, research would take precedence over journalism if the scientists fell behind schedule and a significant number of them failed to get down to the bottom to do their work. But I was hopeful. If lucky, I would climb aboard *Alvin* one day soon to make my first descent into the blackness of inner space.

My optimistic frame of mind was bolstered by the fair day and the general region, which was new and interesting to me. Most of the scientists were from the West Coast—the University of Washington in Seattle, the United States Geological Survey in Menlo Park, California, and the Pacific Marine Environmental Laboratory of NOAA, located in Seattle and Newport, Oregon. Many knew the area and its lore. For me, a Midwesterner by birth, an Easterner by reason of work, it was as invigorating as the ocean air. Behind us lay the rugged coast and mountains, the big vistas and small towns. Astoria, our port at the mouth of the Columbia River, in Oregon, had bustled with rough charm. When all this was wilderness, I was reminded, Lewis and Clark rested here after traveling thousands of miles to explore the continent at the behest of President Jefferson, coming down the Columbia to make winter camp within sight and hearing of the Pacific. And it was here that John Jacob Astor, the New York businessman, set up Fort Astoria less than six years later to further his ambitions in the fur trade and to dream of a Pacific empire.

It was no exaggeration to think that our own voyage might advance this history of exploration and enterprise, though in very different ways. We were to make first-of-a-kind observations of deep volcanism and study how its enormous heats could transform the icy wilderness of the seabed into oases of extravagant life. The hot springs and vents—emitting water rich in microbes and minerals, erecting tall stone chimneys, feeding exotic creatures in densities far greater than anywhere else in the deep—had been a scientific sensation for more than a decade. Yet they were still mysterious. No one knew with any degree of certainty how such oases grew or why they arose where they did. One question was under what conditions the violent heat from a volcanic outburst settled down into the kind of steady flow of hot water that could nourish a virtual jungle of undersea life.

Our chances of making headway seemed reasonably good if the ingenuity shown in the genesis of our voyage was any clue. It began with one of the cold-war spinoffs, the Navy's global network of undersea microphones. After the war, experts at the National Oceanic and Atmospheric Administration, or NOAA, wondered if the spy apparatus had any scientific uses. They quickly learned that the Navy had been filtering out the sounds that geologists found most interesting—the low-frequency vibrations made by seaquakes and undersea volcanoes. Scientists at NOAA's lab in Newport, Oregon, were allowed to examine the signals in 1991 and 1992, finding them hundreds of times more sensitive than land-based apparatus for monitoring deep-sea disturbances. The scientists pressed for more access and won it.

On June 22, 1993, they began getting the data directly so they could track undersea action as it happened. Four days later, in a wonderful stroke of luck, they detected a swarm of seaquakes just off the Oregon coast on the Juan de Fuca Ridge, one of the oceanic spots where the seabed periodically splits open and spews volcanic fire to create new crust. The tremors, slowly moving northward, shook a region some thirty miles in length, going from a northern latitude of 46 degrees, 10 minutes, up to about 35 minutes. The implication was that the quiescent bottom a mile and a half down had come alive with a rush of hot lava. In a first, the site was rapidly reconnoitered by a small flotilla of ships lowering everything from water bottles to a million-dollar robot. What the scientists found was a volcanic gash at least four miles long, and probably much longer, that spewed hot water at one end and molten lava at the other. That August, Stephen R. Hammond, head of the NOAA program that made the discoveries, told reporters in Washington that the Navy gear had opened "a whole new window into a dark ocean."[1]

I had been at the Washington news conference and filed a story, fascinated by the Navy gear and its civilian work. And more was to come. The NOAA experts said scientists in *Alvin* would do the first direct human reconnaissance of the site, probably two months hence, in October. The sub, they added, would be better than robots for sampling and certain kinds of experiments. Eager to join the expedition, I pestered all parties involved in the plan, including the sub's operator, Woods Hole, as well as the two Federal agencies funding the dives, NOAA and the National Science Foundation, both of which had programs to study deep volcanism. The expedition's chief scientist, John R. Delaney, a geologist at the University of Washington and a rising star of oceanography, graciously welcomed my interest but warned that the weather could be rough. Boats smaller than the *Atlantis II* were often lost off the coast in the late season, he said, adding that the expedition was going only because of the possibility to do uncommon science.

I hesitated. Clearly the trip was a gamble that could end with no story. A personal factor also weighed on me. My wife, Tanya, and I had just had a new baby, Juliana, and my leaving home for two weeks would be hard on Tanya and our older children, Max and Isabelle. Even the simple act of calling home daily, easy on most business trips, would be difficult or impossible at sea. But the opportunity seemed too good to pass up. The expedition was a first. My editor at *The New York Times* was supportive. And my wife gave her blessing. So I packed my bags, bringing along books about geology and undersea hot springs, as well as an assortment of pills and potions for seasickness.

My musings on the afterdeck were interrupted by a woman in her thirties with long brown hair. Smiling, clad in jeans and a flannel shirt, she introduced herself as Cindy Van Dover, a biologist at Woods Hole I had talked to on the phone as a reporter. Cindy was a pioneer. Breaking an all-male tradition, she had become the first woman *Alvin* pilot, learning its intricacies and taking the tiny sub down scores of times. Her interest in the deep was so strong that she won a Ph.D. and was now a biologist specializing in seabed creatures. Among her discoveries was that hot vents shine with a diffuse light, too faint for human eyes, that seems to attract some deep fauna, perhaps as a way to find biologically rich areas in the inky darkness.[2]

As people bustled about *Alvin* on the fantail, readying it for two weeks of diving, Cindy explained some of its gear. Most visibly, the sub had two large robot arms made of burnished metal, each some six feet long and tipped with mechanical fingers. The arms were pulled back, elbows high, making the sub look a little like a praying mantis. *Alvin* was distinguished from all regular submarines by its three small and very thick windows, through which passengers and pilot could peer. One window, the pilot's, looked forward, and the others off to the sides. Viewing the deep was aided by more than a dozen lights and cameras clustered on or near the sub's bow. Altogether there were four television cameras linked to internal monitors and tape recorders. Two thirty-five-millimeter still cameras with large film magazines pointed forward, able to take pictures automatically every few seconds. Lights seemed to be everywhere on the sub's front end. High up was one said to be filled with thallium iodide, whose greenish glow was especially good at penetrating deep water.[3]

Low on *Alvin*'s front end, below most of the lights and cameras, was a big metal rack being fitted with instruments, power cables, sampling tubes, and bottles. It was all science gear, I was told. The contents of the rack could readily be changed, allowing each dive to be customized. Remarkably, much of the gear was being lashed down by rubber bands hooked onto plastic milk-carton carriers, which in turn were connected to the metal rack. Though a bit ramshackle, the setup bespoke the universal rule of engineers and tinkerers—whatever works. More than a quarter century of experience and experimentation had clearly given *Alvin* and its keepers a relaxed, unceremonious air that emphasized results over appearances.

Bent over this mass of equipment was a barrel-chested scientist in jeans and sandals, his beard shot with gray. He was introduced to me as Marv Lilley of the University of Washington. A geochemist, he specialized in analyzing seawater. Marv explained the samplers, four of which were so far secured to the basket on *Alvin*'s front end that held science experiments. Each was a titanium bottle topped by a coiled spring that, when triggered, pulled back an internal plunger to suck in hot volcanic fluids, much as a physician's syringe draws in blood or medicine. The bottles were connected by some plumbing to a telescoping tube that extended up to eight feet long. The tube had no ability to move on its own and was instead positioned for sampling by the sub's robot arm.

Over the years, I learned, the tip of that long metallic tube had repeatedly been nudged deep into chimney vents at the bottom of the sea and sampled water as hot as 750 degrees Fahrenheit, or 399 Celsius, a stunning temperature for a region long envisioned as terminally cold. At heats less blistering, the tube had also extracted clouds of microbes from hot vents, with the life forms turning out to be among the planet's most primitive and most closely related to the earliest forms of life.

I stepped back from the sub. No land was visible—just a few birds and foam from the ship's propellers and the restless surface of the sea, alive with light. Far in the distance was the endless horizon, cutting sea and sky in two with remarkable precision.

In contrast, the milling scientists aboard the ship tended to be soft and blurry, like a tribe of aging hippies, a mélange of beards and sandals and T-shirts. Many of the shirts were mementos of earlier voyages, adorned with maps, ships, robots, subs, flags, and creatures. SAVE THE TUBE WORMS, read one. EVENT DETECTION AND RESPONSE, said another. Whimsy tended to mark T-shirt wordplay. Cruise names were often meant to echo the things under study, as with the Ad*vent*ure voyage. Our own cruise was part of the Juan de Fuca E*vent* Response, which had already generated its own T-shirt. Some of the scientists were clearly colleagues from previous campaigns, judging from their similar T-shirts and sloganeering. An anthropologist from another world would no doubt judge oceanographers a distinct race of *Homo sapiens* by virtue of their physical similarities and dress. In contrast to the scientists, members of the *Alvin* support team, who lived aboard ship rather than just visiting for occasional cruises, tended to be lean, mean, and young. Most were clean-shaven.

One of the previous campaigns, I learned, the Adventure voyage, had discovered by accident in April 1991 a recent flow of deep lava. *Alvin* had been diving in the Pacific off Central America in an area of volcanic activity known as Ten North, after its degree of northern latitude. To the surprise of the scientists, fresh lava was found everywhere, apparently having erupted just a few weeks earlier. The ocean floor in some places was a spectacle of death. Previous expeditions to the area had documented rich biological communities on the volcanic rift, including large groups of mussels, clams, and tube worms. But the scientists aboard *Alvin* found that these creatures had either vanished under fresh lava or were still partly visible, charred and lifeless. One site was dubbed the Tube Worm Barbecue.[4]

That discovery off Central America had been followed up carefully but with some difficulty, since the site was far from major oceanographic ports and centers. In contrast, we were speeding toward an eruption of the Juan de Fuca Ridge that was a mere 250 or so miles off the populous West Coast of the United States. Our transit time was estimated at less than twenty-four hours. Since the eruption, that nearness had allowed the site to be probed by a total of five teams, all rerouted from already planned expeditions in the region. Five expeditions in a little more than three months. Ours, the sixth, was the first dedicated to eruption study. The nonstop nature of the analysis was important, given our goals. Ten North had been a killing field where scientists focused on the phenomenon of rebirth. Our site, on the other hand, appeared to have no established oases and offered a chance to study the subtleties of how a lush ecosystem of the deep originated and evolved. We were focused on the riddle of birth.

After lunch the scientists spent a couple of hours stowing gear and preparing equipment. Then they returned to the mess. The expedition leaders gathered where it widened into a lounge with chairs, coffeepot, television, and refrigerator. In the mess proper, available seats quickly filled, so that some of the twenty scientists and ten crew members in attendance for the meeting were forced to stand. The discussion was led by John Delaney, the chief scientist and a natural leader. He was in his early fifties. Tall and self-assured, by turns serious and sarcastic, quick to joke and orate, he began by stressing the importance of the venture, saying we had a historic opportunity to study deep volcanism in action.

A beautiful contour map of the thirty-mile seabed region had been hung on the wall behind him, embellished with false color for easy depth readings. It was a blur of bright oranges, greens, and purples, surrounded by dark blue. The cooler the color, the deeper. The southern part of our study zone began on the flanks of an undersea volcano whose summit lay under a mile or so of water. The study zone's depth increased from 1.3 miles on the volcano's flanks to 1.6 miles at the northern extremity, all in all a gentle downward grade. The underwater volcano was known as Axial, since it was located right on the ridge axis, and the quake region was named Coaxial since it was located adjacent to the volcano on the ridge. The map showed that the Coaxial region had the typical look of a volcanic spreading center—hills and mountains on either side of a narrow rift valley, where lava periodically oozed and erupted.

The quakes had begun in the southern Coaxial section near the volcano and run northward, John said, culminating in a northerly lava flow. The previous expeditions found the lava fresh and glassy and discharging warm water measured at up to 120 degrees Fahrenheit. The lava in places was covered by clumps of white and orangish bacteria. The surrounding water in some areas was fuzzy with bacterial clouds. More southerly, in the central part of the study zone, the last robot dive had found no fresh lava but instead had run into blizzardlike storms of bacteria swirling about, the microbial snow so thick in places that it accumulated in small drifts. Despite this potential feast of nourishment for undersea fauna, the previous expeditions had discovered no animals feeding on the abundant microbes. It appeared to be a clean biological slate. A question for us was whether this outpouring of bacterial material would become the first link of a large food chain, and if so, how fast. John said our overall research plan was to go from north to south along the gash of the spreading center, moving ever closer to the volcano, making dives and observations as we went. This strategy would move us from the known to the unknown, from explored areas to relative wilderness. John stressed that because we were getting into the period of rough weather in the North Pacific, each dive had to be thorough, since it might be the last.

Scientists in their presentations often leave out the most interesting part of the story. This is because the audience knows the general outlines of the research and the shared fascination that drives it. The audience wants specifics. Unstated in John's discussion were things of mystery and wonder at the heart of all our plans and preparations. These were the undersea chimneys, the towering structures that sometimes grew over hot vents to become the hubs of otherworldly ecosystems. Simply put, they were oases of heat and life in a realm of intense cold and darkness.

John's fascination with this world had led him up and down the volcanic ridges off the West Coast. In 1991, while diving in *Alvin* on the Juan de Fuca Ridge west of Seattle, he and his colleagues had stumbled on a chimney fifteen stories high. It was the tallest one ever discovered, a kind of skyscraper of the deep. Godzilla, as the monster was dubbed, was a jumble of stone and rocky appendages. Giant outcroppings known as flanges stuck out its sides for distances up to sixteen feet. A glut of sea creatures covered Godzilla's face.

Prior to the organizational meeting in the mess, John had given me a poster he and his colleagues had drawn up for recruiting graduate students to his oceanography program at the University of Washington. Top to bottom it featured Godzilla, sketched in bright colors on a background of jet black. Hot water was shown streaming from the chimney's flanges and top. Barely noticeable off to one side was *Alvin,* its three humans concealed inside, its mechanical arm reaching out to touch the huge monolith. At the top of the poster was a passage from T. S. Eliot's *Four Quartets.*[5]

GODZILLA. In 1991, scientists discovered a volcanic chimney more than a mile down that rose some fifteen stories high and spewed blistering hot water into icy darkness. The chimney was dubbed Godzilla.

We shall not cease from exploration
And the end of all our exploring
Will be to arrive where we started
And know the place for the first time.
Through the unknown, remembered gate
When the last of earth left to discover
Is that which was the beginning. . . .

John, it seemed to me, was captivated by both sides of the volcanic riddle, by eruptions and their repercussions. In our discussions before the voyage, he had stressed the former more than the latter, saying that the expedition most fundamentally was about understanding the inner workings of the volcanic ridges and their primary processes. He said this with an animation and energy that made it clear he was intoxicated by the thought and the process of explaining it. He called the regular eruptions along the global volcanic rift "the heartbeat of the planet."[6]

This metaphor was too rich for a dry scientific paper, but it drove home the importance of the rhythmic outbursts, which over the eons had been a fundamental part of the process by which planetary crust was formed and continents torn asunder, by which the ocean itself was renewed chemically and circulated. Just as clearly, John was fascinated by the unworldly transformations brought about by the eruptions, by Godzilla and the undersea jungles. Like all the scientists on our voyage, he now hoped to discover the secrets of their formation. John's written proposal to the National Science Foundation almost twitched with eagerness. "We must make every effort to determine if high-temperature vents are present," he wrote, "and sample them if they are found." And his excitement was shared, driving a shipload of scientists to sea at a risky time of year. For many, it was their second round of investigating the eruptive site.[7]

Understanding chimneys and where they came from was like trying to understand how a caterpillar turns into a butterfly. Unless you studied all the different stages and influences, the metamorphosis was impossible to understand. All the scientists on our voyage had studied the volcanic vents for many years, some since the initial discovery more than a decade ago. But they knew only bits and pieces of them. Their knowledge was static and disconnected. Now they fairly vibrated with excitement because, for the first time in their careers, they had an opportunity to try to understand how it all fit together.

The sky that evening was a blaze of reds and oranges. On the fantail people were busy preparing electronic buoys for *Alvin*'s navigational networks. Tomorrow morning the bright yellow balls would be hooked to weights and pushed overboard into the depths of the Juan de Fuca Ridge.

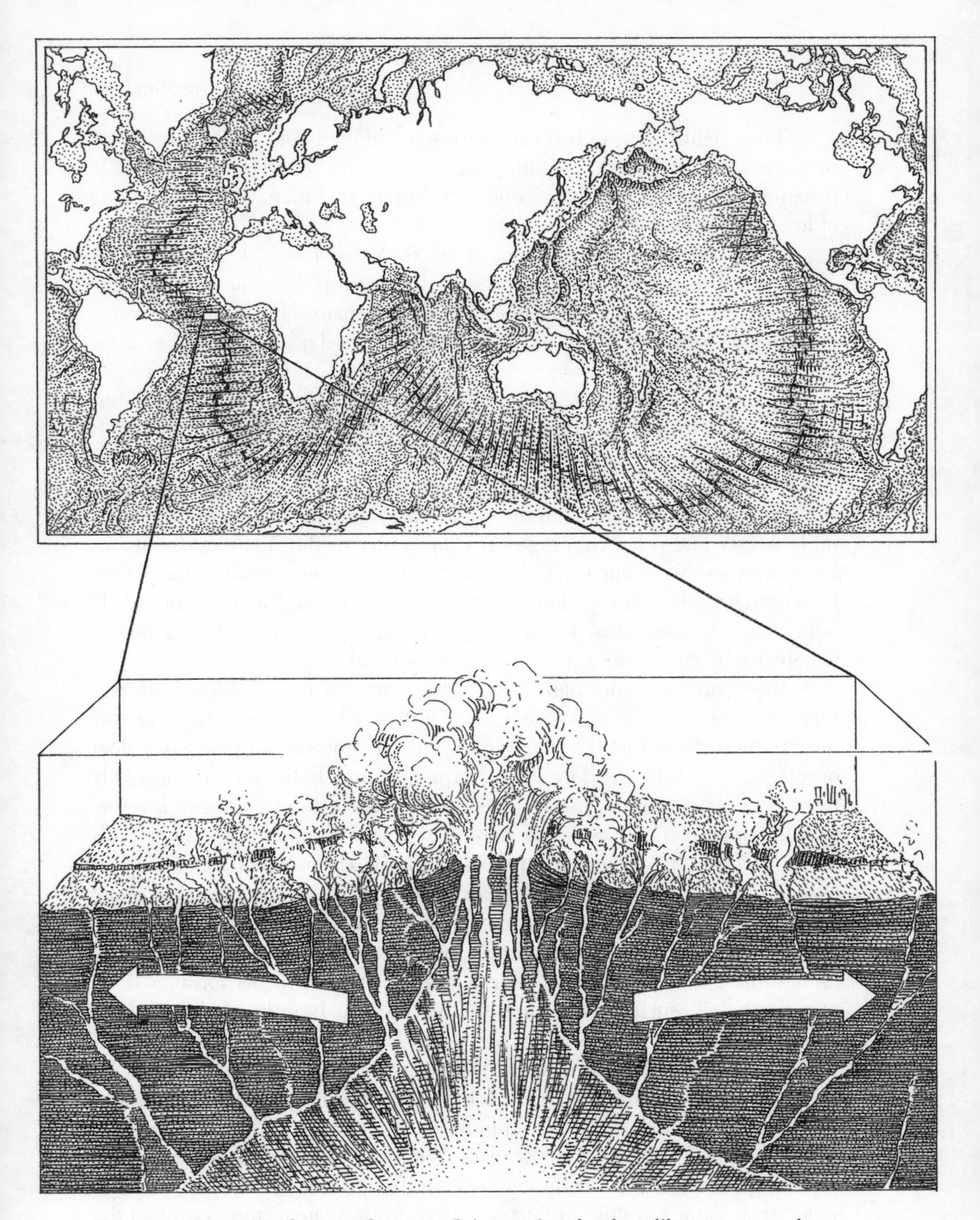

VOLCANIC RIFTS OF GLOBAL SEABED. Crisscrossing the deep like seams on a baseball, chains of undersea mountains mark long rifts where the Earth's tectonic plates diverge, where molten rock wells up to form new crust, where otherworldly animals thrive in pitch darkness, and where life on Earth may have first arisen.

That night I went to bed early with some of the books and articles I had brought along. As I read, I began to see the enormity of our subject. It was as if nature, centuries into the scientific revolution, had managed to save some of her best secrets for last.[8]

The first clues to the existence of this dark world were discovered in the 1950s as oceanographers used echo sounders to map the seafloor with some thoroughness for the first time. They found long chains of underwater mountains that crisscrossed the deep like seams on a baseball. Unknown for ages, first glimpsed by the *Challenger* expedition, the mountain ranges were clearly one of the earth's primary features, dominating the planet, winding for thousands of miles through the darkness of the deep seas. They were mostly linked up, unlike ranges on land, and they were long. All told there were forty-six thousand miles of them. If tied together into a single chain, they could encircle the Moon nearly seven times. And they were high, though not necessarily steep. The ridge running down the center of the Atlantic Ocean was found to rise slowly some two miles, like the Rocky Mountains in some places. Other ridges, like Juan de Fuca, were more modest and akin to the gentle hills of the Appalachians. Importantly, the ridges were discovered to be cut lengthwise by narrow rift valleys shaken by seaquakes.

Based on these and many other clues, scientists in the 1960s proposed that the deep is alive geologically in a remarkable way. According to this theory, the earth's crust is a patchwork of a dozen or so cool plates that float on the planet's hot core. The slow churning of the plastic interior constantly moves the plates around, just as hot water on a stove moves strips of lasagne. Only it happens very slowly. The giant plates move about as fast as fingernails grow, doing so in spasms, shaking violently. Where two plates move apart, molten rock wells up from below to form new crust and, on either side of the cleft, long ranges of low mountains. Such areas are known as volcanic spreading centers. Where two plates come together, one can slip past the other to plunge back into the hot earth for recycling, in the process forming deep ocean trenches, such as the one near the Mariana Islands, as well as high mountain ranges on land, such as the Andes. The Earth's continents are seen as mere playthings in this process, being pushed around and torn apart by the Earth's interior forces, with much of the action taking place under the sea.

After fierce debate, this revolutionary mix of ideas, known informally as continental drift and formally as plate tectonics, won wide acceptance, becoming the established wisdom by the early 1970s. The notion of *terra firma* suddenly became painfully antiquated. In a stroke, plate tectonics unified the field of geology by demonstrating the interrelatedness of all kinds of dissimilar facts, just as genes had done in biology and the big bang theory had done in astronomy. With plate tectonics, geology came of age.

One of its biggest repercussions was a radical shift in the scientific view of the deep's importance. In a generation, it went from being geologically dead and irrelevant to being the most dynamic part of the planet, one marked by a nonstop frenzy of volcanism. Suddenly, its exploration became a very high priority.

Seeing this realm up close was no easy job. The exploratory gear of the 1960s, especially that of civilian researchers, was crude at best. Sonars gave few details of the jagged peaks and gorges that made up the ridge axes. Equipment towed near the bottom for a closer view was damaged and lost at an alarming rate. Things improved in the 1970s with the onset of new gear and cooperation. A French-American team in 1973 dove on the Mid-Atlantic Ridge in tiny submersibles, including *Alvin* and the French sub *Cyana.* Preparations were aided by the American Navy, which swept the area with its high-resolution sonar, whose design was secret. Aid also came from Woods Hole, which for the occasion built a large camera sled, known as *Angus.* Its photos helped locate good dive targets.

As expected, the expedition found a rift valley in the heart of towering mountains. Unexpectedly, nearly everything there was volcanic. What the team anticipated were mainly tectonic features such as folds and faults in the rocky crust, perhaps rent by deep fissures. Instead, lava was found everywhere in a dizzying array of shapes that came to be called pillows. It had obviously welled up from the Earth's interior as the area was torn apart in the spreading process. As the fascinated team probed deeper and deeper, *Alvin* was directed down a narrow crevasse to hunt for warm water, a reasonable thing to do given the volcanic setting, but none was found. The pilot tried to back out. Nothing happened. The sub was stuck. The surrounding rocks held the scientists and *Alvin* in a death grip. With icy calm, the pilot slowly and painstakingly wiggled the craft to safety. The job took two hours.[9]

The discovery of volcanic action at the Mid-Atlantic Ridge fed speculation that such processes were universal and might be fresher at sites where two plates were spreading apart at a faster rate. Attention turned to the eastern Pacific. In 1977 geophysicists and oceanographers took *Alvin* to a mid-ocean ridge near the Galápagos Islands off Ecuador. As in the Atlantic, the Navy mapped the general area with its secret sonar to help the team get its bearings on the rocky bottom. The scientists, diving in waters a mile and a half deep, quickly found fresh lava and the first warm-water springs ever discovered in the abyss. To their great astonishment, they also found lush concentrations of life in a variety of bizarre forms.

"Debra," a geologist aboard *Alvin* called into a hydrophone, "isn't the deep ocean supposed to be like a desert?"

"Yes," came the answer from above.

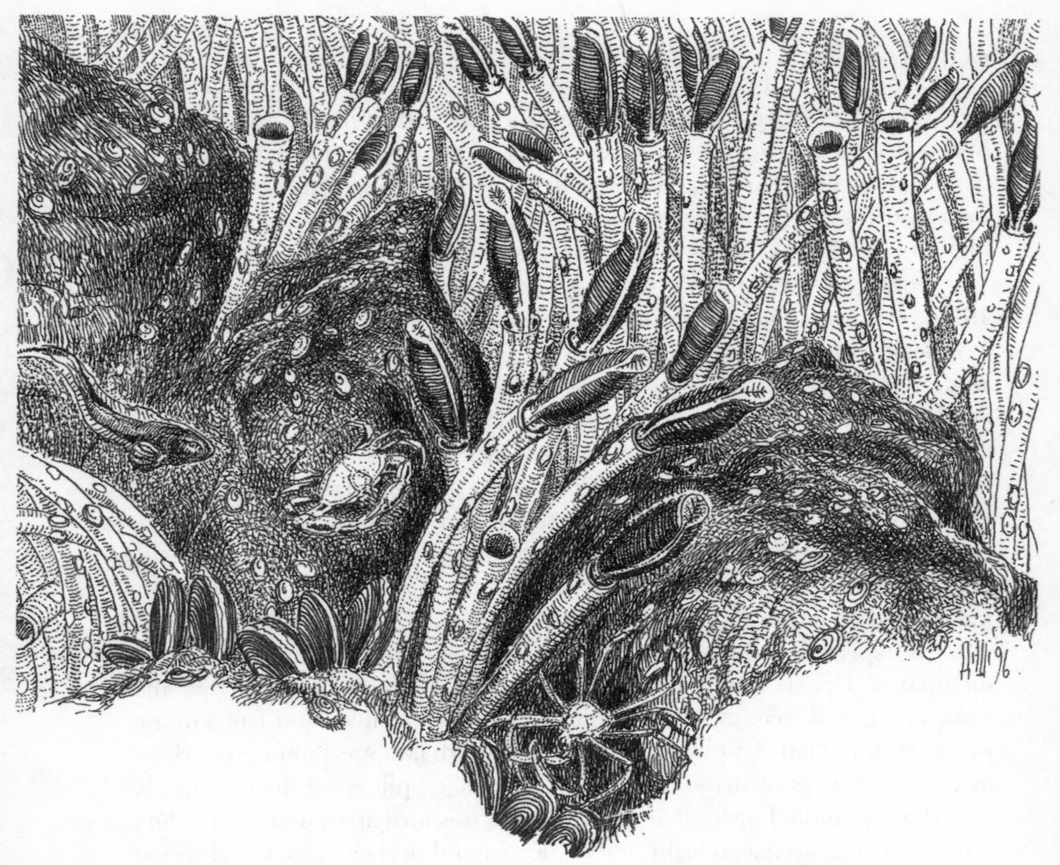

TUBE WORMS. With no mouth and no anus, the tube worms of the deep volcanic rifts grow to lengths of up to nine or ten feet by living in a state of lush symbiosis with tiny microbes that feed on hydrogen sulfide, a compound emitted by the hot springs that is deadly poison to most creatures on land.

"Well, there's all these animals down here."[10]

The deep region was packed with an assortment of creatures in all shapes and sizes. Many were in motion. Bright red shrimp swam by. White crabs lumbered over the lava, as did tiny squat lobsters. Big pink fish undulated around. There were pale anemones and orange balls that looked like dandelions. There were huge clam shells, stark white against the black lava. There were brown mussels, hundreds of them. The warmest areas were choked with life. The mussels were so dense around some vents that their bodies were crowded together to form reefs, channeling the flow of warm water.

The strangest creatures of all were snakelike things that stood upright in long tubes. Reddish tops protruded from the white tubes like lipstick. There were thickets and groves of them. These tube worms had no eyes, no mouths, and no means of locomotion or ingestion. They just stood there, like something from another planet.

The baffled scientists were unaware that they were the first humans to see most of the creatures. After all, the voyage was devoted to the study of geology and had no biologists on board. Eventually, such deep springs would be found to shelter nearly three hundred different animal species, nearly all of them new to science. Some experts came to believe that additional sampling and identification would show that the bizarre ecosystem holds more than one thousand different kinds of previously unknown creatures.

If ignorant of the details, the Galápagos researchers did have some sense of the newness of it all. John Corliss, an expedition leader from Oregon State University, during one *Alvin* dive followed a line of crabs to a new field of warm-water vents teeming with life. He dubbed it Garden of Eden. It was a chaos of crabs, dandelions, and tube worms—the proverbial snake in the garden. All over the area, warm and cold water mixed to form shimmering currents that moved like waves of heat over desert sand. At one point *Alvin* was surrounded by shimmers. Measurements showed that the water there was the warmest of all, sixty-three degrees Fahrenheit, remarkable given the frigid surroundings. It was nearly bathwater. Further inquiry showed that the water was acidic and crawling with bacteria. Most important, all of the animals when brought up to the surface stunk of rotten eggs, a sign of sulfur. That clue spoke volumes, suggesting that the scientists were dealing with a world that was truly alien.

All life needs energy, and most ecosystems on our planet are powered directly or indirectly by energy derived from sunlight and photosynthesis, often to our delight. The warm fire of a midwinter's evening is a study in the slow release of stored sunlight. We tap this solar reservoir more directly when we eat plants, and animals that eat plants. The energies we get from a meal in a good restaurant—from breads and meats and delicate sauces and desserts—all ultimately come to us courtesy of the sun. This solar dependence is true as well for cats and dogs and for most animals of the deep, the sea cucumbers and viper fish and the rest. For nutrition, abyssal fauna mainly depend on food falls from above, from the realms of sunlight and photosynthesis, or prey on one another in the darkness. Whether more or less directly, most familiar life on Earth is energized by the sun.

What was remarkable about the vent communities was that their life-support systems turned out to be so different. They were powered not by sunlight but by the planet's inner heats and energies. The microbes lived off sulfurous compounds emitted by the hot vents, usually oxidizing them, growing and becoming manna for a long line of symbionts and predators, which in turn became nourishment for animals higher up the food chain. The

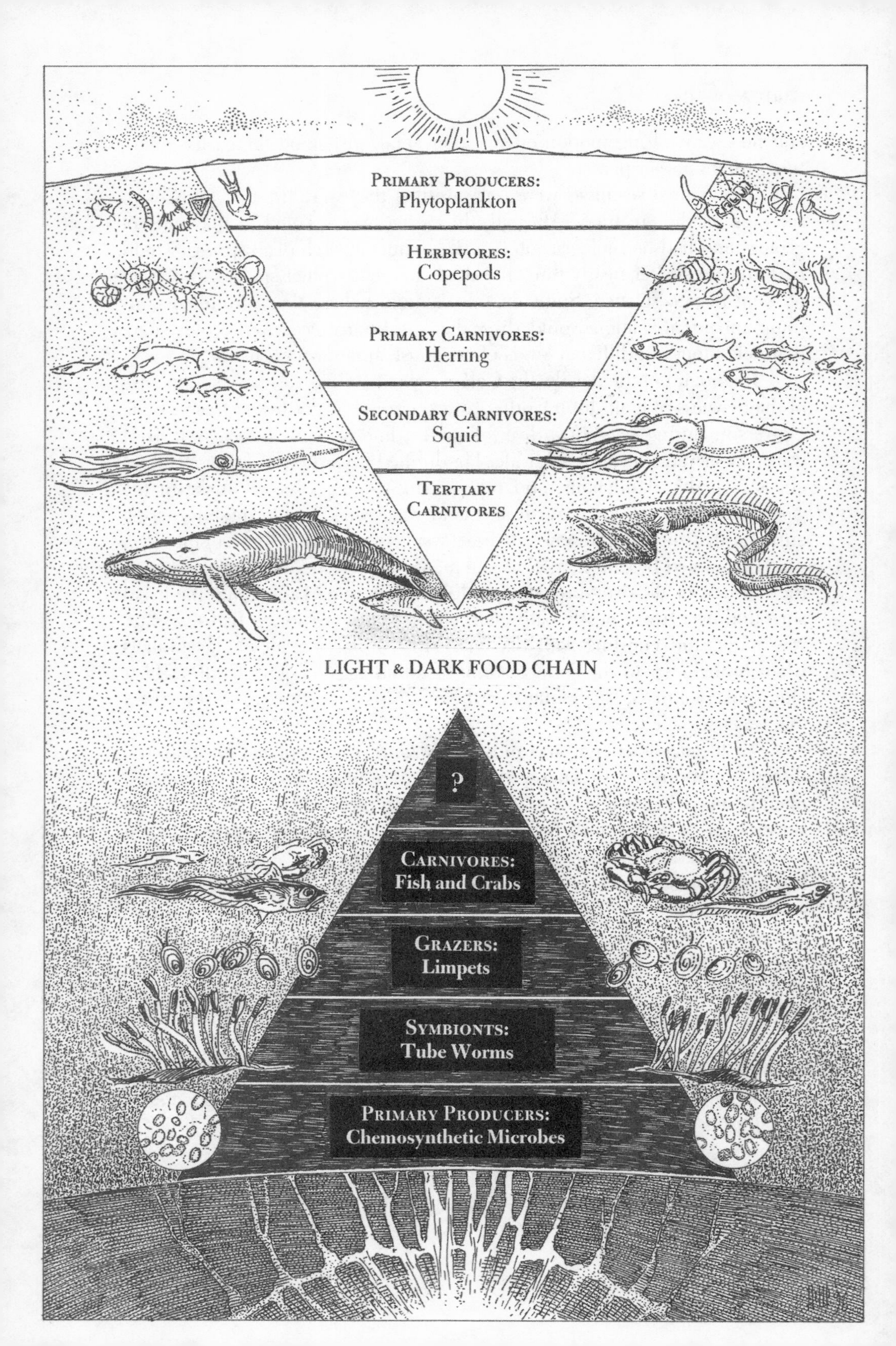

Primary Producers:
Phytoplankton
Herbivores:
Copepods
Primary Carnivores:
Herring
Secondary Carnivores:
Squid
Tertiary
Carnivores
LIGHT & DARK FOOD CHAIN
?
Carnivores:
Fish and Crabs
Grazers:
Limpets
Symbionts:
Tube Worms
Primary Producers:
Chemosynthetic Microbes

microbes were the plants of the deep, the primary producers. The ability of some microbes to live off chemicals rather than light was well known before the discovery of the deep vents, but it was a major revelation to learn that highly complex ecosystems were powered by this principle—that we and all the other light-eaters of Earth shared our planet with an alien horde that thrived in total darkness.

If for no other reason, the discovery was a revelation because the vent world was so inhospitable. These strangers not only were different energetically but lived in an environment that was about as far removed as one could get from the normal prerequisites of life. It was acidic and foul with chemicals. The hydrogen sulfide that provided much of the energy was a strong poison to most creatures on land and a noxious waste product of the mining and power industries. But deep microbes and vent fauna loved it. They thrived in and on substances that others found toxic. Perhaps most fundamentally, their world was inhospitable from the simple standpoint of setting. The violent outbursts associated with volcanism had always been anathema to life. Volcanoes were famous for maiming and killing, not for feeding and nurturing. Pompeii knew the lesson well. Now that logic was suddenly turned on its head. Volcanism was found to support life, and to do so abundantly.[11]

The scientists of the Galápagos voyage had no understanding of such nuances, which would take years to unravel. But they knew instinctively they had stumbled on something different from anything people had seen before.

"We all started jumping up and down," one scientist recalled. "We were dancing off the walls. It was chaos. It was so completely new and unexpected that everyone was fighting to dive. There was so much to learn. It was a discovery cruise. It was like Columbus."[12]

And it got wilder.

As scientists around the world raced one another to probe the new world, expeditions followed the mid-ocean ridge up the Eastern Pacific to a site located at the mouth of the Gulf of California, at about twenty-one degrees north latitude, just off the Mexican resort town of Puerto Vallarta. There, in deep waters, an *Alvin* team in 1979 was startled to see inky-black soot spewing out of something that looked like the smokestack atop a locomotive. Eager to know more, the scientists had the pilot knock off the stack's top. The turbid clouds shot out even faster. The pilot used *Alvin*'s robot arm to push the temperature probe into the smoke. The readout jumped to

LIGHT AND DARK FOOD CHAINS. On land and sea, the sun powers most food chains, the solar energy traveling far as creatures eat plants and one another. In the darkness of the deep sea, however, scientists stumbled upon an unfamiliar world in which food chains are powered by volcanic heat.

ninety-one degrees Fahrenheit. It had to be wrong, the scientists decided. The highest temperature at the Galápagos had been sixty-three degrees. They tried again and got the same reading. Frustrated, the pilot moved *Alvin* into the turbulent smoke. It was impossible to see anything.

Later, up on the surface, the thermometer was closely examined. Curiously, the tip of its support structure was melted. With growing excitement, engineers looked up the melting point of the gray plastic material, polyvinyl chloride. It gave way at 356 degrees Fahrenheit, a stunningly high temperature. It turned out the smoke was nearly twice that hot. Later dives with a recalibrated thermometer found chimneys discharging cloudy fluid with temperatures measured as high as 662 degrees, hot enough to melt lead. Normally, water raised to that temperature boils into superheated steam, but not here. Such a phase transition was impossible because of the deep's crushing pressures.[13]

The chimneys, though wholly unforeseen, once found were plausibly explained. They were envisioned as forming over volcanic rifts as icy seawater trickled down through surrounding fractures and porous rock and came in contact with the heat of molten rock below. The moving water soaked up the heat, leached minerals from the rocks, percolated upward, and then shed minerals as the superhot water flowed back into the icy sea. In retrospect, it was clear that the chimneys bore a family resemblance to geysers on land, some of which are surrounded by low mineralized mounds. It was also clear that heat was the main constructional force, causing the cold water to warm and become more buoyant and rise. Obviously the chimneys grew in height and breadth as minerals precipitated out of solution to form solid deposits, adding layer upon layer of stony material.

Pieces of chimney brought back to the surface from the smoker site were found to be a mix of sulfides and other minerals—including a completely new one, magnesium hydroxysulphate hydrate, which was subsequently dubbed *caminite,* after the Latin *caminus* for chimney. Much of the precipitated sulfur that made up the chimneys was thought to have been leached from the underlying rocks. However it happened, the result was provocative. On land, sulfide ores were the main sources of such valuable metals as copper, zinc, and nickel. Now, oceanographers had stumbled on a feature of the deep that, at least in theory, might produce riches of the same kind.[14]

Soon, expeditions to other mid-ocean ridges were discovering fields of chimneys, many belching smoke. It turned out that the hot emissions were white or gray or black depending on the temperature of the fluid and its mix of minerals. The effluents, striking in volume, were proposed as a feedstock for the potatolike manganese nodules that dotted the floors of all the world's oceans and whose origin had been a mystery ever since they were discovered during the *Challenger* voyage. The flows of hot water through chimneys and ridges were so great that scientists estimated that up to 70 percent of the earth's inner heat radiated through such zones, and that a volume of water equal to all the world's oceans filtered through them once every eight million years or so, a flicker in geological time. If nothing else, the flows clearly played starring roles in the planet's heat and chemical balances.

As explorations widened, it became clear that the chimneys themselves were giant repositories of rocky material. Many had large flanges sticking out of their sides that held inverse pools of hot water, like upside-down bird baths. Hot water would flow through the bath, over the lip of the flange, and rise upward in open seawater, depositing new minerals as it went. The flanges could become so heavy that their weight (perhaps aided by a seaquake or two) caused them to fall off, piling up around a chimney's base in mineral-rich deposits. One chimney field and its umbra of old deposits was the size of the Houston Astrodome. Perhaps the most remarkable discovery of all was that many of the rocky chimneys and underlying deposits of precipitated minerals were laced with such metals as zinc, copper, silver, and gold. People began talking about mining them, though no time soon because of the depth and difficulty of the venture.[15]

Insight grew as scientists realized that this kind of deposit over the ages had been rearranged on the seabed by geological forces and sometimes pushed up onto dry land. Lodes rich in such deposits were known as ophiolites. Geologists had long suspected that these copper-rich sulfide ores were slices of ancient seafloor. Some of these deposits had been mined for millennia, such as the copper lodes of Cyprus, which had been worked during the era of the Greek city-states. Now their genesis was certain. Knowing the secret of their origin caused scientists to look at old mines in a new light. Experts traveled to Oman in the southern Arabian peninsula and examined a copper mine there. It was found to hold fragments of fossilized tube worms. Ninety-five million years ago, in the days of the dinosaurs, the mine had been part of a deep oceanic rift, teeming with life in the darkness of the deep sea.[16]

As I read, the most arresting thing I came across was the suggestion that the deep hot vents not only powered these chimneys and ecosystems but had been around from the start and were responsible for the origin of life on Earth some four billion years ago.

John Corliss, a creative maverick who helped find the deep Galápagos ecosystem, was the idea's main proponent. He had been shaken by the discovery and eventually became convinced he had stumbled on the secret of life. His belief was strengthened when John Baross, a biologist colleague, found that the hot water of some vents and chimneys swarmed with microbes that were biological wonders. The tiny creatures could withstand heats that killed all other forms of life, thriving at temperatures up to 235 degrees Fahrenheit and perhaps briefly surviving heats as high as 700 degrees Fahrenheit. In contrast, most of their terrestrial cousins died long before reaching the boiling point, usually at 212 degrees.

Another piece of the puzzle was supplied by Sarah Hoffman, a graduate student, who learned of experiments in which the building blocks of life—complex proteins and microscopic bubbles looking somewhat like empty cells, without the internal machinery that runs a living cell—could be produced in chemical-rich water if, and only if, the temperatures were above boiling. In 1980 the trio published a paper that put the pieces together. They argued that billions of years ago the chemical precursors of life had been present deep in the interiors of the hot vents and that these substances had been heated and transformed, little by little, into complex, self-replicating molecules. The vents were "ideal reactors" for assembling life, the trio wrote. The first life on the Earth, in their view, was simple, primitive cells that fed off the vent chemicals, slowly reproducing and evolving over millions of years and laying the foundations for all that was to follow.[17]

The unorthodox idea was either ignored or given slight notice by the science community. But its standing grew over the years as favorable evidence began to accumulate. One line of support came from Carl R. Woese of the University of Illinois, whose comparisons of the genetic makeup of all microbes showed that most heat-loving organisms in the hot vents were members of a class that appeared to have undergone less evolutionary change than any other living species on the planet, implying that their ancestors were perhaps the original forms of life. He dubbed these old microbes Archaea, or ancient ones. The Archaea were seen as a third major branch of life, alongside bacteria and all higher organisms. Archaea loved hot environments heavy with sulfur, apparently living unchanged in the hot vents for eons.

Another line of evidence came from the theorizing of Walter Alvarez, a geologist at the University of California at Berkeley. He and his colleagues had found chemical clues that huge rocks from outer space periodically slammed into the earth, raising global palls of dust that blotted out the sun and triggered mass extinctions, including the demise of the dinosaurs. The Earth's surface in its early days was seen increasingly as a hellish place, with a large number of fiery impacts. And the vents looked like a deep refuge.

The idea advanced as scientists sketched out possible chemical steps in the genesis. In 1987, German researcher Günter Wächtershäuser noted that fool's gold, which often appears in chimneys, has a slight positive charge. This, in theory, could have pulled in organic molecules with a negative charge, bringing them in contact and starting complex reactions. The repetition of such chemical intimacies over time, he argued, could have catalyzed the reactions that ultimately led to life. This theory, which gained wide notice because of its elegance and simplicity, had the ironic virtue of suggesting man arose from a glitter he later came to disparage as false.[18]

Greek mythology had the Earth and its inhabitants made by a pantheon of gods, with Zeus often taking the credit or blame for mankind. The Bible laid the beginning to the hand of a single God, with his initial productions being grasses and fruit trees, a scene pastoral enough even for Hollywood. Darwin in his evolutionary theorizing pictured the start in a little warm pond, just as bucolic as the Bible in its own way.

The deep thesis had little room for romance or poetry. Bristling with irreverence, it argued that the agent of life was an acid bath sequestered in perpetual darkness, blistering hot and astir under pressures hard to imagine. As I discovered in my readings, all present and past volcanic springs and their teeming life forms were powered largely by the radioactive breakdown in the Earth's core of the long-lived isotopes of uranium, thorium, and potassium, by the nuclear fires that have stoked the furnace of our planet for ages. Ultimately, the creatures lived off the same heat that melted rock and made lava. What a wonder to think that we, too, might have evolved with the aid of this inner blaze, coming to Earth not on angel wings but on tongues of subterranean fire.

THE SKY WAS STREWN with dark, scattered clouds the next morning as we hovered over the exploration site, having reached our destination in the northern quake zone a little after daybreak. The sea was still friendly. The swells were a few feet high, gently rocking the large ship in a way that was not at all unpleasant. Somewhere below us were the volcanic ridges of Juan de Fuca. The fantail was a blur of activity as the yellow navigational buoys were readied for the plunge. Thick chains on the sides of the fantail held down scores of steel bricks, ready to drag both *Alvin* and the buoys into the depths. The rear of the fantail had no rail but rather a single wire rope stretched across it to keep people from toppling overboard.

The buoys were critical to the success of the voyage, and their positioning would take all day. Known technically as transponders, they were descendants of the devices that the Navy had pioneered after the *Thresher* disaster. The buoys, unlike their familiar cousins that bob in the waves, had no surface duties but instead floated far beneath the sea on tethers, suspended just above the seafloor. Through a dialogue of electronic beeps, they let *Alvin* know its location in the dark. *Alvin* would beep and the transponder buoys would chirp in response, the length of the round trip letting the sub fix its position. And the ship would monitor the sounds as a way to track the sub. The decision of where to put the buoys was important, and had been discussed for days, since they would define the area that could be searched close up with the submersible, and moving them elsewhere would be difficult. The plan was to put them in two places—the northerly lava flow and the central microbe storm, whose flakes in scientific parlance were known as flocculent, since they resembled clumps of sheep's wool. In scientific shorthand the sites had been dubbed Flow and Floc. The Flow site was to get four buoys and the Floc site three, leaving two in reserve for exploratory dives in other areas.

The crew carefully checked them out on the fantail. The yellow balls were about a foot and a half in diameter. The yellow plastic shells were for protection, I was told. Suspended inside were two glass hemispheres sealed together to keep batteries, electronic circuits, and other gear safe from the corrosive effects of seawater and the pressures of the deep. The method of activation was clever. Each yellow ball had a small magnet strapped to its side. Removal of the magnet flipped a magnetic switch inside, powering up the buoys. Another simple trick increased the odds that the costly buoys could be retrieved at the expedition's end. Two light sticks had been tied to each one. As a buoy descended into the depths, the pressure would crush the light sticks, starting their chemical reaction and lighting them up. However, the reaction would soon stop because of the bottom's icy temperatures, which at about thirty-five degrees Fahrenheit were just above freezing. At the end of the mission, the buoys would be released from their deep moorings by a loud buzz from the ship, would rise to the surface, and would light up automatically as they warmed. If the recovery operation had to be carried out at night, the light sticks could make the difference between success and failure.

The first buoy was put near the fantail's rear edge. Its magnet was removed, and the device chirped four times, like a bird, signaling that a rudimentary check of its battery and circuits found them ready to go. The buoy was gently lowered into the water as wire tether was paid out from a large spool. Slowly the yellow ball moved further and further behind the ship, bobbing up and down in the waves.

"One in the water," a crew member said into a walkie-talkie.

The buoy continued to drift back, buoyant. The nerve center for diving operations inside the ship was known as Top Lab because of its location and importance. It was higher than the bridge and pilot house. I was told that Top Lab was now pinging the buoy with an acoustic signal. If it responded with a chirp, the buoy would be judged ready to go.

On the fantail, the buoy's wire tether was attached to a steel brick that weighed perhaps sixty pounds. It was pushed to the fantail's edge. We waited. The radio crackled with Top Lab's okay. The steel brick went overboard with a splash. The buoy remained on the surface for what seemed like a long time and then suddenly disappeared beneath the waves—the first of our party to head for the deep. Its tether was fairly long so that when the undersea buoy eventually reached the seabed it would ride a hundred meters or so above the peaks and valleys of the the rocky terrain, minimizing the chance of acoustic "shadows" where *Alvin* would be unable to interrogate it.

Later I went to Top Lab, three decks up. Its central hub was dominated by a cubicle crowded with electronic gear and monitors. A crew member on a chair scanned the action, aided by two others working on the placement of the navigational buoys. Over a speaker came the sound of buoys already on the bottom pinging away. Like the sound track of an old submarine movie, the din of watery echoes was impressive but hard for the untrained ear to understand. Clearly the *Atlantis II* was interrogating the buoys and they were responding, but the dialogue seemed garbled as it echoed and ricocheted off the uneven bottom.

"You have to do this for a long time before you can start, just by ear, picking out which ones are which," observed Pat Hickey, *Alvin*'s chief pilot, a veteran submariner in his thirties. For the less experienced, glowing monitors automated the job.

Once the transponder buoys were dropped, the ship moved directly over them, one by one, and Top Lab recorded their latitudes and longitudes. The exact position of the ship was constantly displayed on nearby monitors, courtesy of the military's constellation of Global Positioning Satellites in orbit around the Earth, the successor to the old Transit system of navigational satellites. Tall ships once sailed the seas with only the most general sense of their true location, relying on the heavens and magnetic needles, often off course by dozens or even hundreds of miles. In contrast, we knew our place on the globe down to an accuracy of about a hundred feet—less than the length of our ship.

After laying and surveying the net at the Flow site, we headed south to the Floc site. There the process was repeated, only this time I was drafted to help. Journalists like to joke about their best junkets. But this was turning out to be no free ride.

The buoys were prepared by Bill Chadwick, a young geologist with NOAA's lab in Newport, Oregon, and Andra Bobbitt, a colleague from Scripps in La Jolla, California. Bill was an exception to the beard rule. He wore a thick mustache cheek to cheek, perhaps starting a trend for a new generation of oceanographers. Andra wore her long blond hair in a braid. She too represented a new wave. Women on ships used to be taboo. They were considered bad luck and, when aboard, were often thought to stir an overabundance of male attention. But old myths and habits were giving way as women carved out their places on scientific expeditions, as elsewhere. Four of the twenty scientists on board were women.

Being a neophyte in all this, I was given the menial jobs of lifting buoys, dragging steel bricks, and helping pay out line. It was fun despite the hard work. I wore a life jacket and felt the splash of cool water in my face as heavy bricks went overboard. I joked and swapped stories and was silenced by the beauty of the sunset. The first buoy went down that morning at 8:20 A.M. We dropped the last one over the side at 6:30 P.M. It was a long day and we were late for dinner.

In the mess I was surprised to learn that the scientists worked day and night, in shifts. Some were just getting ready to start work. *Alvin*'s dives were usually limited to daylight hours because of the risk that the sub might get lost in the dark and because of the general dangers involved in its launch and recovery. But at night unmanned gear could be towed over the ship's side to scan the ocean floor. I had examined one of these instruments on the fantail —a gorilla cage of white tubular steel about ten feet long and four feet on a side. It turned out to be a camera sled owned by the United States Geological Survey, equipped with thirty-five-millimeter still cameras and eight-millimeter video cameras. Packed with batteries, banishing the dark with lights and strobes, the USGS sled could be towed back and forth to survey a general region of the seabed. Later, its films and tapes could be studied for clues about where to send *Alvin* for follow-up inquiries.

Having no experience in ocean expeditions, I was beginning to see that the round-the-clock rhythms might tax a journalist. Even the setup in the mess hinted at the possibility of upsets to come. All the table tops had wooden dividers that marked off individual eating areas and a central condiment zone, which was packed tight with bottles. The dividers, I was told, kept things from sliding around in bad weather. And when the seas got really high, the table tops were equipped with special rubber mats that held on to cups and dishes like glue. Then again, the scientists said, there were days when nobody wanted to eat or even think about food.

THE OCEAN in the dark before dawn was hard to see. The slight glow on the horizon revealed little. No whitecaps seemed to be out there, though it was hard to be sure.

The fantail throbbed with quiet activity and *Alvin*'s hangar was lit up nearly bright as day as the submersible was readied for its first dive. Its robot arms were put through their paces. They moved up and down, back and forth, all with cool precision. Pincers opened and closed and robot hands rotated on wrists in ways no human could mimic, circling around more than 360 degrees—just the thing for turning a screwdriver. A steady stream of scientists made adjustments to scientific gear in the front basket. Crew members spoke into walkie-talkies, in communion with Top Lab. A few onlookers shuffled about, sipping coffee, waiting for the start of the dive.

In theory the tiny submersible was the same as when it debuted nearly three decades earlier, in 1964, though in fact it had been renovated so many times that few if any of the original parts remained. Its aluminum structural frame had been replaced with one made of titanium, the superstrong metal that is lighter than steel. A new titanium T-bar on top allowed the eighteen-ton sub to be picked up from a single point and placed ever so gingerly in the water. Previously the sub rode to and from the sea in a temperamental elevator that often broke down and once accidentally dropped *Alvin* into the deep, where it lay empty and lifeless for nearly a year before rescue. Of all the titanium upgrades, the most important centered on the seven-foot personnel ball, which originally had been made of steel. In a stroke, the titanium sphere doubled the sub's operating depth to nearly two and a half miles, opening up large new areas of the ocean floor to study. At the sub's stern, the single large propeller had been replaced with four small thrusters, cutting overall speed but improving control. And most conspicuously, the front end of the sub, once fairly streamlined, had exploded in a welter of gear as the vehicle had gained prominence as a tool for oceanographic research.

All this revamping had come at a price. *Alvin* over the years cost some $50 million to build and rebuild, and the complexity of the craft meant its operations were anything but cheap, currently running about $25,000 a day. The last and least conspicuous of the changes was how the sub's general costs were borne. Written on *Alvin*'s red sail in small white letters was NSF ONR NOAA, for the National Science Foundation, the Office of Naval Research, and the National Oceanic and Atmospheric Administration. What was once a military tool had become a national resource, at least in terms of operational funding. The Navy still owned the little white submersible.

Undergoing almost no change over the decades was the sub's general layout. The batteries, ballast tanks, and main items for propulsion were still located toward the rear, as were sheets of syntactic foam that helped give the heavy sub its buoyancy. The seven-foot personnel ball, also surrounded by thick foam, was still located at the sub's front end. It carried not only the pilot and two observers but the vessel's life-support gear and some electrical equipment. The hatch for the personnel ball was still atop almost everything else, located at the base of a small tower meant to prevent waves from sloshing in. The method of descent was also the same. When placed in the ocean and ready to go, *Alvin*'s main ballast tanks would flood with water, making the sub heavy and sending it down. At the bottom, one set of steel weights would be dropped to make the sub neutrally buoyant. For the return, the pilot would drop the rest of the weights located on either side of the sub, lightening it for the long journey to the surface.

Dawn finally broke under a uniformly gray sky, and all was deemed ready. Pat Hickey, the sub's pilot of the day, wearing a bright red T-shirt, went over a checklist one last time. Then the sub was slowly pulled straight back from the hangar on rails by a motor-driven chain as some two dozen onlookers milled about the fantail. It was clearly a ritual. The passengers for the dive were Marv Lilley, wearing a SAVE THE TUBE WORMS T-shirt, and Bob Embley, a senior NOAA scientist, who helped organize the voyage and pinpoint the eruptive site at sea. They climbed up some steps, over a short bridge, and down into *Alvin,* joining Pat the pilot. Atop the sub in wet suits were two divers, one a woman.

Rising some forty or fifty feet over the fantail was a large steel structure known as an A-frame because it looked like the letter *A.* Dangling from its center strut was a heavy-duty rope about four inches thick, whose end formed a loop. The divers put it around *Alvin*'s T-bar. Then the real action began. Roaring and whining, powered by engines and giant hydraulic rams, the A-frame came alive. A winch at its top reeled in the thick rope and lifted the heavy sub and then the whole huge frame began to pivot on its hinged base. The top of the gargantuan *A* was tipping back toward the sea. As it did the sub and the divers were swept in a long, slow, graceful arc over the back of the ship and down into the water. *Alvin* bobbed up and down. Swells moved the ship and sub to very different rhythms, and it was easy to envision them colliding in a sickening crunch. After the two pulled apart a safe distance, the divers plunged into the water to inspect *Alvin*'s submerged science basket and such critical gear as the steel weights. Eventually the divers were picked up by an inflatable boat. For a long time *Alvin*'s sail bobbed above the waves, a red dot in a sea of gray. Then, in a blink, it was gone, headed into the depths to investigate the Flow site.

It was 8:30 in the morning.

Later I went up to Top Lab, joining a small group gathered about the command station. Over the speakers came a strange chorus as each transponder on the seafloor sang its characteristic note, the pitch rising progressively higher before the pattern was repeated. Somewhere in the watery song was *Alvin*'s chirp. As evidence, the automatic plotter on a nearby table came alive every so often and marked a new position for the sub. The drawn line grew at a snail's pace. *Alvin*'s top speed on the bottom was less than two miles per hour. Though often described as flying, its movements were really more like a crawl. The sub's position was marked in red on the electronic plotter and the ship's in green.

For a long time no voice from down below came over the hydrophone.

"When it's quiet it means they're working," observed Cindy, the former *Alvin* pilot, adding that the flow of real news would begin only after the two scientists were safely back on deck. A timer went off, apparently to prompt a routine check. The console operator signaled the sub. Seconds passed. From the bottom of the Pacific, Pat, the pilot, replied by giving *Alvin*'s depth, 2,394 meters, about a mile and a half down. The operator went on to ask about the performance of a particular piece of equipment. An echo of his query came back before the answer. It was ghostly.

"Yup," Pat replied from the bottom, "works fine, but most of the other science gear is not doing so good today." People around the console glanced at one another.

At 5:30 P.M. *Alvin* bobbed to the surface. The mechanical ballet of its lowering was run in reverse. The trickiest part came at the beginning when a diver atop the sub reached up to grab the thick rope hanging down from the A-frame. He almost lost his balance as the sub lolled back and forth in the waves like a sick whale. With *Alvin* safely on board, a crowd of scientists gathered around to hear the informal report of the two weary travelers, who went down in T-shirts but now wore flannel shirts and sweatshirts to fight the deep chill. The lava was still glassy and fresh, they reported. Shimmers were all about, with one vent spewing water whose temperature was measured at ninety-seven degrees Fahrenheit, up from thirty-five degrees in the surrounding area. But there were no signs of hot vents or young chimneys. Moreover, the scientific gear had been balky.

Obviously tired, his hair askew in the wind, Bob Embley ticked off a list of equipment troubles—electronic cameras that shorted out, water bottles that failed to fill, a temperature probe that broke down.

"We had all the first-dive kinds of problems," said Marv, seeming to take the day in stride.

Though no large animals were found, Marv said they had seen many thick deposits of microbes on the lava, some in piles of loose flocculent-type material and others linked in strands, tendrils, and an unusual kind of velvety fuzz. "It was really funny," he said, stroking his beard. "You're going from black, clean pillows, and move into a zone where they look velvety, almost like deer antlers. Sometimes the pillows are completely covered—like this thick." He held his fingers three or four inches apart.

"There's nothing dining on it yet?" asked John Delaney, the chief scientist. No, came the answer.

"It probably doesn't taste very good," Bob Embley observed.

Cindy, eager for details about the area's biology, pressed Bob about a reported sighting of a tube worm half hidden in a drift of bacterial floc. His description was austere.

"It was down in the hollow of one of these areas," he said. "It was just, you know, something down there. It looked like a tube with a little red tip. I mean, I saw it for about three seconds."

He threw up his hands in frustration, a day of equipment failures capped by an enigmatic sighting.

"That's pretty convincing to me," Cindy said with a smile, promoting a round of laughter from the assembled group.

"Can you publish it?" John teased.

Taking charge and changing focus, John asked for a list of what instruments needed to be repaired. "The sooner we fix those, the better off we are," he remarked. Discussion ensued of what could be done overnight before the next day's dive and what might take longer.

Alvin was breaking down—and so was I. After dinner and a group meeting, I went to bed early, tired, fighting a cold, missing my family. I wondered if my gamble was going to pay off, whether this voyage would prove to be a boon, a waste of time, or perhaps even dangerous. Before leaving home I had repeatedly reassured Tanya, and myself, that everything about it was very safe. After all, *Alvin* had been doing this stuff for years, even decades. Now I meditated on the flip side of that fact. The sub was aged, perhaps in some respects well past its prime. It was a patchwork of old and new parts. Sure, safety was obviously right at the top of the agenda out here. And the things that failed were scientific instruments rather than life-support gear. Still, the problems hinted at the possibility of unforeseen trouble. The question was how extensive it might become—what systems might break down and when. I took little comfort in such thoughts as I tossed in my bunk, trying to go to sleep.

THE NEWS AT BREAKFAST was good. An important discovery had been made last night with a NOAA device I knew of only sketchily and to my knowledge had never seen. So much for my reportorial skills. It was a system towed over the ship's side on a long tether, like the USGS sled, only instead of taking pictures it was like a miniature laboratory that scanned the water for chemicals, temperatures, and particles. And instead of being towed evenly over the bottom, the device was moved up and down a thousand feet or so through the water above the seafloor in a sawtooth pattern, not unlike a yo-yo, hence the name of the operation, Tow-Yo.

It sounded prosaic, but the results were stunning. The method turned out to be ideal for finding plumes from vents and chimneys. Large areas could be examined relatively easily, and clues of activity were often conspicuous. Like a smoky pall over a factory town, a warm plume laden with microbes and minerals was a strong indication that venting, and perhaps chimneys and

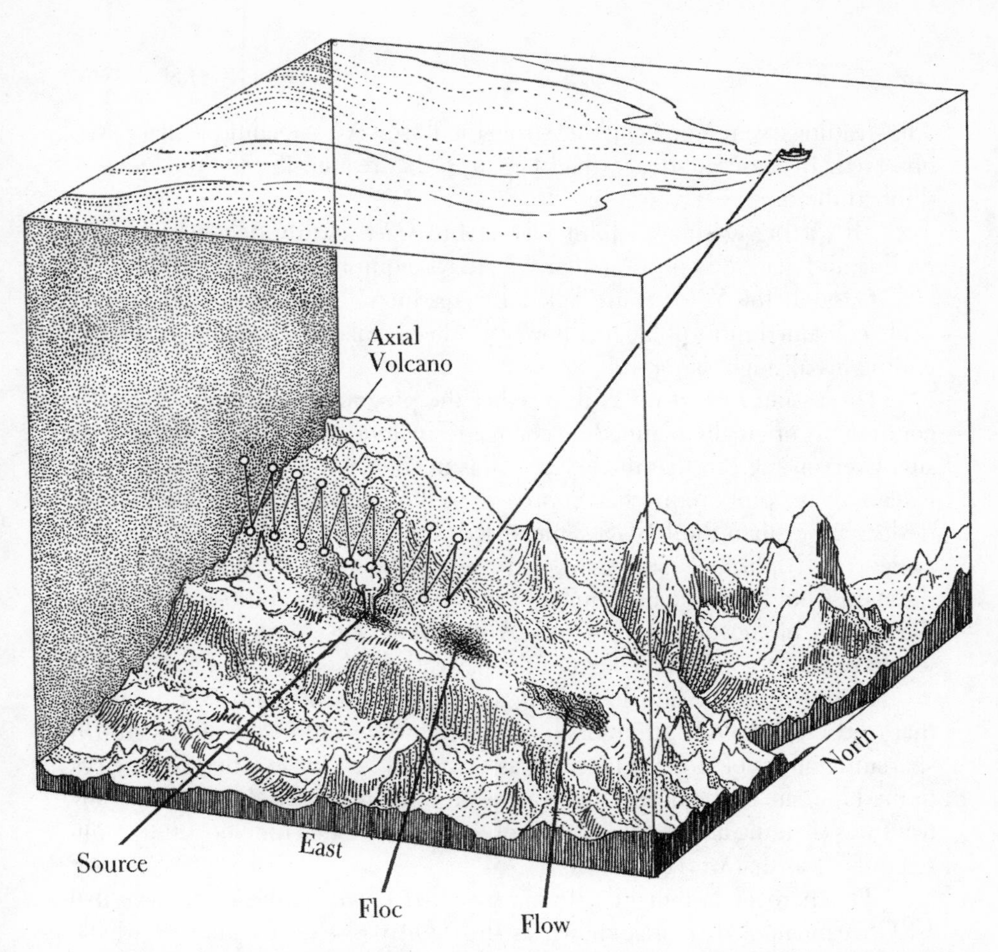

JUAN DE FUCA HOT SPRING. Like smoke from a factory town, a plume of warm water laden with microbes and minerals rose from a deep volcanic eruption on the Pacific seabed, its position helping scientists locate hot chimneys in the icy darkness.

black smokers, were somewhere in the vicinity. So it was last night. Strikingly, the breakthrough was in an area previous expeditions had judged uninteresting. Two NOAA scientists, Ed Baker, inventor of the Tow-Yo technique, and his colleague, Gary Massoth, had begun the tow along the ridge axis at the very southern end of the study area, the opposite extreme from yesterday's dive. It began at a latitude of 46 degrees, 11 minutes, close to where the quakes started. Their goal was to sweep northward in search of venting as far as the central microbial Floc site, which at 19 minutes was to be the tow's end. Since the Floc site was one of two sites designated for *Alvin* dives, the areas around it were being thoroughly studied at night. But to everyone's surprise, the start of the tow produced the strongest readings of all, indicating very warm water and heavy particulate matter in the area around 11 minutes.

The readings were some of the strongest the Tow-Yo technique had ever observed. Because the quakes had begun in that region, the site was quickly dubbed the Source.

"It's a big anomaly," John said at breakfast, clearly delighted by the finding and glad to be upbeat after yesterday's equipment woes.

Gary, a Tow-Yo veteran, called the readings "really fascinating." His smile contained hints of fatigue, however. After breakfast he would go to bed, ending his all-night labor.

Discussion picked up as to whether the program of *Alvin* dives should continue as originally planned or change to include new ones at the Source site. Everyone agreed that the first step was to confirm the readings and try to extend them, pinpointing the Source as well as possible. Thus the NOAA device, rather than the USGS camera sled, would be towed overboard again tonight. If confirmed, strong readings would be hard to ignore, and the question would become how quickly plans could be rearranged to search for the origins of the tantalizing clues. A changed plan, among other things, would call for the rearrangement of the navigational buoys.

I headed for the fantail, noticing on my way out a sign over a doorway that read, "PB4UGO," an injunction to those who would work all day in the spartan confines of *Alvin.* The sub was being readied for another dive, again to the Flow site in the north. Since any shift in planning would take days, the team was pressing ahead with studies of the lava flow, this time hopefully with scientific gear that worked.

The hero of last night's drama was tied down to the deck near the USGS camera sled. It was known as the NOAA rosette. I could see why I had overlooked the device, industrial in appearance if not downright ugly. It was about five feet high, its petals made up of ten large gray bottles arranged around a circular frame. The bottles could apparently be closed to sample large quantities of deep water. At the frame's bottom was a small suite of sensors. Their drab appearance gave no hint of their power. Unlike the USGS sled, the NOAA device carried no large batteries and instead got its power down a tether from the ship and sent up most of its readings the same way. And unlike *Alvin,* the rosette had no engines or contoured surfaces. It was zigged and zagged through the water by the ship's winch, which dealt only in brute force. A curious parallel to the rosette's utilitarianism was the fact that its masters, Ed and Gary, wore no facial hair. It was as if they and their machine were all business, with no concessions to the romance of the sea.[19]

Awakened to the importance of the Tow-Yo technique, I went back inside the ship and looked with new respect at a homely device on Gary's workbench that he had been fiddling with ever since we left port. Known as the Scanner, it was a miniature chemical lab the size of a large fish tank that could be carried by the NOAA rosette to sample minute amounts of seawater and quickly analyze them for the presence of various metals. It was a dense maze of small parts. Notable amid its tangle of tiny plastic tubes and wires were baby bottles that held various solutions to aid the chemical analyses. The whole assembly, contained in a small metal frame, was custom made by Gary and colleagues. Its computerized routine was to take readings every five seconds for up to four thousand observations per session. It had been carried by robots and *Alvin.* Tonight it would be strapped to the rosette and plunged into the deep sea, sniffing for whiffs of iron and manganese, hunting for metallic evidence of a strange world.

Many hours later, after dinner, Cindy and Bill Chadwick reported no striking findings from their *Alvin* dive today, just more lava and shimmering water and a tepid high temperature of forty-four degrees.

But at night the Tow-Yo again found a very strong signal at the Source site, pinpointing its center to 9.5 minutes, right atop the southern terminus of the quake zone. The intense, symmetrical plume was measured at about five miles in length, and Gary's Scanner picked up high levels of dissolved iron and manganese. It seemed almost certain that a field of high-temperature vents was chugging away somewhere down in the darkness. Now the question was whether we could find them.

WE MADE a total of three dives to the lavas of the Flow site, none of which produced much to get excited about. The weather was good, the waves far from unruly. *Alvin* behaved. Basic data were gathered that one day would help scientists paint a detailed portrait of a deep eruption's aftermath. But the region to all appearances was unspectacular. We moved slowly south, disciplining ourselves to be thorough and ignoring for the moment the allure of the Source. On Thursday we dove for the first time on the Floc site, which we knew from previous expeditions and our own measurements to be quite active. Indeed, Marv and Randy Koski of the United States Geological Survey found a microbial world that was very rich.

Videos of the region taken by *Alvin*'s cameras were extraordinary, as Randy showed me Friday. The area was made of old volcanic rock rent by deep fissures, some discharging warm water because of the recent flow of molten rock far below. Cracks between lava pillows were packed with bacterial mats, dark, bushy, and motionless. One vent was painted in bacterial oranges and yellows with a trim of white, as if tinged with frost on a cold winter morning. Another vent blew white clouds of microbial floc up into the water column with such vigor it looked like a snowstorm working in reverse. I began to appreciate how clouds of such particulate matter would make great targets for the Tow-Yo apparatus.

The final section of the video was eerie. It showed a jungle of microbial slime undulating out of a vent, with hundreds of long tendrils swaying back and forth in the currents. It was hypnotic. Some of the filaments were long and slender, like blades of wild grass. Others were thicker, like kelp. All were mixed together in a jumble of motion. The most peculiar thing about them was their suppleness. An undulation would begin at the bottom of a filament and work its way upward so that the top zigged in one direction while the bottom zagged in another. It was as if tendrils were made of thin strips of jelly. For some reason the wiggles running up the different filaments and tendrils often moved to their own rhythms in the vent currents, making the scene somehow chaotic and soothing at the same time. To all this writhing was added a continuous flow of particulate matter, some small, like the floc seen earlier, and some quite large, apparently bits of tendrils and related matter that broke off to join the upward swirling storm. It was a very strange environment, especially considering that it was so extraordinarily remote from the domains of everyday life.[20]

"I've never seen anything like it," said Randy, a veteran of such deep explorations.

With seeming nonchalance, Randy also told me of how his dive in the tiny submersible had suffered a blackout on the bottom when all the power went out temporarily. It was just one of those things that apparently happened to *Alvin* once in a while. Randy said it was kind of spooky.

"When an airplane suddenly drops you can always look to the flight attendants to see if they're worried," he said. "But here it was utter blackness."

I wandered up a few decks to talk to Jim Holden, a graduate student in biological oceanography at the University of Washington. Boyish and bereft of facial hair, Jim spent his day isolating microbes from samples of vent water that *Alvin* kept bringing to the surface, helping analyze the general microfauna and the makeup of the microbial jungles.

It turned out that both warm and hot vents harbored swarms of heat-loving microbes known as hyperthermophiles, a kind of Archaea, the likely ancestors of all life. These creatures were dormant at temperatures hot enough to kill most other forms of life. The temperatures they really enjoyed, the heats at which they woke up to eat and swim and reproduce, were even higher. At sea level the boiling point of water is 100 degrees Celsius, or 212 degrees Fahrenheit. Hyperthermophiles had been found flourishing at temperatures hotter than that—up to 113 degrees Celsius, or 235 degrees Fahrenheit, and perhaps briefly surviving temperatures hundreds of degrees hotter. It was an extraordinary feat. Such talents were unquestionably an oddity on the planet today but would have been quite handy during the Earth's early days, when temperatures on the surface were probably 100 degrees Celsius or hotter. Among other things, understanding these ancient forms of life was helping to answer all kinds of evolutionary riddles, indeed, to illuminate the secret of the origin of life.

In his small lab, Jim painstakingly re-created the hot world of the hyperthermophiles, taking them out of their slumber. He took vent samples, placed them in nutrient-laden test tubes, and then slowly raised the temperature—moving from 80 degrees Celsius to 90 degrees, to 100 degrees and higher. His aim was to slowly weed out the terrestrial organisms that invariably contaminated all laboratory materials and to isolate the deep-sea microbes.

"Historically, every time we've isolated a hyperthermophile, it's been a new species," Jim told me. This voyage, he added, might discover a dozen or more of the organisms, all new to science. Jim reassured me that the microbes posed no threat to man. They were creatures of extremes and dormant at the ho-hum temperatures enjoyed by humans and most of the rest of creation.

Friday's dive was the first to the Source. With high hopes for discovery, John and Bob left in *Alvin,* as usual, at daybreak. But their trek was cut short by electrical failure—an ominous continuation of Randy's woe. The sub's voltage converters had overheated, creating a "brownout" during the dive, in which power was sufficient only to begin the return trip. The dive was aborted at 10:45 A.M. It was a bad ending to a week that was anything but untroubled. Back aboard the *Atlantis II,* John with his usual pithiness summed up the trip as a hundred minutes down, ten minutes on the bottom, and a hundred minutes up.

"We were within three hundred meters of the target," he rued. "That's what was so frustrating." With typical aplomb he declared that things would move forward swiftly as soon as the submarine was fixed. To all appearances the abort left him unshaken. Cindy, the calm voice of *Alvin* experience, said that kind of total power outage happened about once a year.

One and all, the scientists seemed untroubled by the woes, which to me seemed to be getting worse. Having no basis for judgment, no experience in such matters, all I could do was watch and wait and wonder what would happen when and if my own turn came.

ON SATURDAY we had a cookout on deck, featuring grilled tuna steaks and mahi-mahi. The food, as usual, was great, a real morale builder. In a stunning light show, the cloudy horizon at sunset lit up with a rainbow. But there was no luck for Bob and John, who returned from *Alvin*'s second Source dive to report finding no evidence of smokers or hot vents after traveling like crazy all over the bottom. On Sunday we temporarily retreated from our southward march of seven days, sending *Alvin* back to the Floc site. The divers, while finding warm waters of seventy-four degrees Fahrenheit and a vent gushing so much floc that it was dubbed Snow Blower, discovered no hot vents or young chimneys. That night the NOAA rosette was again lowered to try to zero in on the smoldering fires of the Source.

I WENT TO BED early Sunday evening, trying to shake my lingering cold. There was a knock at the door. It was John, saying it was time for my *Alvin* checkout. He emphasized that there was no guarantee I would dive, but the voyage was half over and it was time to prepare.

Dazed and disoriented, I got dressed and readied to be initiated into the cult of the tiny submarine, about which I had developed a number of decidedly mixed feelings. In charge was Dave Lovalvo, my roommate and an *Alvin* pilot-in-training. Also going through the ritual as a refresher was Russ McDuff, a University of Washington scientist, who was to dive the next day. To begin, Dave put us on a scale and noted our weights, which were important to take into consideration so *Alvin* could be correctly adjusted to sink and rise on command. Then, standing beside the sub, he explained how its port and starboard lights could be controlled by scientific viewers. Over each observation window were strobe lights we were able to flash while taking pictures with thirty-five-millimeter cameras. Finally, we climbed up some stairs and took off our shoes. Standing atop the sub, Dave pointed out that the topsail was fragile. No banging, please. He further warned us to be careful while going through the hatch to avoid scratching its rim, which could weaken the seal. No belt buckles would be allowed on dive day, he said.

One by one we went down a metal ladder into the tiny sphere and sat on its carpeted floor. All around us were racks and monitors and electronic gear and what seemed like hundreds of switches. Calling it cramped was charitable. Yet, surprisingly, I found myself reassured. It was warm and womblike. There were soft lights. Padding lined the wall near my observation port. All the racks and electronic arrays looked sturdy and thoughtfully laid out. It was like a college dorm packed with stereos and high-tech gear. The space, though cramped, was far more comfortable than I had expected. My long limbs would surely get stiff over the length of a day, but it seemed like that would be relatively easy to handle. More challenging would be keeping an eye to the observation port, which was about level with my belly button. To peer out I had to bend down quite far.

Dave began the inner tour with a life-support survey. Oxygen would be slowly and continuously released from bottles and, as we breathed, our exhaled carbon dioxide would be removed from the air by cylinders filled with lithium hydroxide, the same absorbent used by astronauts in space. If the whole system failed, there were face masks. Two separate ways were used to test the cabin's atmosphere for its oxygen and carbon-dioxide levels, the redundancy being a safety feature. In a pinch, Dave said, there was enough oxygen on board for three days of breathing. And there were sleeping bags for warmth, in addition to whatever hats and sweaters or shirts we brought along. It was clear that *Alvin* had no heaters.

We moved on to electronics. Dave reviewed the operation of the video monitors and switches that were clustered next to each observer's window. They gave us a personal view of the surrounding sea through the sub's cameras and thus helped establish a context for anything we might glimpse outside the viewing ports. We could select any of *Alvin*'s television cameras, and whatever we chose would automatically be recorded on two high-quality videotape machines. Another set of switches ruled the lights outside each port. In general, they were to be switched off unless needed for viewing, since the heavy drain on the sub's batteries could cut a dive short. The thick window itself, five or six inches in diameter, was plastic and scratchable, Dave warned. It could be cleaned only with special wipes, and no glasses or camera lenses should be moved across its surface. Above the window was a small pocket holding a hand-held tape recorder for oral notes about observations. Except for flashlights, he said, the only illumination inside the sphere during the dive would be the glow from instrument panels and light from the screens of video monitors.

Next, Dave went through the sub's communications gear—a radio for the surface, a special system for talking with divers, and a hydrophone for when we were submerged on the bottom.

Lastly, and of much interest given my jitters, he described how to handle various kinds of crises. There were a number of urine bottles, which had to be carefully sealed after any use. There were fire extinguishers. And there were a whole raft of procedures for when things really got grim. The number-one rule in a serious emergency was to communicate with the ship for guidance, if possible. In case the hydrophone failed, and the pilot was unconscious or dead, Dave showed us how to switch on the emergency power. This would be sufficient to extract *Alvin* from any number of ticklish situations, allowing entangled robot arms to be jettisoned, an overloaded science basket to be ditched, and, most important, the steel weights to be dropped so the homeward journey could begin.

For the most desperate emergencies, when all hope of escape had vanished, when fading flashlights were the only source of power and oxygen was running low, Dave showed us the system whereby the seven-foot titanium ball could be separated from the rest of the sub. It was a big red handle buried in the floor. This was a drastic step, he cautioned, implying that there was considerable doubt in the technical community over whether anyone would be able to survive the ride. The method had never been used or tested. Stripped of the sub's heavy batteries and other gear, the sphere would rise to the surface like a rocket. If it ever came to that, Dave advised, the best thing to do was to wrap your head in a sleeping bag at the start of the ascent and hope for the best.

The checkout was over. But going back to bed was the farthest thing from my mind. I was wide awake.

Around midnight I ended up in the lounge area of the mess, which, as usual at almost any hour, was busy if not crowded with activity. My brain was reeling with extremes, with new kinds of life and death, with discovery and danger.

Sipping juice, courting sleepiness, I sat down next to Marv, who radiated a feeling of calmness. We talked. Little by little, in a gentle voice, the bearded oceanographer with his well-worn sandals sketched out his microbial vision. It was somehow comforting, like a glass of warm milk, even though his ideas in fact were rather startling.

He spoke fondly of "the bugs" that we had been seeing on the ocean floor. These colonies and tendrils and storms of microbial floc, he said, were only the most superficial expression of a deep, hot world of unfamiliar life that lined the fissures and hollows of the world's oceanic crust. It had to be, Marv said. The ejecta were too great in volume to have been created at or near the crust's surface. The habitat probably went down miles. In places, he said, there were undoubtedly big undersea caverns where lava once flowed upward and then receded and now held only lakes of warm seawater percolating with thick clouds of microbial life. It was another world, he said. By viewing the vents and chimneys, by tracking the plumes, by exploring the ocean ridges, we were only now beginning to get a glimpse of this ancient world.

I went back to bed, in dreamland.

ON MONDAY I took a close look at a paper John had given me earlier on the voyage. Back then it seemed like a distraction. Now it looked more like required reading.

The paper went on at some length about the possible dimensions of the hidden microbial realm around the globe. It was written by Thomas Gold, a Cornell University theorist, who over a long career had distinguished himself for unfailing boldness. His main achievement had been finding out what powered the throbbing radio stars known as pulsars. But his restless mind had also applied itself to large questions of cosmology and fine points of physiology, to astrophysics and geophysics, to lunar and planetary science.

His ideas were often contrary to the conventional wisdom of the day and, even when proved wrong, were always creative and provocative.

Gold in his paper suggested, like Marv, that our planet harbors a biosphere of deep microbial life that goes down miles. Amazingly, he argued that its total mass might rival or exceed that of all surface life. From this perspective, all the world's people, mammals, birds, amphibians, reptiles, fish, insects, invertebrates, sea creatures, seaweeds, plants, forests, gardens, and jungles were mere icing on the cake of life. Heat-loving microbes were possibly not only the Earth's first inhabitants, in Gold's view, but its dominant ones as well.

As evidence, Gold described how microbes had been found in deep-sea vents, terrestrial hot springs, and other kinds of deep, hot places. The clues were few but tantalizing. Then Gold took a leap of logic that only further research, perhaps decades of it, could prove right or wrong. From the sketchy evidence, he suggested that microbial life might be widespread throughout the upper few miles of the Earth's crust, inhabiting fluid-filled pores, cracks, and interstices of rocks while living off the Earth's inner heat and chemicals. Ignorant of sunlight, the deep microbes would constitute a kingdom independent of the Earth's surface and its diverse inhabitants, powered by the planet's inner energies, thriving in darkness.

In his paper, Gold conducted a remarkable thought experiment based on common knowledge about the porosity of the Earth's crust. He assumed that microbes would be found that could thrive at temperatures up to 150 degrees Celsius, or about 300 degrees Fahrenheit (a fairly common assumption among microbiologists), and thus could live at a maximum depth of three to six miles. Life beyond that point was seen as unlikely, since the Earth's increasingly hot interior would kill the organisms. Conservatively assuming that three miles was the limit, and that only one percent of the available pore space in rocks was occupied by microbes, and that a band of concealed subterranean life encircled the globe, Gold calculated that the total mass of this deep life would be about two hundred trillion tons. If that amount of animate goo was spread over all the planet's land surfaces, Gold wrote, the layer would be nearly five feet thick.

"This would indeed be more than the existing surface flora and fauna," he noted. At the very least, he added, this undiscovered band of subterranean life in theory is "comparable to all the living mass at the surface."[21]

Even more intriguing, he speculated that subterranean life may dot the heavens, hidden beneath the surfaces of planets and moons and energized by geological processes. In contrast to everything science had assumed for so long, he wrote, life on other planets might have no need of the warming radiations of nearby stars. Indeed, the surfaces of these planets could be quite hostile to life, though their interiors might not be. Our solar system alone, he wrote, might harbor ten planetary bodies with deep biospheres, some quite distant from the Sun. Throughout the firmament, the interiors of many planets were likely to possess the right temperatures and pressures for the maintenance of liquid water, he wrote, making them ideal breeding grounds for alien microbes.

"Such life," he wrote, "may be widely disseminated in the universe."

BACK ON EARTH, I finally got serious about souvenirs. Late Monday, with the advice and encouragement of a number of seasoned hands, I prepared four foam coffee cups with drawings and messages for my wife and kids. Then early Tuesday we put them into a bag hung on *Alvin*'s side, along with those of crew members and scientists. Down they went during the dive and up they came at the end of the day, transformed. The intense pressure at the bottom of the sea squeezed them down relentlessly into something smaller than shot glasses. They were still recognizable as cups, however, since the pressure had been so uniform. And the writing was still clear, if tiny. Interestingly, the shrunken cups were rock hard. All the bubbles that give plastic foam its springiness were gone.

SIGNS OF THE SOURCE as revealed by the Tow-Yo apparatus faded early in the week, and Tuesday yet another Source dive failed to locate the fire behind the erratic heat. The tantalizing target had all but vanished. While the NOAA rosette still found warm plumes in the area, they were weak and their centers were hard to pin down. None of the signals were as strong as the ones we had first detected.

On Wednesday we deployed a battery-powered camera and temperature sensor at the Floc site near a likely-looking vent, hoping that a year of time-lapse photography would reveal chimney growth. The camera was taken down by *Alvin* and the large temperature probe by a transportation device known as an elevator. The elevator hit bottom and rose back up before *Alvin* had a chance to remove the probe. A second try at the operation was successful. A warm vent had been chosen to monitor since our team of fifty-two scientists and experts and aides and hangers-on had failed to find any hot ones.

The ocean was like glass, its smoothness eerie, given all the warnings about bad weather. I was allowed to leave our ship in the inflatable boat to photograph *Alvin*'s return from the deployment mission. The experience was great fun. I was liberated from the small confines of the *Atlantis II,* out and about in the wide Pacific. The whine of the boat's outboard motor and the hubbub of the sub's recovery did little to dispel the sense of solitude at sea.

It was all sky and sea and that remarkable, razor-sharp line in the distance where they met. Humans and their activities seemed like a very small part of the picture.

That evening a science meeting turned into a philosophical clash. At odds were those who wanted to continue pursuing the Source and those who wanted to shift the focus back to comprehensive studies of the Floc and Flow sites. The debate was complex, but it boiled down to a conflict between safety and risk. John, the chief scientist, displaying no small amount of fervor, said our scientific duty was to study the whole ridge and as many of its changes as possible. The Source was the biggest potential shift brought about by the volcanic eruption, he argued, and therefore it had to be tracked down and explored. The forest could not be ignored for the trees. Dissenters said that opportunities for good science, perhaps great science, were slipping away amid the rush after a phantom. Incremental studies of the Flow and Floc sites might not be sexy, they argued, but the workings of the ridges were so mysterious that such efforts in the long run could provide the most important insights of all.

John, a man with no lack of charm and rhetorical skill, a conciliator who usually sought to promote discussion and consensus, made it clear in a gentle way that we were going after the Source.

NATURE CAST HER VOTE, just hours after the debate on Wednesday night over what to do. The period of exceptional calm ended. Having never been at sea in a storm, I now learned what happens when big waves toss people, equipment, and ships around like toys.

Last night's Source tow was cut short. By Thursday morning things had settled down enough so that Bob and Randy were able to take *Alvin* to the bottom on a Source hunt. But they were called back early as the wind and waves grew dangerously high.

The sky was dark gray. Whitecaps were everywhere. Just off the fantail, *Alvin* sloshed and rolled and heaved amid the swells and breaking waves. Two divers in dark wet suits clung fast to the sub's topsail, water surging between their legs as they and *Alvin* pitched up and down. After a few tries, the divers managed to grab the big rope hanging down from the A-frame. For a tense moment they stretched like rubber bands as the rope and *Alvin* heaved in dissimilar rhythms. Then, just as the sub rose to the crest of a wave, they gracefully slipped the loop around the T-bar. It was done. The divers plunged into the boiling water as the sub was hauled up. Its eighteen tons swung back and forth in a threatening arc before being pulled snugly into the A-frame's grippers.

• •

THE DIVE PLANNED for Friday was canceled by the continued run of dirty weather. It was the second week to end on a dark note.

I got little or no sleep the previous night as the ship crashed through the waves. At one point some of the books I had brought along went flying across the room. Ian Jonasson, a Canadian geologist on board, looking rested and chipper as usual, told me the secret of a successful night's sleep under such conditions. Rather than lying *on* the mattress and struggling to maintain that position, you get *under* the mattress, preferably slightly off to one side, so you are pinned between it and a wall.

The good news was that signs of the Source had come back, with a vengeance. Though *Alvin* was out of commission, the NOAA rosette was successfully towed over the site, finding the most intense plume yet since the original discovery eleven days ago, pinpointed with new precision. The target was estimated to lie at a latitude of 46 degrees, 9.27 minutes, and a longitude of 129 degrees, 48.48 minutes. And then all further work ground to a halt because of the storm.

At our evening science meeting in the mess, John said everybody had agreed that I should have the next dive if the weather broke. Jokes were made about John's zeal to go after the Source and about who would inherit my laptop computer. Time was clearly running out. We were scheduled to leave for Astoria on Sunday.

I WAS UP EARLY Saturday morning, waiting, hoping, going up to the bridge, peering out its windows into the darkness. No luck. Rain hammered us. The needle of the ship's barometer headed for the floor. By 10:30 A.M. it was official. There would be no dive.

Then, as fate would have it, the weather began to improve. The barometer shot skyward. The rain let up. The waves were high but diminishing. All in all it seemed like a bad movie. Nothing this dramatic was supposed to happen in real life. The improving weather prompted John to get in touch with Woods Hole by radio telephone. Yes, they said, the voyage could be extended for one day. John made the announcement and basically went into hiding. It was no small thing to rearrange fifty-two lives, fifty-two schedules, fifty-two rendezvous with colleagues and friends and lovers and spouses and families. The ship's fax machine and satellite antenna went into overdrive. Another burden for John, and Woods Hole, was that extending the voyage would delay those groups waiting patiently in line to use *Alvin* and the *Atlantis II,* surely inconveniencing them and perhaps winning their ire. Jokes began to circulate about John's mad hunt for Moby Vent, the great white chimney. There was a whiff of mutiny in the air, just a whiff.

• •

IT LOOKED LIKE pea soup outside the bridge early Sunday morning. Then, ever so slowly, my eyes adjusted to the darkness. Peering through its wide windows, I saw stars. Lots of them.

I stepped onto the flying bridge and looked up, stunned by the sight of so many points of light swishing back and forth in the dark as the ship rolled lazily in the waves. Orion the hunter shone bright. Even the fuzzy patch in his dagger—the great nebula of Orion, a vast cloud of incandescent gas and dust lit from within by hot blue stars—was unusually luminous. A falling star flashed through the constellation. It seemed like a good sign. The hunter was accompanied by a celestial guide. The ship's first mate, who was manning the bridge, confirmed that conditions were improving. He said the swells were running to a height of ten or twelve feet, down from twenty or twenty-five. That wasn't too bad. It looked like an acceptable day for a dive, he added with a smile.

Preparations were a blur. Suddenly I found myself on the fantail high atop the sub, looking out at distant clouds shot with gold from the rising sun. I had no thoughts of *Alvin*'s woes, of dives cut short, of risk and danger. Instead, I felt extraordinarily calm. I went down the ladder, followed by John. The hatch closed behind us and, before I knew it, silvery bubbles were swirling outside my observation window, as we prepared to dive to a depth of more than a thousand fathoms.[22]

Our pilot was Bob Grieve, an ex-Navy submersible pilot with a strong jaw and a relaxed, confident air. He was in his late twenties. He wore sweatpants, a sweatshirt, and furry slippers tipped with fake animal claws. His ear was pierced by a small gold loop. Playful and irreverent, good at his job, he was a great companion to have on a trek to the bottom of the Pacific. Around 8:10 A.M. we got permission to dive.

"May the force be with you," the controller in Top Lab said over the hydrophone as we began our descent.

The daylight outside my viewing port slowly began to fade. Then came darkness, deeper and deeper. Before long I was transfixed by the flashing lights and glows and glimmers of an endless procession of gelatinous creatures, some quite small, others seemingly a foot or two in length. Ripples of light would pass along the bodies of the larger ones as we swept past. Smaller ones would explode in a luminescent flash. The overall density of the life was amazing. It seemed as if every cubic foot of seawater held at least one thing that was phosphorescent. At any moment there could be dozens of lights in view. Sometimes four or five would explode at once in a blaze of bioluminescent glory. All the living lights appeared to be streaking upward as we moved relentlessly down. Someone once remarked that it was like falling through stars. So it seemed to me.

"Snot," Bob said derisively of the light show. "It's biosnot."

"It's your ancestor," John needled him.

Well, not quite.

John and Bob busied themselves with all kinds of preparations. We also had time to chat. John, always eager to poke fun, went on at some length about how death on the bottom of the sea would be a great way to go. If the sphere had a structural failure, it would be instantaneous. One minute we'd be busy doing our work and the next, whaaaaaamm, flat as a pancake. His banter got me thinking. I told him I had recently written about six volcanologists who died while investigating Galeras volcano, a smoky giant in the Colombian Andes. Its violent eruption had been totally unexpected. How likely, I asked, was an outburst in our dive area? He thought for a moment. There was a very small but finite chance of an eruption, John said, leaving it at that.

We fell for more than a hour, nearing the bottom around 9:30 A.M. Bob dropped some of *Alvin*'s steel weights, lightening us to near neutral buoyancy. The sound of their impact on the bottom reverberated through the sub's hydrophone speaker. Bob made some fine adjustments to our buoyancy with the ballast tanks and switched on the main lights.

All around us in every direction were humps and lumps of dark, bulbous lava. It was unbelievably spooky. The water was murky and filled with drifting debris, clouding our vision. Our lights were strong enough to illuminate a small area right around the sub. But the wider region was only vaguely discernible, a blur of indistinct shapes and dim shadows. The Empire State Building could have been standing at the edge of our vision and been easy to overlook. Overall, there was no way to judge whether we were in a deep fissure or high atop a ridge. The only things we could see clearly were right in our hemisphere of light.

Close by, the giant globules of lava were deeply gnarled and cut with grooves. They were clearly old. Some were powdered with light-gray sediment, which also lined their cracks and the interstices between pillows. Such mounds were thought to form during an eruption when the outer surface of molten lava was frozen fast by icy seawater until increasing pressure inside began to squeeze the liquid rock through a weak spot, the red-hot lava oozing out like toothpaste from a tube. Repetition of the freeze-flow cycle made the terrain look quite lumpy. The term *pillow,* though adequate, failed to do the globules justice. They were an alien material that had been twisted into an alien landscape. The variety of bulbous turns and bends and curls was enormous and odd. Moreover, the word *pillow* had associations that were too soft and dreamy. This stuff was rigid and hard. And though the mounds and globules were clearly old they still conveyed a strong feeling of pent-up energy, as if their violent birth had been cut short.

Suddenly a large, reddish shrimp swam past my window, its antennae fluid and willowy, each perhaps a foot long.

We had landed relatively close to our first target, according to a message from the ship, which was closely monitoring our position through the chorus of transponder chirps. In theory the target lay just 250 feet away at a bearing of 140 degrees, or to the southeast. (In compass degrees, north is 000, east 090, south 180, and west 270.) The targets had been picked on the basis of rosette findings and topographical analysis. But we knew the exactitude was illusory. At best our targets were educated guesses and at worst convenient fictions. Four other *Alvin* dives to the Source had failed to find any hot spots amid a clutter of targets. Now it was our turn.

"It's a downhill shot," Bob said of the terrain in the direction of the target.

"If we're not successful," John grumbled, half serious, "it's going to be downhill for the rest of my career."

We moved slowly over the pillows, *Alvin* flying perhaps fifteen feet above the volcanic rock. Soon, the water began to clear up. John said that was a good sign and that we would keep moving in and out of clouds of particulate matter that had accumulated as a result of hot venting. Bob noted that, in good conditions, we could hope to see thirty or forty feet. As we moved forward, the lava visible out my viewing port, on the starboard side, suddenly began to rise upward to become a wall, then opened into a deep fissure about two yards across. John asked me to note the time, so later back on the surface he could make a detailed map of the area's geology. Bob remarked that we were about a hundred feet from the target.

John took the opportunity to lower expectations. "Remember," he said, "this is just a place to begin. We're not going to stumble on some giant, massive industrial smokestack here."

A big fish, perhaps two or three feet long, swam by my window, its tail slowly undulating back and forth with a surprising degree of suppleness. And the water began to get hazy again, raising my expectations despite John's warning. Wads of floc floated by. Suddenly the ground on my side gave way. The area seemed almost bottomless. We knew we were next to a fissure, since neither Bob nor John experienced similar falloffs. The murky water out my window began to writhe with tiny bits of floc and small insectlike creatures. The critter index was high. "The water is just crawling with things," I said, peering into the gloom.

"That's good," said John, again trying to lower expectations for what

SKATE ON PILLOW LAVA. A big skate, perhaps five or six feet long, lay motionless atop gnarled lava a mile and a half down on the Pacific seabed, its mass remarkable for a domain once thought to be devoid of life.

could be a very long day. "That sort of comes and goes. That's not a positive indicator," but rather a necessary prerequisite.

Indeed, the critter index began to go down. I felt a knot of disappointment. I could almost sense the chimneys around me. We seemed to get close to them, to detect some life in their umbra, and then to pass them by in ignorance, unable to comprehend our surroundings. On the other hand, it was still early in the morning, only 9:50 or so. We had been exploring the bottom for all of twenty minutes. The reasonable thing to do was relax and get back to work.

We went up a small ridge and came to an area of heavy sediment. It was almost like a sandy beach, with dark rocks protruding around its edges. As we moved back over solid lava we passed a large animal. It was a big skate, perhaps two feet wide across its winglike fins and five or six feet long from snout to long, thin tail. It lay perfectly still atop the dark lava, an eerie mass of whitish flesh.

The pillows became interminable, punctuated only by occasional ridges and fissures. Slowly we seemed to be climbing out of a valley. My head swam with the blur of volcanic images. At a high point we paused to change course to a heading of 240, or west southwest, going back across the valley. As we took off, the murk index started to rise and the water was again filled with floc and critters, including small, fast-moving fish with wonderfully flexible tails, little insectlike creatures that moved by rapidly snapping some body part, and sinuous blurs that moved like snakes across sand. Then the water became extraordinarily clear.

It went like that, back and forth, back and forth. My fatigue increased. We reached another high point and turned eastward back across the valley. Again we entered an area where the critter index was high, with fleshy sea cucumbers lying on the rocks and small fish darting about. The murk was relatively heavy. Then, out of the blue, we made the first major discovery of our voyage.

Recall the context. Since the eruption in June, this general area of the Juan de Fuca Ridge had been studied by no fewer than five expeditions, ours being the sixth. To make our trip possible, a number of Federal agencies had rapidly and closely cooperated, no small thing in itself. Our team of prominent experts had been assembled from across North America, Cindy flying in from Woods Hole on Cape Cod, Ian from Ottawa, a dozen scientists from up and down the West Coast. At our helm was John Delaney, a rising star of oceanography. We had risked bad weather late in the working season and been at sea for fifteen hard days, tantalized and increasingly frustrated as clues of major activity came and went, the target always eluding us. John and Woods Hole had gambled and extended the voyage to continue the hunt, risking no little professional embarrassment. In all, *Alvin* had gone down a dozen times. Then, in primal darkness at the bottom of the Pacific, nearly a mile and half down, in a place no human had ever laid eyes on before, we made our first big discovery.

A tennis shoe.

"No," said John, his voice low and incredulous.

It was a Reebok, sitting upright on the sediment, bright white and seemingly in perfect condition. Framing it were pods and curls of swollen lava, dark and gnarled, poised threateningly. If the Reebok company had paid millions of dollars to have some agency on Madison Avenue do this scene up as an advertisement, it couldn't have looked better.

Bob reached out with one of the sub's robotic arms to retrieve this sunken bit of somebody's vacation.

John, considering the stress he was under, handled the situation well, joking about how the video footage would make a good commercial.

"Pick up a basalt while you're at it," he told Bob, referring to the volcanic rock that was all around us.

It was 10:30 A.M. We had been submerged for more than two hours and had found nothing of significance. To me, it felt as if we were starting to run out of time.

Alvin lifted up and we began moving again. The areas of murkiness came and went, as did the critters. My stomach started to get queasy with the endless viewing of lava. Around 11:10 A.M. we paused on the bottom to take stock of our situation. Bob started eating a sandwich. I didn't. Our tiny sphere was growing colder. Water vapor had begun to condense on exposed parts of the titanium hull, forming a glistening sheen. Bob pointed up. The bottom of the hatch was covered with droplets of moisture, looking as if they might soon start dribbling down. I put on a sweater.

John was sketching a map of the area's topography when Top Lab called down. "*Alvin,* people on the surface are curious about whether you plan to visit the max anomaly target, over."

This was curious. We thought we had traversed that spot. But, according to Top Lab, we had yet to visit the area that ahead of time had been determined to be the core of the heat anomaly—in effect, had failed to play our top card. I found this discrepancy jarring. Despite the huge investment of time and energy in navigation and getting our bearings, in transponder buoys and all the rest of the electronic gear, in whole decades of military effort and development, we still seemed to be lost in the dark at the bottom of the sea, or something close to it.

John calculated that our current position was near the point where we had originally come down, our route having been roughly a triangle across a central valley. Going to the site recommended by Top Lab would be easy, though, in his judgment, it also might be repetitious. The site lay some five hundred feet away at a bearing of 030, slightly east of due north. Since John was ready to begin a pattern of north-south traverses to track fissures identified in our survey, it was more or less in his intended path. Munching on a candy bar, John sent us off in that direction. *Alvin* lifted up and began to move again over endless piles of pillows. Hungry, I began to eat a sandwich, my stomach settled.

Rather quickly, the flat bottom fell away abruptly on the left, out John's window. We were atop a deep fissure. John called up to Top Lab that we would continue to follow it northward.

"For the people who are worrying and wondering," he said into the hydrophone, "we're basically trying to identify what the cross-structures are, then we're planning to follow those." The assumption, he added, was that the strong plume found by the NOAA rosette was askew in the water column because of the currents and that the real action "is liable to come from fissures and fractures, over." It seemed like a gentle rebuke. The scientists up top had a theoretical plot, while we were learning the lay of the deep terrain firsthand, carefully, painstakingly. We had the facts.

We followed the wide fissure northward. Nothing or little was visible out

John's window. Near mine lay the fissure's upper edge. All the areas, as usual, were gnarled with lava. On my television monitor, via the sub's top camera, I could just make out the fissure's rocky floor, which meant it was perhaps thirty or forty feet deep, given the limits on visibility down here. Bob moved *Alvin* more directly over the fissure, so its rocky face slid past my observation port like a wall. The altimeter in the sub's tail, whose readings were displayed on my monitor, indicated the fissure's bottom was down twelve meters, or thirty-nine feet. We passed a huge lobe of lava that had begun to drip into the fissure before freezing stiff.

The murk index went up, as did the critter count. On the wall moving past my window I began to see occasional sea fans and sea anemones. We passed a big sluglike creature that had a soft cloud of gelatinous tentacles radiating from its body, probably some kind of sea cucumber. Top Lab called down to say we were in the main anomaly area. Bob remarked that it was getting murkier and murkier. I saw small fishes in the wall's cracks. Most interesting of all, as we continued northward along the fissure I began to see brownish growths all over the rocks. They were different from anything I had seen during our dive or videos of other dives. John asked if the growths looked like bacteria.

"Yes," I replied, "standing up an inch or two all over the rocks, getting bigger as we move forward, maybe two or three inches high now.

"More of it, lots more," I said. "Really fuzzy rocks. It's like they're all growing hair." It was starting to look like a beard, I said.

We proceeded along the living wall, moving forward perhaps another hundred feet.

Then Bob sang out, his voice tense.

"I've got a spire."

"A what?"

"I have a chimney. I've got a spire."

It materialized slowly out of the gloom before us, increasingly clear as we moved forward through the murky water. The chimney was poised on the fissure's rim. It was a patchwork of dark and light colors, framed by the dark surrounding lava. Its bottom was thick and craggy, perhaps ten or twelve feet across. Its midsection slowly tapered upward, like a spire. The image of the monolith overflowed my monitor as we drew closer.

"See anything moving?" John asked as we advanced.

"Nothing yet," Bob replied, peering out his port. He quickly updated his report. "Hydrothermal sulfide chimney," he cooed. "It's hot. It's hot, bud. It's got tube worms all over it."

I let out a whoop.

Now that we were closer, we could see that the craggy tower was spewing water that shimmered with motion. And its surface was thick with animals and microbial life.

John, knowing it was around 11:30 A.M. and that time was indeed running out, immediately called Top Lab, telling them of the discovery and requesting that they prepare to send a time-lapse camera down on the elevator. This procedure would take at least an hour and perhaps longer if we had difficulty finding the elevator on the bottom. It had to be dropped soon if we were going to have a chance to recover it and set it up.

Bob pulled *Alvin* close to the monolith and eyed it. He had seen a lot of chimneys in his day but this one was odd. "It's covered with weird-looking stuff," he said. "It's got worms on it but they're unusual-looking. I've never seen them before."

Tube worms can grow in length up to ten feet, and sometimes do. These seemed to be four or five inches long.

Alvin moved slowly up the spire's side. Its rocky top gushed no hot water, no black smoke. There was some venting, but most of the structure's shimmers seemed to come from cracks on its flanks. We were up eleven meters, or thirty-six feet, according to the sub's altimeter. The bottom of the seafloor near the chimney was uneven, and this measurement seemed the best we were likely to get. The chimney was roughly the height of a building three or four stories tall.

We slid down the side and moved in close for a detailed survey, panning a color camera over its surface. The thing was a tangle of life. All the strands and tentacles and filaments of the various creatures and growths seemed to be knotted together into a dense carpet. There were red-tipped tube worms emerging from brown outer cases, tiny white crabs, long white microbial tendrils, and many small creatures that were hard to identify, such as little strings that looked like Chinese noodles. Perhaps most prevalent was brownish and whitish fuzz similar to what I observed on the fissure's wall. And this, by such happenstance, was how the chimney got its name—Beard.

Many of the animals had tendrils or arms that waved gently in the currents or swayed in the simmers of hot water. Overall, the chimney's lumpy surface was separated into different zones, where one or another fauna would predominate, giving that zone a dominant texture and color. That mottling had given the chimney its patchwork appearance from a distance. Brown beard would predominate in one area, white noodles in another, and scores of tiny crabs somewhere else. These were just the dominant creatures, each zone also having a host of less-flashy inhabitants. Visible in a few places beneath this knot of biology was the rocky surface of the chimney, dappled with whites, grays, blues, and mysterious patches of red and orange.

Adding to the strangeness of this overall scene was the frequent flow of shimmering water over the chimney's surface, making the image of whatever we were looking at ripple at times like a mirage.

Off to the side of the chimney was a haze of floc and swirling debris. The upward flow was heavy, if not a storm.

One vent we saw during our brief survey had been gushing thin gray smoke. Now we went back and took its temperature. It was near the chimney's top. Bob directed one of *Alvin*'s mechanical arms to grab the temperature sensor in the science basket and push it into the vent. The sensor was long and thin, like the wire in a metal coat hanger. But its thick handle was easy for the robot claw to grab.

The temperature readings on our monitors shot up to 209 degrees Celsius before dropping off as some of the rocky orifice broke and partly clogged the vent. The body of the chimney was very crumbly. A cloud of debris rose in the water, mixing with the shimmers. Bob pulled back to make fine adjustments to the sub's buoyancy and try again. Hovering in the icy water, we had to be as steady as possible for such delicate work. Thinking ahead, John remarked that it was unclear whether this vent would be the hottest and that we needed to measure it carefully and survey the general area before making decisions about where to take water samples. We carried four metal bottles that could hold high-temperature water, and we had to husband them carefully, being sure to get the hottest water possible for later analysis.

"We have about two hours before we have to pull the plug," John added.

Though our temperature work was momentarily frustrated, I was amazed. Reading about chimneys had somehow left me unprepared for the reality of high heats. It was like the difference between reading about camping and having the flames of a fire warm you. This was hot. And it was nearby. A Celsius reading of 209 degrees was equal to 408 Fahrenheit. That was hot enough to melt many modern materials, to sear flesh, to cook a pizza. In sufficient quantities it could heat homes, power turbines, light cities.

Our adjustments done, we pulled close again to Beard. His hands on the robot controls, Bob reached out with the gripper, placed it around the top of the vent, and squeezed. The rocky material burst into a blizzard of debris. Amid the swirl were tube worms and a small crab. Bob picked up the temperature probe and placed it down the widened vent. The readings shot up to 259 degrees Celsius, or 498 degrees Fahrenheit. Happily for us, things were getting hotter.

John said he was encouraged with our progress but wanted to press ahead so we could check out the wider area. As we had approached Beard he had seen other chimneys down in the fissure, and wanted to explore them and any of their cousins that might be around.

First we did a television survey of Beard's base, which John said would help in the determination of its age. "It's easy for something like this to grow in the time it's had," he said, "if it's related to this event."

We focused on the border where the chimney grew out of the rocky pillows. The area was painted in dim reds and oranges. There was no large buildup of debris or sulfides around the base of the chimney, as can happen in old fields as flanges and chimneys clog up, fall, decompose, and new ones arise in numerous cycles. It grew straight up from the rocks. The camera also found animals galore. Snaking through the rocks around the chimney's bottom were a number of large rattail fish, their thin, sinuous tails wiggling slowly back and forth. Also poking around were assorted but unidentifiable snouts. Higher up on the chimney proper were thickets of sessile animals whose numerous spines made them look like bottle brushes. Each of these white animals was perhaps eight inches long and had hundreds of thin barbs up and down its body. Bob reached out with the robot arm and grabbed a handful for analysis up top. Tangled beneath and around the bottle brushes was the brownish fuzz, and crawling through this jungle were waves of the tiny crabs.

Bob said they looked very small.

"Smaller than normal?" John asked.

"Yeah."

"That's interesting."

Out my port in the rocks near the monolith I could see big orangish-brown spider crabs lurking in the shadows, trying to hide from the lights. They had big claws. These deep-sea crabs were different from the tiny ones on the chimney and were common inhabitants of the deep ocean, though not in the concentrations we saw. Atop the nearby rocks were large sea anemones with thin, graceful arms that extended like flower petals. Most interesting of all on the rocks were large networks of small pinkish organisms full of delicate tentacles. The colonies were cemented together on patches of whitish goo over the rocks. Rising a few inches from this gooey substructure were hundreds of individual stalks, each exploding at its top in starbursts of radiating arms colored in pinks and oranges. It was like a field of flowers. Both the anemone and starburst arms were generally reaching out in the direction of the chimney, perhaps to feed on the swirling debris in the water. Visually it was striking. All their attentions seemed focused on the monolith.

As we surveyed Beard's lower reaches, Top Lab called down to ask us about the camera deployment. John asked the opinion of the people up top. "I don't think it should go," Pat, the senior *Alvin* pilot, replied. "You have to sample, find the thing, get it back to the site. I personally don't think you've got enough time." John said he was inclined to agree, adding that we had no clear target for a time-lapsed observation. This chimney was already large and colonized. "It's not clear to me that they're three-month-old items," John said of the chimneys. "We'll do the best we can with video. But in terms of a yearlong time-lapse colonization study, it's not clear to me that's something we can accomplish easily in the two hours we have left."

Pat, eager to begin preparations for heading back to Astoria, told John he wanted to take the ship northward to pick up navigational transponders and flow meters. "Go for it," John replied.

This gave me pause. We were about to be temporarily abandoned by the ship, our ultimate lifeline. We were to be on our own at the bottom of the sea, our batteries running low. In two weeks of dives and troubles, no similar situation had come up.

We pulled *Alvin* back from Beard. Perhaps ten feet northward on the edge of the fissure were two small chimneys side by side, and beyond that a squat, larger one. We moved forward. The bigger one was apparently clogged and emitting no water. The two small chimneys were perhaps three and six feet high, obviously young and fresh. It was too late now, but they seemed to be ideal candidates for a time-lapse study.

Suddenly a loud, deep buzzing noise reverberated through the sphere. It was like a fire alarm gone haywire. I had no idea what was going on and tensed up, ready for emergency action.

DEEP ANEMONE. In frigid darkness, a sea anemone held its tentacles toward a volcanic chimney whose temperature was measured as more than twice as hot as boiling water.

John, continuing his observations without pause, calmly explained that it was just the ship blasting coded signals through the water to release the deep transponders from their bottom moorings. My heartbeat slowly returned to normal.

The smaller chimneys were joined at the bottom and alive with shimmers. Currents of hot water scintillated all over their surfaces rather than occurring in patches, as on Beard. The chimneys had no coating of creatures. The only life visible was a few microbial strands stirring in the currents. Their surfaces were mostly smooth and cream-colored. John said these small spires would probably produce the hottest water, and suggested that Bob knock the top off the tallest one to increase its flow. The robotic claw reached out, its fingers wide, and squeezed. Rather than just cracking or breaking a bit, the chimney collapsed in an explosion of swirling dust.

"It's all anhydrite," Bob said. "It just crumbled."

Anhydrite is a main building block of chimneys. The granular mineral is made of calcium sulfate. Forming only at high temperatures, and requiring no minerals for its origin other than those normally found in the seawater, it dissolves back into the sea when temperatures go down. Since it dissolves quickly, extinct chimneys and the cool surfaces of live ones usually have no anhydrite. Conversely, thin, young chimneys like those we were now examining were made principally of the crumbly stuff.

We waited for the murky water to clear. After a couple of minutes we could see that the two small chimneys were now roughly the same height. Pure gray smoke came out of the vent. Bob lowered the temperature probe down the widened throat. The readings shot up to 276 degrees Celsius, or 529 Fahrenheit, then rapidly fell off as the chimney's throat became clogged with debris. Things were getting hotter still. The young chimney was roughly 30 degrees hotter than Beard.

Bob knocked over the smaller chimney with the robot arm, revealing a nice, clean hole any plumber would have been happy to plumb. It gushed clouds of hot gray water. He put down the temperature probe and the readings soared to 280 degrees Celsius, or 536 Fahrenheit, seven degrees hotter than before. It was water-sample time, given the high temperature and heavy flow. Bob grabbed the sampler arm connected to the titanium bottles and moved it over to the flowing vent. Temperature readings from the sensor in its tip quickly climbed into the same range, at which point John flipped a switch to begin to pump water through the circuit to the sampling bottles, which were to be filled when the temperatures through the pipe were uniform.

Up to this point, our dive had been going smoothly from a mechanical point of view. No major systems had malfunctioned or failed. But the curse of oceanographers now descended on us.

As John began the pumping operation, the temperature sensor at the tube's rear exit showed dropping temperatures—just the opposite of what should have occurred as hot water flowed through its length. It was baffling. And time was precious. We needed to move on and survey the rest of the dark region. But good water samples ranked very high among the expedition's many goals. If nothing else, they might hold new kinds of hyperthermophiles, which would be a biological windfall. So John entered into a detailed discussion with the distant ship about how to solve the problem. For the moment, the northward-moving *Atlantis II* was still within hydrophone range. John speculated that the wires for the pump had somehow been reversed, and that water was moving the wrong way through the circuit. Top Lab concurred and advised that the flush pump should be turned off before the samples were taken. Since the flushing mechanism removed only a small amount of residual water from the pipe, the majority of the sample would still be water from the hot vent.

"Okay," John told the ship, "I'll wait until the input temperature is stabilized and then trigger the bottle, is that correct?"

"That's affirmative," said Gary, who had been called into Top Lab to help out with the problem. His voice through the speaker was weak, the ship increasingly distant.

Our detour had taken perhaps twenty minutes, not too bad considering the importance of the sampling. But things on the seafloor had deteriorated during that time. Bob was no longer getting high temperatures from the vent, try as he might to jam the temperature sensor down its throat. The problem was that new anhydrite kept clogging the throat—amazingly, the chimney was trying to grow right before our eyes! The chalky material was also forming on the temperature probe, threatening to cement it into the body of the growing chimney.

"We'll be here the rest of the night," John joked.

Bob kept shaking the robot arm to rapidly break down the new encrustations. Finally, clouds of gray water from the vent billowed up. Even then it was slow going, since any movement at all within the sub caused the sampling arm to shift and lose its high readings. I started doing breathing exercises, relaxing, willing myself to ignore my cramped leg muscles and a persistent itch. Finally, the temperature readings shot back up, and John called for a spring-loaded bottle to be triggered.

"Okay," said Bob, "number four is coming up." Our instructions were to start with the last bottle. Bob could see its plunger rising steadily, meaning the bottle was sucking in hot water. The fill took about thirty seconds. During that time we all remained perfectly still. The time was 12:38 P.M. After more than two weeks at sea, we had succeeded in taking the expedition's first sampling of hot water.

Bob managed to get high temperature readings once again, and we immediately filled bottle number three. Before we left the site, Bob used the robot arm to grab a chunk of sulfide from the nearby base of Beard. Since we were running out of room for samples, he put it in the "critter box" on the science basket. John also for the first time saw the gooey colonial organisms that had so captivated me earlier. Neither he nor Bob had seen anything like them. As gently as possible, using the robot arm, Bob scooped some of the creatures up for examination back on the surface.

As we prepared to leave Beard and survey more of the region, John said that one of our remaining goals was to look for age-related clues. "The real question," he said, "is whether or not these things are new."

"They don't look new to me," Bob remarked.

"No, they don't."

"This one looks pretty old, as a matter of fact."

While agreeing with Bob's general impression, John emphasized that the understanding of chimney evolution was immature, and that the life cycle of the monoliths was largely a mystery.

"We don't know how fast they evolve," he said.

"That's true," Bob agreed.

"And the basalt around it is really fresh," John added, "and there is no debris on the basalt."

There was also the apparent smallness of the creatures. Many seemed to be juveniles. One possibility, it seemed to me, was that this was an old field that had been reactivated by the heat of the recent eruption. That might explain the mix of small and large chimneys and the diminutive size of many of the creatures. Perhaps some had arrived only in the past few months. Perhaps this field wasn't so much a nursery as a condominium with a mix of young and old tenants.

"Let's go," said Bob. "I want to see what else is around here."

We moved northward along the edge of the fissure and down into it a little. Right away, Bob saw a large chimney perched on the rim, which he said looked like a church. To one side it had a pointed spire that looked very much like a steeple.

"This thing looks much hotter, much hotter," he said. "It's not quite as big as the last one but it's definitely hotter, shimmering, smoking."

We circled Church. Its hot flanks were almost barren of life, its surface a mottle of mineral whites and grays and oranges. Its upper reaches looked like fresh anhydrite, and its sides shimmered with thick waves of hot water. Many of the vents emitted dense gray smoke, indicating high temperatures. As we circled, John saw a chimney lower down in the fissure, and directed

Bob to go there. John was curious to see if elevation in relation to the fissure made a difference.

We dropped perhaps forty or fifty feet into the gloom. The lower chimney turned out to be very large and gnarled and covered with dense clusters of creatures. It had a couple of dead flanges that stuck out horizontally. All in all it was perhaps forty feet tall. Its head was tumescent and covered with life. No water came from its top, only shimmers up its sides. Its base was surrounded by rubble and sediment and brownish and orangish debris, unlike our first chimney. Surprisingly, it also had a number of large cracks on its flanks, which were edged with whitish stains and percolating with hot water.

"It's hard to believe this could have gotten started in three months," said John.

"No way," said Bob.

The microbial tendrils on its sides were longer and plumper than on the first big chimney, and packed into dense thickets. They undulated back and forth hypnotically in the shimmering currents, similar to the ones I had seen with Randy on the video aboard ship. Their contrasts and colors were vivid, bright whites intermingled with soft browns and oranges. At John's direction, Bob had a robot arm take out of the science basket a device called a slurp gun, which, like a vacuum cleaner, could suck up small things. Bob put it to work, zigging it back and forth through the microbial thicket.

"It's vacuum Bob," said John in the exaggerated voice of a television announcer.

"Why go for those cheap models when you can have the Electrolux," Bob joked in the same voice, while moving the chrome tube through the soft tendrils.

Ever since the start of the voyage I had wondered about a possible implication of the existence of all this life down at the bottom of the sea in the middle of nowhere. Now the thought came to me again as I eyed the darkness around chimneys that were dense with life. There was so much raw biological mass all around, so much material for the formation of long food chains, that it seemed to me not entirely farfetched to imagine that the ends of the chains down here in the depths harbored predators sufficiently big and ugly to warrant the designation *sea monsters.*

I asked John what he thought.

"We're here in the sub," he said, laughing.

John had us cruise up to Church, eager to sample water that promised to be very hot. As we settled down next to the chimney, Bob moved the sampler arm into a thick flow of gray smoke, and the readings started to soar. As Bob worked, the speaker came to life with an unintelligible crackle.

"They're calling us," John remarked.

"Tell them to switch ducers," said Bob, thinking a transducer on a different frequency might work better.

John called into the hydrophone: "Switch ducers, please."

There was a pause. The watery echo of his message came through the speakers: *"Switch . . . ducers . . . please."* But nothing else that was intelligible, only garble from the ship.

"I can't quite hear what you're saying," John said into the hydrophone. "We're sitting at the base of a structure using the manifold sampler and understand that we need to leave soon."

More garble from the ship.

John gave up trying to communicate and turned his attention back to the sampling work. I was uneasy but said nothing. It felt lonely to be so deep, to have your support ship calling, and to have no idea whether the message was routine or urgent. Intellectually I knew the chances of the latter were extraordinarily slim. Nonetheless, my heart rate had risen. Again, I told myself to calm down.

As Bob had predicted, the temperatures of Church were in fact quite high. He deftly positioned the sampler arm and tripped bottles two and one, both of which worked perfectly. Then he took out the dedicated temperature probe to measure the vent's exact temperature. The readings quickly rose to 284 degrees Celsius, or 543 degrees Fahrenheit. It was the hottest water yet —more than a hundred degrees hotter by the Fahrenheit scale than our first sample. Church was percolating. Its blistering heats could melt lots of elements and a host of modern plastics. I had no idea whether it could melt *Alvin*'s plastic windows, and didn't particularly want to know.

Static came through the speakers. This time the fuzziness formed itself into an indistinct but audible voice.

We hailed the ship.

"Read you much better now," Top Lab crackled in response. "Transponder A is now aboard. We are proceeding down in your area, will get good nav and a positive fix on your location again. Then we'll determine whether to pick up the C transponder or you guys first."

John said aloud, but not into the microphone, that the ship should take the C transponder first.

Bob rejected the idea.

"We're going to have to leave here," he said. "I don't have enough power to drive to the surface." Since ascent with the aid of the sub's propellers was faster, a return could be somewhat delayed, but our low battery level ended that option. Soon we would have to drop our remaining steel weights and begin our slow rise to the surface by means of buoyancy alone.

Bob stowed the sampling arm. A camera pan showed a juvenile chimney at the base of Church that was really chugging, emitting dense clouds of gray smoke.

"Jesus, look at that," said John, obviously impressed and probably frustrated that we had no time to sample its waters.

Our final act was difficult. Although we had earlier decided to pass up the time-lapse camera, we still carried in our science basket a large temperature sensor powered by an automobile battery. In theory it could work for a year or more tracking the changes in a chimney's temperature—changes that might reveal some of the subtleties of chimney evolution, especially because this one was so young and hot.

John asked if the temperature probe could be put in the vent we had just sampled.

No problem, Bob replied.

The trouble was getting good video of the setup, which was important for eventual analysis of the data. After all, the chimney might change and grow dramatically during its year of monitoring. In lieu of a deployed camera, we needed a good visual fix on where the data initially came from.

Bob said that as soon as the eighty-pound battery was dropped, *Alvin* would start to rise, making it difficult to survey the setup on video. Normally as a countermeasure he could take on eighty pounds of water in the ballast tanks to compensate for the dropped load, thus remaining neutrally buoyant. But in an active chimney field the water was too acidic and corrosive for that. It could damage the tanks, he said. So deploying the gear had to be our last substantive act.

Bob pushed the probe down the throat of the hot vent, grabbed a yellow rope handle that looped around the battery, gently lifted the heavy mass, and swung it over next to Church.

All this activity stirred up muck in the water as *Alvin*'s thrusters worked to keep us down. Bob panned the camera over the site as we started to rise. I noted the time, 1:56 P.M., some six hours since we had entered the sub. The battery's position next to the chimney was evident through the murky water, but the probe's was not. Frustrated, John directed various lights to be turned on and off, and rotated his video selector switch through different channels. He gazed out his window. We were moving up slowly, with Bob slowing the ascent by means of the sub's thrusters, so that John could make last-minute observations.

"I guess I do see it," said John. "Can you drive down? I can kind of make it out, Bob, but can't quite . . ."

"Oh, shit."

"What?"

"We're over the top of something hot."

Temperature sensors in *Alvin*'s skin showed the hull was starting to heat up, a situation that was potentially deadly. Rapidly Bob swung the sub to one side, off the threat.

John, ever the scientist, ever the oceanographic crusader in search of Moby Vent, was absolutely nonchalant and kept directing all his energies to getting the best possible video of the experimental setup. Without pause he directed Bob to churn the thrusters to swing *Alvin* around the chimney and down low for better viewing.

"I can't see where I am," Bob grumbled as the thrusters raised clouds of muck.

John, with clearer vision out his port, became Bob's eyes, telling him to turn to the right to bring the chimney into full view. "Swing it past my window and then let's get out of here. I just need to get a visual image of the whole thing."

Finally, he could see where the temperature probe was placed.

"God, that's perfect," said John. "Let's get out of here."

It was 2:00 P.M. We started to rise as Bob requested permission from the ship to surface. Unexpectedly, Top Lab said to stand by and hold position.

"Ship traffic," it gave as the reason.

We waited, suspended between worlds.

"Okay, *Alvin*," Top Lab called down, "you're clear to surface."

Out my observation port I watched the bottom fade from view. Somehow it now looked beautiful, not eerie. I could see no chimneys, even though the ones we had just investigated were quite close by.

On the way up, I watched the flashes of living light and wondered what else lay out there.

FOUR

Lost Worlds

ANGRA BAY IN THE AZORES is hardly ever visited today. It's a backwater, an out-of-the-way pinprick on the map of the North Atlantic. But during the age of discovery, starting late in the fifteenth century, it was a major crossroads of the maritime world.

The sheltered cove was visited by thousands of ships returning to Europe from the Orient and the Americas, their holds full of gold and silver, silks and spices, pearls and diamonds, porcelains and fine steels. To weary seamen and colonists alike, Angra and its environs on the small isle of Terceira were an oasis of earthly delights. The climate was mild. The springs were sweet. The soil was volcanic and fertile, perfect for fattening cattle and growing grains and citrus and grapes. Mariners were gladly accommodated by the taverns and merchants of the city of Angra, who turned a tidy profit and multiplied to fill the growing demand.

Jan Huygen van Linschoten, a Dutch traveler who lived in Angra for a few years late in the sixteenth century, drew a panoramic overview of the place, which showed orchards, rolling fields, and a tidy city of Renaissancian order set by a bay where galleons rested at anchor and tall ships came and went, their sails full. His picture overstated the prominence of the harbor, underscoring its psychological significance. But what ultimately made the area so important was not the anchorage or the bounty of the nearby land. It was the wind.

NORTH AMERICA
ANGRA
PORTUGAL
AFRICA
SOUTH AMERICA
N

The North Atlantic has a system of fairly steady breezes that blow past one another in opposite directions as part of a large clockwise rotation of winds and currents. Through trial and error, the navigators of the age, beginning with Columbus, discovered that their sails could catch easterly winds in the lower latitudes near the Canary Islands off the African coast and ride the breezes across the Atlantic to the Caribbean and the Americas. To return home, they simply moved northward along the coast of Florida to catch the westerlies that blew them back toward Europe. Their return was aided by the Gulf Stream, which nudged the wooden ships gently northward and then eastward. The winds and currents almost invariably took the mariners into the Azores, an archipelago of nine volcanic isles spread generously over the sea and hard to miss. A thousand or so miles west of Lisbon and politically part of the Portuguese state, the Azores were an obligatory last stop for returning ships. And the isle of Terceira ("third" in Portuguese, for its third place in the order of discovery, in 1427) had a protected harbor that mariners eagerly sought out.

Each year large fleets of Portuguese and Spanish ships coming from the New World would stop at Angra (which means "bay" in Portuguese), particularly between 1580 and 1640, when the two Iberian states were politically one. Adding to the influx over the centuries were convoys from Asia. In the wake of Vasco da Gama's discoveries, so-called East Indiamen sailed out of Europe and slipped around the southern tip of Africa to trade with India, China, and Japan. They also sailed through Indonesian waters to the Spice Islands. On their long return voyages, these ships rode far out into the Atlantic to catch the prevailing winds and currents, which, as with the New World fleets, took them into the Azores.[1]

The wealth that flowed through Angra was immense. The city was a stepping stone to empire. Spain and Portugal in the century after the opening of the sea lanes enjoyed a monopoly in the often grim art of global conquest, ruling over the world's first dominion on which the sun never set. Their acquisitions stretched from Macao in China to Potosí in Bolivia and produced a river of riches from trade and plunder that flowed in staggering quantities back through the Azores. Some of this wealth—over time, much of it—inadvertently stayed in the isles.

As a rule, the incoming ships tended to be in bad shape. Hulls after so many months and years at sea were often dense with barnacles and riddled

TREASURE WINDS OF AZORES. Early explorers in the age of discovery found a clockwise whirl of winds and currents that ultimately carried tall ships into the Azores. Hundreds of ships headed for Europe eventually sank there, littering the seabed with gold and silver, pearls and diamonds, porcelain and fine china.

with worm holes, which slowed travel. Men were weak and crews thin, owing to fever and scurvy, death and desertion. Exhausted sailors often faced extreme danger. Hurricanes lashed the isles every few years, sinking many ships each time. Another foe was fog and sudden shifts in the trade winds, which made navigation perilous along volcanic coastlines, where pinnacles of rock stuck out like jagged knives. One of the greatest dangers was men. Privateers and pirates lurked about the isles to prey upon returning treasure ships, as did French, English, and Dutch warships. The Spanish were known to set their ships ablaze rather than let them fall into enemy hands. The builders of Angra endowed it with massive fortifications, which bristled with cannon and succeeded in thwarting many aggressors. Sir Francis Drake attacked in 1589 and the Earl of Essex in 1597. Both were successfully repulsed.

Angra Bay, despite its beauty and that of the surrounding isle, was often a deadly place. One way or another, by acts of God or man, hundreds of ships and men that sailed for the sanctuary of its waters instead ended up far beneath the waves, the voyages unfinished, the dreams unrealized.

One of the most famous episodes featured the destructiveness of both man and sea. The English, emboldened by their rout of the Spanish Armada, sailed on the Azores in 1591 with a squadron of thirty vessels, aiming to seize a treasure fleet. But the Spanish rallied and surprised the English with a force of fifty or so galleons and warships. The two sides exchanged broadsides. With great daring or foolishness, depending on your point of view, an English commander by the name of Sir Richard Grenville attempted to scatter the Spanish by sailing the *Revenge,* a celebrated warship, into the heart of the enemy formation. His ship was quickly becalmed by the big Spanish sails, seized by grappling hooks, and boarded. Hand-to-hand fighting raged through the night, the decks slippery with blood. (Not a few works of fiction are based on the battle.) Grenville urged his men onward, everywhere at once, somehow surviving amid the growing carnage. But finally he collapsed, bleeding from many wounds.

The next day, the dying Grenville surrendered to save his surviving men —to no avail. As the Spanish threw his body overboard and sailed for shore, a great hurricane descended on them, sending dozens of ships to the bottom, including what was left of the *Revenge* as well as such Spanish galleons as *Santa Maria Del Puerto, San Juan Baptista,* and *Nuestra Señora Del Rosario.* Thousands of bodies washed ashore. The islanders came to believe that Grenville was in league with the devil, who took revenge on them all. The British came to view Grenville as a great hero.[2]

Tennyson bemoaned the loss in a long poem, "The Revenge," concluding in its final lines:

And or ever that evening ended a great gale blew,
And a wave like the wave that is raised by an earthquake
grew,
Till it smote on their hulls and their sails and their
masts and their flags,
And the whole sea plunged and fell on the shot-shatter'd
navy of Spain,
And the little Revenge herself went down by the island
crags
To be lost evermore in the main.

Or so it seemed.

I arrived on Terceira early one morning late in the summer of 1995, tired and edgy after days of hard travel. In Angra, tall churches and houses with red tile roofs were packed close together on narrow streets paved in dark volcanic stone. It was beautiful, but I was too distracted to play tourist. In my hotel room, I opened the curtains and caught my breath. The bay was spread out before me, its periphery rough with black rocks, its surface glistening. I threw open the window and the balcony door. The waves rolling onto the beach filled the air with a continuous sigh. Scattered clouds swept the sky from the west.

The genius of the bay was evident at a glance. Rising on its western rim was an extinct volcano that formed a huge natural jetty, its slopes a peninsula that connected the volcano to the mainland. It obviously sheltered the southward-looking bay from the westerlies and the storms carried by the trade winds. The peak was also an ideal place to spy incoming ships. The whole base of the volcano had been transformed into a dense stone fortification that climbed to complex ramparts dating to the sixteenth century. The steep stone walls seemed unscalable from the sea. Closer to the city and my hotel, at the volcano's base, was a huge old fort, which a brochure said had once kept treasures safe while galleons were at anchor. The bay's waters held no cruise ships, no sailboats, no moving craft of any kind. It was dead in terms of tourists and water sports. Nobody was on the beach and its dark volcanic sand. A half-dozen fishing boats were tied up at the municipal pier on the cove's eastern edge, sheltered beneath another old fort. Angra Bay was beautiful and nearly lifeless. And its depths seemed to interest no one, an impression I knew was remarkably wrong.

For decades quiet battles had been waged over the fate of the precious rubble that was strewn across the bottom of the bay and around the whole isle of Terceira. By some estimates, it was the world's greatest concentration of treasure wrecks. The salvage war began in the 1950s and '60s as divers used newly available scuba gear to find, in the bay's relatively shallow waters, any number of lost ships. The clue was usually a jumble of cannons. Some of the adventurers stole artifacts. Others sought governmental permission to open the area around Terceira to organized recovery. Lisbon over the years took no action other than to keep the area officially off limits. The Navy raised some cannons. Bureaucrats debated some issues. But in the end, mainly by default, it was decided that the treasures should remain for the time being in the safekeeping of the sea.

The issue went far beyond gold. For archaeologists and their study of bygone peoples, lost ships in general were great prizes because they could inform about whole civilizations, not just leaders eulogized in tombs and monuments on land. They were cities in miniature, cross-sections of society frozen in time. Their concentrated nature was suggested by their nickname, time capsules. A good example was the *Mary Rose,* a flagship of Henry the Eighth that sank in 1545 in shallow waters off Portsmouth, England, as the king watched in horror. Treasure hunters might have ransacked the wreck for gold coins. Instead, the ship, after its discovery in 1971, was painstakingly unearthed by marine archaeologists. Up from the mud came cannons, shot, gun carriages, longbows, arrows, wicker baskets, barrels, jars, bottles, bowls, razors, mallets, peppercorns, chessboards, leather purses, shoes, flutes, dice, and even hay from the mattresses of sailors. In short, it was a remarkable look at Tudor life.[3]

Most shipwrecks around the Azores were safe from plunder and study because the waters are so deep compared to most continental shores. There is no shelf. Instead, the islands are old volcanoes whose slopes quickly drop off a mile or more into abyssal darkness. This put the majority of the wooden shipwrecks out of reach. It also enhanced their appeal. Through trial and error, scuba divers over the decades had learned that wrecks in shallow waters were often a mess. Around Florida and the Caribbean, treasure ships caught in hurricanes had frequently been torn apart by coral reefs and their riches scattered over miles of seabed, making comprehensive recovery nearly impossible, especially after shifting sands and corals buried the gold and silver. But many of the wrecks around the Azores, at least in theory, were sitting on the bottom right where they had gone down centuries earlier, with the deepest ones perhaps in unusually good condition because of the icy temperatures.

The impossibility of rescuing these lost ships began to change in the late 1980s and early '90s as a wave of advanced technologies opened the deep to civilians. Quite suddenly, the riches around the Azores were in play.

The allure was especially great for a soldier of fortune by the name of Robert F. Marx. A treasure hunter and self-taught archaeologist, he had first glimpsed the Azorean graveyard in the fifties as a military diver traveling with the U.S. Navy. Ever since, while finding hundreds of other shipwrecks around the world, the stocky ex-marine had unsuccessfully pestered Lisbon for salvage rights. Now, in his sixties, he pictured the Azores and its hundreds of wrecks as the crowning glory of his career.[4]

Marx threw himself into fathering a Portuguese law that was up to his ambitions, and succeeded in having one written that reversed the presumption of state ownership over old artifacts found in Portugal's territorial waters and opened the way for free enterprise. It was tempting. After all, Portugal after its early glory had become the poor sister of Western Europe and had virtually no funds for underwater recovery. Now the country was facing a rising wave of high-tech piracy as salvors and robots dove surreptitiously to the seabed to scavenge for loot. Better to exploit such forces rather than trying in vain to suppress them, went the logic. The law passed without serious opposition in July 1993. Under its provisions, the salvor was awarded up to 70 percent of the wreck's value, depending on the difficulty of the job and the estimated value of the finds. If Portugal judged any artifacts to be particularly attractive, it had the right to buy them back. Exploration licenses were to cover up to one hundred square miles.[5]

Marx requested one for Angra Bay and a large swath of sea around Terceira, claiming his interests were mainly historical.

"I've found so much gold and silver in my life that I'm sick of it," Marx told me on the phone from Lisbon as he lobbied for an exploration license. "This is different. It's my childhood dream come true. It's the only place I know where there's hundreds and hundreds of intact ships."[6]

In readying for the job, Marx was like a consumer in a supermarket. He was overwhelmed by choices. Makers of cold-war gear were almost giving it away. Westinghouse promised to loan him a laser line scanner so he could probe the Azorean depths with its concentrated beams of light, sweeping the darkness for old wrecks. Graham Hawkes, the San Francisco inventor, wanted to join the hunt and eagerly sized up military gear, including a big Navy research ship, for deep exploration. The Russians were ready to go with the *Keldysh* and the ship's *Mir* submersibles, which could easily plunge to any Azorean depths. In addition to all this gear, Marx had lined up a group of wealthy investors who wanted to finance the hunt.[7]

A number of Portuguese companies and groups were also galvanized by the new law. But in general, they had little money for probing the deep and were forced to limit their hopes to shallow water, usually planning to work with scuba divers.

Many Portuguese citizens and scientists appeared to hate the new law, seeing it as sanctioning the theft of the state's cultural patrimony. Its implementation soon bogged down in fierce debate as its pros and cons were argued in scores if not hundreds of articles and television shows. The dispute was usually portrayed as scientific virtue versus commercial evil. And Marx, to be sure, was cast as no angel.[8]

The bane of the treasure hunters was Francisco Alves, the outspoken director of Portugal's National Museum of Archaeology and a member of the Shipwreck Commission, a government body that was administering the law. Almost single-handedly, Alves had succeeded in weakening the act and slowing its implementation, even while denouncing Marx as a pirate.

I had come to Terceira knowing that Alves was on the isle to teach a crash course in underwater archaeology for young Azorians, the archaeologist being eager to create a political and technical counterpoint to the treasure hunters. The fieldwork was being done with scuba gear, though it was unclear to me whether the group was planning to dive in the bay. On that chance, however, I had brought along my fins, mask, and wet suit.

Stiff from an early-morning flight, I donned my jogging shorts to search out Alves. Most houses in the city seemed to have balconies above and shops below. Moving past the old cathedral and the main square, I headed up a hill toward the city museum, where Alves had been holding the classroom part of his archaeology course. The museum was housed in a former church that was very old. On Vasco da Gama's return voyage from India, he had stopped in the Azores and buried his brother Paulo inside the church, a common practice in those days.

A plaza beside the museum held dozens of big cannons from sunken ships, many of them made of iron and heavily pitted from the salty waters of the bay. Inside, Alves was nowhere to be found, but evidence of his presence was scattered throughout a wing of the old church that was obviously being used as a staging area for the students. Amid old stonework and ancient carvings were air bottles, red-and-yellow surveying rods, ropes, measuring tools, numbered weights, and a chalkboard. It was a striking mix of old and new. The equipment was clearly for helping the students learn how to survey and document old artifacts found beneath the sea, a venture not without risk. A bright red sticker on an air compressor advertised the telephone number of DAN, the Divers Alert Network, a group that seeks to prevent and treat such scuba emergencies as decompression illness, the dangerous malady otherwise known as the bends.

Inside the old church was an exhibition of artwork and instruments from the golden age of Portuguese exploration, with emphasis on the role of Angra. Maps and paintings showed the bay and old city in various phases of their power and glory. The exhibit was a remembrance of things past, and perhaps an unobtrusive call to arms. At the front desk, the museum sold large prints of van Linschoten's sixteenth-century drawing of Angra Bay, an artful reminder of early renown.

After a shower at the hotel, I happened upon Alves in a restaurant, and we headed back to the museum together. The man was fiftyish and balding. Puffing a pipe, bespectacled, he wore jeans, T-shirt, and fanny pack—sort of a cross between a scholar and a tourist.

"We are not some undeveloped country," he told me as we crossed the city square. "Marx is wrong if he thinks he can come in here and dictate his agenda. We have history. We have pride."[9]

Despite his informal dress and manner, Alves was a power to be reckoned with in Portugal. His address in Lisbon was one of the best—the Jerónimos monastery and church, a sprawling late-Gothic masterpiece of spires, vaults, and ornamentation on the banks of the Tagus River, built beginning in 1501 to honor Vasco da Gama's discoveries. By the middle of the sixteenth century, Lisbon was the richest city in Europe, brimming with confidence, filled with buildings and monuments worthy of an imperial power. Jerónimos was the epitome of that pride. One of its palatial wings housed the National Museum of Archaeology, where Alves was installed as director in 1980 by the socialists. After Marx's successful legislative push, Alves lobbied against the law as the museum's director, as a member of the Shipwreck Commission, and as president of the Archaeonautica Studies Center, a nonprofit group associated with the museum. Its 250 members include archaeologists, geologists, engineers, jurists, architects, students, and scuba divers. Archaeonautica ran the classes in underwater archaeology, which began in Lisbon in October 1994 after the law's passage and were intended to help block the rush for sunken treasure.[10]

Rumor had it that Alves was about to quit the museum altogether and go full time into archaeological fieldwork, seeking to make Angra Bay a model of integrity in the art and science of recovering old artifacts from beneath the waves.

Sitting by the cannons saved from the sea, Alves said four dozen people from the Azores had just finished the course and would constitute an early-warning network to monitor the archipelago's waters for pirates and new discoveries, which he said were often made accidentally by fishermen hauling nets over the bottom. He disparaged deep-water exploration and its costly gear, saying the depths around Terceira were so rich in wrecks that relatively shallow waters reachable by scuba divers would present sufficient challenge for generations of scientists. Galleons in the deep would keep just fine, he said. Bringing them up now and struggling to preserve them on land would drain scarce archaeological funds from the serious business of doing wide surveys of the cultural riches hidden in the sea.

"The treasure hunter's point of view is quantitative, of wanting more and more," Alves said. "The archaeologist's is qualitative. You must dig as little as possible—that's an archaeological law. To dig is to destroy. So you measure. You make a grid. You observe every detail before removing anything. It's like Sherlock Holmes. And it assumes the superimportance of documentation, of recording your observations, of taking notes and photos. With treasure hunters, it's always dig, dig, dig. 'This one has no treasures, let's go to the next.' They do double damage. They have not studied what they have discovered and they have exposed it to the elements."

Happily, I learned that Alves and the instructors were planning to dive in Angra Bay that afternoon. Yes, of course I could go along. They would find me a tank, regulator, and weight belt. For the instructors, it was a reward for two weeks of teaching scuba-based archaeology at a remote beach. The dive was to be the group's first exploration of the bay.

We walked through the old church, stepping on old gravestones and climbing a dark stairway to a room used by the archaeology class. Its wooden ceiling was heavily beamed. On a chair was a nautical chart of the Azores in which the islands were encircled by masses of colored pins. According to an index, each pin represented a wreck—yellow for the twentieth century, green for the nineteenth, black for the eighteenth, blue for the seventeenth, red for the sixteenth. No wrecks dotted the northern coast of Terceira, only the southern half of the island around Angra Bay.

At a glance, the map showed that Terceira's sunken ships were far and away the oldest and most abundant of the archipelago, as indicated by a dense crowd of blue and red pins. The chart also showed that the waters just south of the pin-denoted shipwrecks reached a depth of 1,573 meters, or about a mile. The second-greatest concentration of wrecks lay around the isle of São Miguel, the third around Faial, the fourth around Flores. Clearly, someone had been doing his or her homework.

Alves introduced me to Paulo Monteiro, a young man who was vice president of the Friends of the Museum in Angra, which had been founded in 1994 as the controversy over the law heated up. The group was helping to sponsor the archaeology course.

Paulo told me, "Up till now, there've been 535 registered shipwrecks, the oldest from 1515. We have 138 documented in Angra Bay alone. A main reason for them was shifting winds that drove ships onto the rocks. They called them carpenter winds because of all the wood.

"In a way," he added, "we're in debt to the treasure hunters. We're Portuguese. We tend to do nothing unless there is grave danger. Now we have to run hard to counteract them."

The dive party would split in two, I was told. Alves and Paulo were to set out from the eastern side of the bay to survey its central area with dive scooters—small motors that would help them cover a bigger area than otherwise possible. The larger portion of the party would head out from the opposite shore, on the western side of the bay, under the old fortifications and volcano. That was my group.

I picked up my dive gear at the hotel and walked to the waterfront. The bay was small, less than a kilometer wide. At the base of the old fortifications on the western rim was a modern concrete landing ringed by a railing and lined with a dozen or so traditional Portuguese fishing boats, bottoms red, tops white. At the far end of the walkway, twenty or so concrete steps led down to the bay's waters. I had looked at the area in the morning during my jog. The water then was gin clear, and the bright sunlight had made it easy to see to the rocky bottom. Now the water was murky. The wind had shifted to the south and the sky had clouded up. A light rain fell. It looked as if nature was going to be no help.

The old drawing of Angra by Linschoten had depicted the west side of the bay as the main anchorage, with galleon after galleon lined up in a neat row. That was our dive site. Though excited to get in the water, I worked to lower my expectations. Old wrecks that have lain at the bottom of the sea for centuries are invariably in bad shape, rotted and covered with sand and sediment. There are no treasure chests with lids ajar, spilling jewels and doubloons. In fact, little wreckage is usually visible, especially in shallow waters, where animals and warm temperatures and surface agitation from storms speed burial and decomposition. Modern undersea recovery is largely the art of digging very carefully into chaotic heaps to see what may have survived the ravages of time. An unknown at this site was the extent to which such arts might already have been practiced on the sly.

Eventually the group showed up, a dozen people in all, almost evenly divided between men and women. They were mostly in their twenties and thirties, about half instructors from the mainland. A prominent exception was Valdemar Reis, a portly man who lived in Angra and dove for a living in Terceira's waters to hunt octopi, which he sold to restaurants. Valdemar was an expert on the bay's undersea features. He was our guide and clearly a serious diver. His face was ruddy, except for where the hood of his wet suit fit around his neck, cheeks, and forehead.

Everybody suited up. I was given some dive gear, for which I was grateful yet about which I was anxious. I had recently won my advanced certification in the States and had learned to emphasize safety above all else. Scuba diving can be tricky. The pressures mount rapidly as you descend and can do great physiological damage, at times life-threatening. So I was surprised to find that the air regulator loaned me had no built-in depth gauge, usually a standard item. And it had no hose for inflating a buoyancy vest. Worst of all, the rig had no pressure gauge to warn when the tank was running low on air. Most modern regulators have four or more hoses radiating from the air tank, each doing a different job that bears on safety. Mine had a single hose leading to a mouthpiece.

My diving buddy was Fernando Rodrigues, a short man with a commanding air who was also the dive leader. He said if I ran out of air we would share his. As we reviewed the hand signals for air cutoff, I silently vowed to stay close. My second dive buddy, Tom Spiker, a lanky Dutchman who lived on Terceira, also offered to share air with me if I needed it.

Fernando briefed us on the plan in the drizzle. It was to be no pleasure dive but an application of the principles the instructors had helped Alves teach for two weeks. Carrying various measuring tools, the divers would fan out and, upon finding an artifact, go through a routine of measuring it, photographing it, and then sweeping around it in a circle to hunt the vicinity for other items. Fernando had a Nikon underwater camera, a yellow rope for the circular hunts, and a lot of other gear. I was to help him by carrying a white plastic slate on which he could take notes.

Scuba divers when suited up in all their gear carry a lot of weight, making it hard to move around out of the water. Clumsily, and very carefully, we lumbered down the cement steps, which were slippery near the bottom. One by one we plunged into the sea.

Still on the surface, we swam as a group out a hundred or so meters from shore. The waves were less than a foot high, so the going was easy. The old volcano loomed above us. Luckily, the rain was letting up and the sky clearing, giving us a little more daylight for bottom illumination. Fernando, clearly in charge, slowly made eye contact with everyone in the group and verified that all was well. Then, at his signal, we all went under.

Before long the bottom came into view. Large boulders were all about. Scattered among them were hundreds of rounded rocks a foot or so across. Rumor had it that they were all ballast stones from centuries of shipwrecks. It seemed hard to believe, given their number. They stretched as far as the eye could see. In theory, the stones could be cracked open and examined to determine their place of origin, local or foreign, to see if they were made of lava or continental granite.

At the bottom, Fernando proceeded to get tangled in his gear, the rope particularly. The simplest task on land can be diabolically complex under water. It took a while to undo the mess, but Fernando did so calmly and then gave me the rope, silently acknowledging that he had tried to carry too much equipment himself. We glided over the bottom. The large boulders gradually diminished in number and then ceased altogether. Gone too were the other divers, who had disappeared into the haze. The three of us—Fernando, Tom, and I—were on our own.

We came upon a long straight object. Tom and Fernando swam up to it, pulled out their knives, and cut into the encrustations, apparently thinking it might be a cannon. Their investigation left the issue in doubt. We swam on.

Eventually we came upon a group of divers clustered about something on the bottom, working away feverishly with their hands to remove sand and sediment. It turned out to be the wooden ribs of an old ship, rotten but still clearly recognizable, the ribs all lined up in parallel rows. The area to one side was sandy, while the other side, from which the ribs were protruding, was dense with rounded stones.

Fernando signaled all the divers to pull back. A collapsible ruler was unfolded and laid down, after which he photographed the site. It all happened in slow motion, like a ballet. Afterward, Fernando motioned to me, got the slate, and took some notes.

Our team swam away on the periphery of the stone field, letting the other divers continue to excavate the ribs. We quickly came upon a pile of yellow and reddish ceramic pieces, some glazed on one side, clearly remnants of ancient jars that had been shattered. No piece was much larger than a hand. Tom and Fernando gathered some of the larger ones into a pile atop a rock. Fernando swam upward, paused, and then hung motionless as he pointed his camera downward. Afterward, the pieces were returned to their resting place.

Swimming back, we passed the excavating divers, and further on we came to another place where a ship's ribs were exposed, as well as pieces of planking beneath them. Again, the area from which the woodwork protruded was dense with the rounded stones.

At this point, the immensity of it all began to hit home. I had been absorbed by the details, the rocks and wood and shards. But this was much more than that. It was a shipwreck, the remnants of a city of the sea that centuries ago had teemed with activity and human drama. Perhaps the ship had sailed through the Straits of Banda and Malacca to carry spices from distant Pacific isles. Perhaps it had been heavy with Inca gold.

I began to agonize over whether to take a bit of the wreckage, to capture part of this lost world. On land, I had engaged in no discussions with Fernando or anybody else about picking things up. It never occurred to me. I had no idea what the group's ethic was on recoveries, or whether they had instructions on the topic from Alves. But I was seized with an urge to capture a bit of this bygone world. I also realized, however dimly, that the impulse was destined to achieve little. I nonetheless decided to follow it.

The wood was in terrible shape, rotten and full of worm holes. I pulled on an edge of a rib and it easily gave way. I put the small piece, no bigger than a large coin, under the sleeve of my glove for safekeeping.

Near the exposed ribs was another pile of broken pottery. I lifted a small piece and held it in my hand amid the coils of yellow rope, having nowhere else to put it.

Tom and Fernando were surveying the edges of the wreck, making measurements and taking photographs. I decided to do my own survey of the wider area. Somehow, my earlier fears about running out of air had vanished.

I passed over a sandy region in a direction I judged to be away from the shore and out to sea. Rather quickly, I came upon another old wreck. Parts of its wooden skeleton stuck out from the bottom even higher than the first wreck and were in a better state of preservation. Excited, I swam back to get Tom and Fernando. At the new site, Tom pulled out his knife and took a sample of the wood.

We swam back, passing again over an area of dense stones. Suddenly, not far from a group of giant boulders, Fernando swam down to the bottom and began to carefully examine a long object. It was heavily encrusted, and at one end it was hollowed out. As Fernando had done before me, I put my fingers carefully inside, feeling around the circular hole. No doubt about it. The object was a cannon.

Fernando laid down a plastic measuring tape along the cannon's length, but currents kept pushing it off. After several false starts, we finally succeeded in an arrangement whereby Tom held one end of the tape and I held the other. Fernando then photographed the cannon.

Tom swam off with Valdemar, the octopus man, and Fernando and I ventured further out to sea. Repeatedly we passed over large sandy areas interspersed with piles of ballast stones. Finally, we came across the ghostly wreck of a large modern ship, far deteriorated, its steel ribs sticking high out of the sand. Swimming through its skeleton was a large school of jack mackerels, a bluish fish. Fernando, clearly happy with the dive, threw his arms open wide to embrace them.

We surfaced and slowly swam back to the cement stairs, having explored the bottom of the bay for more than an hour. The sky had cleared up. Scattered clouds were soft and low as the sun began to set. It was beautiful. We climbed the long steps and took off our heavy equipment.

Nowhere to be found was the tiny bit of wood I had broken off, despite all my agonizing. The same was true of the ceramic shard. Both were gone, kept by the sea.

Little by little, the other divers came back to shore, elated by the adventure. Dripping wet, smiling and laughing, they compared notes and enthused about the richness of the shipwreck cemetery. More than one person remarked that there were centuries of archaeological work to be done at the bottom of Angra Bay.

Valdemar, the octopus man, the hood of his wet suit pulled back to reveal graying hair, seemed more relaxed than the other divers, having seen much of the bottom before. He said he had found two large cannons at a different spot from ours and knew that there was much more to be discovered down below. Once while hunting octopi, he said, he had stumbled on a wreck that quickly yielded gold and silver coins.

"It has taken many years for people to realize how important this is," he said, leaning on the rail, gazing at the water.

Further along the landing, Fernando and other divers were comparing notes. Miguel Aleluia, a tall man with a cigarette dangling out of his mouth, who lived in Porto on the mainland, held a large slate on which he was drawing a composite picture of our investigations, based on all the individual reports.

Like a puzzle, the emerging picture had many parts—the cannon, the pottery shards, a loose bit of wood, the two exposed-rib sites, plus a third one I never saw. What I hadn't appreciated on the bottom was that many of the individual sites were tied together by a single array of ballast stones, indicating a single wreck. Lying amidships was the cannon Fernando had found. Miguel said no overall measurements of the site had been taken but the shipwreck debris seemed to have a width of about twelve meters (or forty feet) and a length of about twenty meters (or sixty-five feet).

"The problem is too much to do and not enough time," Miguel said, sighing as he sketched his picture.

Even so, in an hour the divers had captured an impressive amount of detail. At the second of the two rib sites, careful measurements had been taken of all the exposed wood, down to the gaps between the ribs (four centimeters), the width of ribs (fifteen and sixteen centimeters), and the width of exterior planking (twenty-eight centimeters). Miguel said analysis of such measurements could reveal much about the size, age, and provenance of a ship. It appeared, he added, that our shipwreck was from the sixteenth or early seventeenth century.

The divers had also measured the distance between the wreck and the large neighboring boulders (eight meters, or twenty-six feet). In their informal analysis, they agreed that the ship's end quite possibly had come as violent winds, carpenter winds, threw it against the boulders and crushed its side.

We had found no gold, but a dozen people on a short dive to the bottom of Angra Bay had come back with the bones of history, something that is perhaps more valuable.

It was dark outside and getting late. The group got ready to head back to its quarters at a youth hostel outside Angra. Yes, of course I could come along. Miguel, the man with the dangling cigarette, and his girlfriend, Margarida, agreed to ride in my rental car to show me the way.

We drove over roads that were incredibly bumpy. The hostel was about four kilometers outside of Angra, isolated on a majestic point of land. It was a modern structure with big windows.

After showers, the instructors gathered in a large living room and examined three or four bathymetric maps of the waters around Terceira, their minds clearly ablaze with visions of shipwrecks in waters far deeper than those we had just explored.

"That's why it's a paradise," Rita Cortez, one of the instructors, said of the deep wrecks as she inspected the maps. "Nobody can get down there with scuba equipment."

We drove to the fishing village of São Mateus da Calheta for dinner at a seaside restaurant. Alves joined us. The place was empty because it was so late. We arranged the tables family style and had a good meal, marked by much joking and laughter, as well as continuing analysis of the dive. Alves agreed with those who said that all the round stones we had seen were ship ballast. Puffing on his pipe, he theorized that they were strewn on the bottom not only because of sinkings but also because of the rebalancing of anchored ships. During such work centuries ago, in the age of discovery, ballast stones were sometimes thrown overboard and replaced with such heavy objects as cannons.

During their own dive to the middle of the bay, Alves and Paulo had come upon a rubble pile in which they were able to discern a dozen or more bronze cannons. That shipwreck, further out from shore and deeper than ours, had obviously gotten little human attention. In general, bronze cannons held up much better in seawater than iron ones and were far more valuable as artillery pieces and antiques. Most cannons were dated at the time of manufacture, but when immersed in seawater for centuries, these marks eventually disappeared on iron cannons as oxidation ate away the exterior. On bronze ones the marks were usually undiminished.

Next morning, the instructors gathered at the church-museum to pack for their return to Lisbon. Alves was in shorts and a T-shirt, directing things, ashes falling from his pipe.

I asked if it was possible to recover artifacts in a way that had archaeological integrity, but still sell them.

"I'm not a fundamentalist," he answered. "The only problem is that experience shows that archaeologists must return once a decade to their records and revise old conclusions. Anything that doesn't allow complete study and revision—because you've sold things at Christie's or to private collectors—doesn't respect the completeness of archaeology. You've lost the possibility of global study."[11]

Couldn't commercial work generate fees that would nourish worthwhile research?

"Underwater archaeology doesn't need a lot of money," he replied. "It's a false question sown by the treasure hunters."

Alves went on to say that he was at a crossroads in his professional life, possibly ending his directorship of the Archaeology Museum and going full time into teaching and undersea fieldwork, trying to move it in the direction of scientific virtue. "I think this is the moment to fight until the end, whether I win or lose," he said, sitting near ecclesiastical stonework that probably dated to the sixteenth century.

I had lunch at a restaurant across the street with Fernando, my dive buddy. Whereas Alves insisted that deep gear was superfluous, Fernando was open to its use in the investigation of deep wrecks—but not now, and not in the hands of foreigners.

"We should leave it to the next generation when we have the techniques, the means, and the instruction," he said, his eyes steely.

The next day, I visited a man that Alves and the group had mentioned over and over. He had an incredibly rich collection of historical artifacts, they reported, rolling their eyes. And they hinted that he had worked with undersea salvors, who over the years had quietly gathered in some of the accessible bounty in the waters surrounding Terceira.

The man was Francisco Ernesto de Oliveira Martins. His home was on the outskirts of Angra.

The undistinguished building was typical of homes in the city—two stories high with narrow balconies above, probably built in the eighteenth or nineteenth century. I rang the doorbell. After a minute or so, a gray-haired man in his sixties stuck his head out a balcony door and buzzed me in. With an unceremonious air, he gave me a tour of his home, which was indeed filled with four hundred years of painting, pottery, carving, and sculpture. Despite my high expectations, it was quite impressive. With a flourish, he opened old cases of teak and mahogany to reveal exquisitely carved jade and ivory, which he said the Metropolitan Museum of Art in New York City wanted to acquire.[12]

I explained why I was on Terceira and he proceeded to point out things "from the sea"—pottery, candlestick holders, an ornate silver incense burner, a giant vase, perhaps four feet high and three feet wide, made in China during the fifteenth century by Ming Dynasty craftsmen.

Arching his eyebrows with a conspiratorial air, he produced a key from a pot and opened an old wooden cabinet.

Velvety boxes inside were filled with many rows of gold and silver coins.

"From the sea," he said, the words solemn.

I asked his opinion about the debate over what to do with the undersea treasures. He dismissed it all as an unseemly squabble.

"Things now are not good," he said. "We must change. I want those things safe for study and enjoyment. I don't care if they go to New York or Paris or the Azores. The pieces are living. They need to be respected and put into a good place. They don't belong to the Azores. They belong to mankind. The gold of the Incas was taken by the Spanish and lost in the Azores. It belongs to the world."

I flew to Lisbon, curious to know the wider context of the war between the salvors and archaeologists. On the city's high-rise periphery, I found the administrative offices of Expo '98, the last world's fair of the century. Its chosen theme was "The Oceans, a Heritage for the Future."

Expo was undoubtedly a factor in the law's passage. The Portuguese saw the exposition as a way to highlight their maritime past (the opening was set for the quincentennial of da Gama's reaching India) and hoped that the law might allow a largely intact Portuguese ship from the age of discovery to be brought up from the depths for display. I went to Expo's offices, wondering if the fair's organizers had grown disenchanted with the recovery idea in the wake of the bitter public debate.

Its administrator, António Mega Ferreira, was a sharp dresser in a dark suit, cordial and well spoken. As we talked, it became clear that he was still eager to acquire an old ship, saying the debate had only kindled Portuguese interest in what lay beneath the sea.

"Ten or fifteen years ago, people didn't know or care," he said of shipwreck exploration. "Now it's becoming more and more a part of our culture and our heritage to understand what's possible. By picking this kind of exhibit, the major attraction of the Portuguese pavilion, we want to state that there's a lot down there to be discovered. Look, there's something like four or five thousand shipwreck sites in Portuguese waters alone.[13]

"This field is going to take off," Ferreira added, "if not prior to Expo, then after. Millions of people will realize that you cannot postpone this situation any longer."

Later, closer to the heart of Lisbon, I found Marx at the Holiday Inn, his usual haunt. He bore no resemblance to a man who was losing a battle he had waged for nearly four decades. Instead, he looked like a fox who had just been handed the keys to the chicken coop, all smiles, puffing a big cigar.

To see Marx is to understand immediately one of his main disadvantages with detractors—he looks like a pirate. He has no wooden leg, but he does have a paunch, a mustache, and a scar under one eye. It's easy to picture him with a cutlass between his teeth. Gruff and sarcastic, he apparently likes to trade insults the way other people exchange pleasantries. Largely self-educated, an adventurer at heart, he sneers at most archaeologists as timid do-nothings more interested in protecting academic turf than in saving important aspects of the past.[14]

From recent talks in Lisbon, he felt that his great foe, Alves, had been effectively neutralized—either out of his job and influence as director of the Archaeology Museum or, if he stayed on, muzzled by new rules just adopted by the Shipwreck Commission. More important, Marx felt that his own permit to probe the waters around Terceira was as good as signed.

"Today my whole life changed," he enthused.[15]

One reason for the shift was that the Shipwreck Commission was worried that all the big players who had applied for permits were about to pull out and that Portugal would thus lose anticipated revenues after having weathered the storm of controversy over the new law. "The greatest embarrassment for them would be going through all this abuse, hundreds and hundreds of articles, and then having nobody left," Marx said.

I told him of my experiences on Terceira and of the charges by Alves and the young people. He grew angry.

"If I wanted to do this for my own enrichment, to simply bring up treasure, I'd do it off Florida or in international waters," he fumed.

"By going after intact ships, I'm doing the opposite. By working in the Azores, I'm going against all my principles. I usually get seventy percent off the top and the government can take the rest. Here I can't bring up anything without permission. It's ridiculous. If there were anyplace else in the world where I could find intact ships, I wouldn't put up with this for a minute. If you're looking for treasure, it's better not to get intact ships. Then it's just lying there.

"This is my obsession, my dream," he continued. "It has nothing to do with money. There are so many intact ships down there it's like toys at the bottom of a swimming pool. The first thing I'd like to find is an early caravel from the time of discovery, because nothing is known about their construction." The caravels, sturdy little craft that could navigate in a cross-wind, were important innovations that helped beget the age of discovery. They included the *Niña, Pinta,* and *Santa Maria.*

"All the people I'm involved with are multimillionaires," Marx added. "I only pick people that can afford to live a fantasy. Sure, we'll make money in the end, but it won't be right off the bat."

Marx accused his detractors of having twisted his record into an unrecognizable caricature.

"Every time I find something, it's always treasure. They don't like the fact that I do good archaeology, that I've taught it in seminars around the world. All they see is gold bars.

"I don't give a damn about gold. Treasure is trouble—the more treasure, the more trouble.

"This is my last big bang," he added, still upset. "It's Marx landing on the Moon. I've done everything else. Since 1965, I've expounded on deep shipwrecks. It was considered heresy. My friends used to laugh. Now everything has changed. It's happening. There's nothing that we can't find and nothing that we can't bring up.

"The Azores are the only place we can find deepwater colonial wrecks, East Indiamen, Spanish galleons. It's the only place on Earth like this. Other than going after those wrecks, there's nothing left I want to do."

THE SEA on the evening of Sunday, April 14, 1912, had been unusually calm and the night bitterly cold when the *Titanic,* the largest and most luxurious ship then afloat, an icon of the Edwardian era rich in pearls and mahogany, socialites and industrialists, plowed into a small mountain of ice. The luxury liner had been sailing on its inaugural voyage from Southampton in England to New York City and carried more than twenty-two hundred people, its passenger list a *Who's Who* of the period. The *Titanic* was said to be unsinkable. Yet the vessel, nearly nine hundred feet long, had been cut open while speeding through icy waters 380 miles off Newfoundland and had slid beneath the waves in a little more than two and a half hours, taking the lives of more than fifteen hundred men, women, and children. It was the greatest maritime disaster of the day.

The sinking also left a huge dent in the human ego, in the notion of advancement at a time when that idea was at its zenith. Ships back then were society's most advanced technical arts. After the loss of the greatest ship of them all, a true leviathan, the word *progress* never had quite the same ring. Humanity never had quite the same confidence. Four centuries after the start of the scientific revolution, the sinking of the *Titanic* marked the beginning of a series of technological upsets, of wars and disasters, that shook the twentieth century and drove home the realization that the price of progress is often extraordinarily high.

We were over the grave. Beneath us, down nearly two and a half miles, lay the bones of the great liner.

It was August 1996 and we had arrived over the site to join an ongoing expedition that was praised by some people for rescuing history and condemned by others for dishonoring it.

Riding the waves a half mile or so away from us was the French ship *Nadir.* In both French and English, *nadir* means the lowest point. The ship is one of the handful of craft in the world that are in the business of launching and recovering submersibles, its large A-frame similar to the one that sends *Alvin* into the depths, only the French submersible can go deeper. Each morning it dove down to the wreck and each evening it came back, often bearing some of the *Titanic*'s remains.[16]

We rode our inflatable boat over to *Nadir*, fighting the waves. On board we watched the sea as the sun inched toward the horizon. Finally, the yellow submersible was sighted, bobbing in the waves. It had a prize, a rare one, yet a tiny fraction of the thousands of resurrections to date.

The telegraph from the *Titanic*'s decks once relayed commands between the ship's officers and crew, between bridge and engine room. About four feet tall, it was central to the ship's nervous system, helping set the liner on her fateful course. Now it lay on *Nadir*'s gently rocking deck in two parts, its cylindrical head and tubular pedestal, both removed from the *Titanic*'s debris field after eighty-four years of icy solitude.

The old artifact was wet. Its greenish patina, probably a mix of microbes and oxidized metal, darkened my fingers. The frigid temperatures of the deep had kept the device in a remarkable state of preservation. Its overall shape and contours were wonderfully intact and evocative of its relay function. You could see how the head fit atop the pedestal and held the device at just the right height for a *Titanic* officer to send signals through the ship's internal system of pulleys and sprockets. The metal indicator arms ("full ahead") were intact, though their wooden handles were gone. You could almost hear the bell ringing to announce the receipt of a new command.

But the telegraph also bore hints of tragedy. Around the pedestal's base were empty holes where heavy bolts once held the device fast. Most likely, the device had been ripped from the *Titanic*'s deck by the blow of the ship's collision with the seabed. Whatever tore it loose also bent the pedestal's rim.

The effects of long abandonment were also visible. The places where the telegraph's head and pedestal had rested in the frigid ooze were corroded and eaten away, the metal puckered and pitted. In spots, the metal was gone entirely.

The gold and porcelain lost to the depths over the ages may be intact and lying on the bottom, strong enough to withstand the assault of briny corrosiveness. Perhaps whole galleons sit there as well, their sails and beams saved from marine worms by pockets on the bottom that lack oxygen. No one knows for sure. The fate of the millions of ships lost over the centuries, especially in deep water, is for the most part unaddressed and unknown.

With the *Titanic,* however, it is clear that some of its artifacts are beginning to go, their turn-of-the-century materials and workmanship perhaps contributing to their demise. By some estimates, parts of the wreck that are now structurally intact might collapse sometime in the next century, the thick plates of mediocre steel melted into rivers of rust.

Had the passing submersible cast no beam of light on the telegraph and made no retrieval, the decomposition of the device would have continued slowly, and in time, perhaps decades, perhaps centuries, it would have disappeared entirely. But now the telegraph was headed for a new life.

First it was given a number—96/0023—a way of tracking it through a maze of recoveries. Then the device was hustled down the fantail to the waiting Zodiac, the inflatable boat rising and falling in the waves like an unruly elevator.

With a roar of twin outboard motors, we were off, pounding across the waves, airborne at times, the driver for some reason fixated on speed, his passengers clutching ropes and grips tightly. Spray wet my face. The unshackled telegraph lifted off the Zodiac's floor, pounding up and down, threatening to fly skyward.

After a trip that was doubtless shorter than it seemed, we pulled up to the ship in which we had arrived that morning, *Ocean Voyager,* which had sailed from Saint John's, Newfoundland. The telegraph was carefully lifted out of the Zodiac and into the hands of waiting conservators, proven experts at restoring *Titanic* artifacts. In time, the telegraph would join thousands of other artifacts saved from the wreckage and painstakingly restored to a semblance of their former grandeur, eventually to go on public display as representatives of a lost age.[17]

Beside me was the man responsible for all the activity. George Tulloch was short and wiry, in his fifties but boyish with his blond hair. A cigarette bobbed in his mouth. He crackled with energy, talking and gesturing, smiling and scowling, looking and questioning, fueled, it seemed, by little more than cans of cola and clouds of smoke.

"We have a good collection," he told a French conservator who was new at the job. "So unless it's a brilliant piece of glass or an object that somehow epitomizes the ship, we don't need to pick it up."

To date, he had retrieved from the pitch darkness at the bottom of the sea about four thousand of the liner's artifacts.

Tulloch was working hard to turn a profit for his investors after sinking roughly $20 million into various kinds of *Titanic* ventures and recoveries. He and his allies, including a number of historians and scholars, argued that the work was a cultural windfall and no different in principle from the restorations of Colonial Williamsburg outside Washington or any number of ancient tombs, which people pay considerable sums to view and tour.

Critics vigorously disagreed, calling the work an unseemly mix of greed and sacrilege, embodying the worst aspects of free enterprise. It was a recent grave, they said in disgust, not the sarcophagus of a lost pharaoh. Many of the critics would bar recoveries altogether and leave the wreck untouched, its own best memorial.

Tulloch was undaunted. He was preparing to put three hundred artifacts on display at the Wonders exhibition center in Memphis, the show opening eight months hence, in April 1997, the eighty-fifth anniversary of the sinking. It was to be the first major American exhibition and the largest anywhere to date. On display would be such objects as gold coins, silver dinnerware, fine china, and a bronze cherub that once graced a first-class staircase. Surprisingly, the deep's frigid temperatures and inky darkness had preserved many of the wreck's delicate paper objects, and some of these were to be displayed as well, including stock certificates, playing cards, and love letters.

To people unopposed in principle to the recovery of the ship's artifacts, the main questions often came down to whether the work could be done with care and dignity. Eight survivors of the tragedy were still alive, though none were said to recall the events of that frigid night, being either too young then or too old now. Even so, more than eight decades later, some experts wondered if enough time had passed for the crossing of the blurry line that can separate the violation of a graveyard from the legitimate pursuits of history and archaeology. The situation called for an extra measure of discretion.

On Monday, August 12, 1996, the first day of a week's visit to the monthlong expedition, the prospects for tact already seemed mixed, as suggested by the handling of the telegraph. And although a number of scholars were associated with the expedition, it was clear that none oversaw the recovery work at sea. Instead, it was ruled over by Tulloch. The expedition's ground rules, he said, were to leave the main wreckage untouched, to gather artifacts from the debris field, and to display them in public collections rather than selling them. Beyond that, he was clearly eager to gain a financial edge in most any way he could. His merchandising blitz included the sale of *Titanic* coal at twenty-five dollars a lump ("Authentic Anthracite From The 1912 Maiden Voyage"), each encased in a plastic box.

Before our arrival, Bass Ale, an expedition sponsor, had sailed a ship out here. On board were ten winners of a sweepstakes who had come to witness the retrieval of some of the thousands of bottles of ale that went down with the *Titanic.* A handful were picked up, the company's sponsorship netting Tulloch a quarter million dollars.

A main goal of our leg of the expedition was the deep lighting and filming of the wreck. The work was to be done by Tulloch's company for a movie to be shown at artifact exhibitions and by the Discovery Channel, which, along with its European partners, had paid Tulloch nearly $3 million to come along on the voyage. The channel planned a one-hour special on the recovery work and a two-hour special on how and why the *Titanic* sank.

After we left, two cruise ships, one from Boston and one from New York, were to show up for the raising of a piece of hull section. The size of a big living room, it was some twenty feet wide and twenty-four feet long, bearing four portholes and traces of four others, the largest block of *Titanic* debris ever slated for recovery. ("You can witness history when for the first and only time . . . ," Tulloch's newspaper ads declared.) The recovered hull fragment was to be the centerpiece of a museum, perhaps a traveling one on a special ship. And the audience at sea was to net Tulloch's organization a few million dollars, depending on how many people showed up. The cruise ships in total had room for thirty-five hundred customers and cabins that went for up to seven thousand dollars. *Titanic* lectures and movies were to occur throughout the cruises, while gambling at casinos and slot machines was to stop while the ships circled over the grave. More than anything else, the cruise ships had reignited disputes over the recovery work, with some critics saying they represented a new low for Tulloch as a peddler of disaster cachet.[18]

It was not just a question of taste. Bets here were divided on whether Tulloch could pull off the hull raising and make good on his sales pitch to the cruise-ship customers. The raising was a major endeavor, involving a small flotilla of lifting bags and boats, drivers and winches. Nothing like it had ever been tried before.

And, in truth, the issue was not just Tulloch. The commercial appeal of the wreckage down below was so great that a war had broken out between him and rivals seeking to cash in on the old disaster. Today the French submersible lifted from the depths not only the old telegraph but very fresh wreckage from a Russian *Mir* submersible that was found scattered on the *Titanic*'s deck. The tattered pieces of Russian fiberglass—roughly the size and shape of a child's snow sled—apparently broke off during inadvertent collisions last year, damaging some parts of the *Titanic*. With dispatch, Tulloch had now seized upon the *Mir* debris as a sign of disrespect for the sunken ship and planned to submit it in ongoing court proceedings meant to block rivals from visiting the wreck.

"I felt like a cop," Tulloch said of his own inspection dive earlier in the expedition, upset by the damage but happy in the role of a good guy.

Nearly two weeks into the monthlong venture, Tulloch was still highly energetic but also surrounded by a palpable air of overextension. He had few lieutenants and handled most of the details himself in an operation with scores of people and agendas. His son Matt assisted with the artifacts, helping to conserve and catalog them.

The question was how long Tulloch could keep it up. If the past was any indication, perhaps quite a while. As I learned to my surprise, he and his rivals had been battling over the artifacts for more than a decade. The history of activity was so extensive that this expedition would mark the hundredth dive to the resting place of the vanished ship.

Quiet searches began as early as 1953, fueled in part by unsubstantiated rumors that fortunes in gold, silver, and jewels had gone down with the liner. Where other hunts failed, Ballard, the Woods Hole explorer, succeeded in 1985 while towing over the seabed a fifteen-foot robot named *Argo* which he had recently built for the Navy. More than two miles down, bearing sensitive cameras, beaming up images over fiber-optic lines, *Argo* came upon a ghost ship seemingly frozen in time.[19]

Triumphant, Ballard returned in 1986 with *Alvin,* the Navy's adept submersible. He found the liner's bow remarkably intact, its railings undamaged. But overall, most surviving parts of the ship were covered in thousands of reddish-brown stalactites of rust that hung down as much as several feet, produced by bacteria and looking very much like icicles. Ballard dubbed them rusticles.

It was evident that the giant ship had split in two during the sinking. The bow and stern now rested nearly a half mile apart. Between them in the ooze lay thousands of artifacts spilled from the ship's decks and innards, many in a remarkable state of repair. Ballard in his richly illustrated *Titanic* book showed boilers, telegraphs, steam valves, floor tiles, wine bottles, cooking pots, serving bowls, bed frames, spittoons, safes, champagne bottles, lifeboat davits, and, hauntingly, pairs of shoes lying side by side.[20]

At first, in 1985, Ballard called for the recovery of artifacts from the debris field, arguing that they should be saved and placed in a special museum as a memorial to the dead, even as intact parts of the ship and its contents were left undisturbed. Deep submersibles, Ballard told Congress, should be put to work "carefully recording and recovering those delicate items lying outside the hull." Later, he reversed himself and advertised his opposition to salvage, criticizing any recovery as plunder and desecration. Such pleas multiplied. Some survivors called it grave robbing, as did a number of *Titanic* societies around the world. On the issue, however, Ballard stood above all others in terms of visibility. Eventually, despite his early stance, he became synonymous in the public mind with the preservation of the hulk and its artifacts, with leaving them untouched.[21]

The French, Ballard's partners in the 1985 discovery, had no such reservations and joined with various businessmen to fund a program of *Titanic* recoveries. Other motives simmered as well. The French felt Ballard had stolen undue credit for the discovery (which Ballard denied) and were happy to oppose him. Wanting an American participant with financial skills, they sought out Tulloch, a car salesman in Greenwich, Connecticut, who had French ties and European business experience and had built up the largest BMW dealership in the United States. Tulloch was interested. Car sales looked far less interesting than an opportunity to help recover what remained of the world's most famous shipwreck.[22]

The French team arrived here in 1987, with Tulloch so rushed he barely had time to pack. He was, as he puts it, the bride at the wedding. The recovery work was done then as it is now, by IFREMER (for Institut Français de Recherche pour l'Exploitation de la Mer), a French governmental unit for oceanography and deep inquiry that was Ballard's partner in the original discovery and has close ties to the French Navy. Its workhorse is *Nautile,* an updated version of *Alvin.* The French submersible is named after the nautilus, the ancient cephalopod that uses its spiral shell as a float to rise and fall through the sea. During the expedition, artifacts were brought up by *Nautile* as well as by lifting devices. Under international law (basically a refined version of "finders keepers, losers weepers"), the team won salvage rights to the vessel by virtue of being the first to recover artifacts.

In thirty-two dives that summer, the team probed the length and breadth of the debris field and recovered some eighteen hundred objects, including a safe, a porthole, a ship's bell, a compass, a gilt chandelier, and a leather bag that turned out to hold jewelry. After some of the artifacts were refurbished, small exhibitions of them were held in Paris, Stockholm, Göteborg, Oslo, and Malmö near Copenhagen, with viewers in most cases being charged an entrance fee.[23]

Meanwhile at sea, the famous wreck had no visitors during the final years of the cold war, 1988, 1989, and 1990.

After that, it suddenly faced a rush, at least compared to the solitude of the previous decades. The burst of activity was based on exploratory gear developed during the cold war that afterward was redirected to commercial investigations of the deep. Now, each season, without fail, some of that gear showed up at the site of the sinking.

The Russians led the parade. Their socialistic empire in ruins, newly poor and desirous of cash and recognition as deep experts, they arrived in *Akademik Keldysh,* mother ship to the *Mir* submersibles. Moscow made the pair of teardrop-shaped subs in the late 1980s as part of a belated effort to catch up with the West in deep exploration. Though built to Russian specifications, the subs were designed and manufactured in Finland, famous for its shipbuilding. Each cost $25 million and each dove as deep as six kilometers, or nearly four miles. Eager to publicize their deep abilities and drum up commercial business, the Russians in 1991 illuminated and filmed large parts of the sunken liner for the IMAX Corporation of Toronto, which used the footage to make *Titanica,* a movie for its big-screen movie theaters. In one eerie scene after another, the film brought the ghost ship back to life. But the costars were clearly the *Mir* submersibles and their crew members, who were cast as approachable and easy to work with. It was a model advertisement that generated much business.[24]

The next year, 1992, Herbert Humphreys, Jr., a Tennessee entrepreneur and treasure hunter, set his sights on *Titanic* artifacts. His partner was Jack Grimm, a Texas oilman who in the early 1980s had tried and failed three times to find the lost wreck. For recovery gear, the team lined up one of the deep-diving *Magellan* robots of Oceaneering Technologies in Maryland, the American Navy contractor that was then diversifying its clientele. Humphreys sent a ship to the wreck site, awaiting the outcome of a court ruling over who legally owned the wreck, ready at a moment's notice to lower the robot into the darkness. The season was late. The ship tossed in rough seas. Finally, a federal judge in Norfolk, Virginia, ruled against the team and for Tulloch. The ship and her frustrated crew turned back to port.[25]

After the court challenge, Tulloch and his French partners mounted back-to-back expeditions in 1993 and 1994 that raised thousands of artifacts, including such large pieces as the ship's whistles and a huge davit for lowering lifeboats. The work was financed partly through a stock offering. Tulloch was now president of RMS Titanic, a company based in Manhattan. In 1994 and 1995 it won a new degree of respectability when it sponsored an exhibition of artifacts at the National Maritime Museum of Britain, outside London in Greenwich, a distinguished institution housed in elegant old buildings. And the museum agreed to become a guiding hand in gathering artifacts, helping determine which ones were needed to round out the collection.[26]

The Russians returned to the site in 1995 to film the ship's remains again, only this time for Hollywood director James Cameron, maker of such action hits as *Aliens, The Abyss,* and the *Terminator* films. More than two miles down, Cameron gathered images for a *Titanic* thriller reportedly being made at a cost of more than $100 million.[27]

Tulloch and the French sailed back to the North Atlantic in 1996, amid rumors that the Russians, while ferrying moviemakers through the darkness, had suffered one or more serious accidents, not life-threatening but enough to have shaken them up. Indeed, the French team found much evidence of discord down below.

"I still find the ship very beautiful," Charles A. Haas said at breakfast on Tuesday morning, refreshed after a hard day in the depths and happy to talk about his dive, including finding the telegraph and the *Mir* wreckage.

Haas was a main scholar of the expedition, the coauthor of *Titanic:*

Triumph and Tragedy, Titanic: Destination Disaster, and *Falling Star: Misadventures of White Star Line Ships.* When not writing books or editing a journal devoted to old liners, he was a high-school teacher in Randolph, New Jersey. A *Titanic* buff since he was a kid, Haas probably knew as much about the ship's history as anybody and now had the added advantage of having toured the wreckage twice. Tulloch had invited Haas and his coauthor, John P. Eaton, once vocal critics of the recovery work, to join the 1993 voyage as participants. On separate dives the two toured the aging wreckage and helped the recovery team identify artifacts. Afterward, they wrote glowing accounts.

After breakfast we discussed his latest dive and observations. Haas was considered an expert witness on the *Titanic* and had been asked by Tulloch to draw up an affidavit about the *Mir* damage and debris.

The current theory of the accident, or accidents, was that a Russian submersible had tried a risky maneuver and hit its tail (a notorious blind spot) on the *Titanic,* smashing the sub's cowling, which encircles the rear propeller like a doughnut for protection and thrust control. The spinning propeller then proceeded to tear the bent cowling to bits.

One damaged area that Haas talked about was the *Titanic*'s Marconi room, where the wireless operator that night had been so busy sending messages for passengers that he ignored ice warnings. The wooden frame of its skylight had clearly been pulled out of its mounting and shoved aside, Haas said, adding that French pilots had earlier retrieved *Mir* debris there.

Moving past the accident site during his own dive, Haas watched in bewilderment as the pilots sought to extract from an officer's room a mysterious two-foot-long object.

"It was black or dark gray, and I was rather surprised by that because I had not seen those colors in my travels around the *Titanic,*" he said. "It was foreign." Only later did Haas discover that it was yet another piece of *Mir* cowling, bringing to seven the total number of fragments of black fiberglass hauled up from the deep Atlantic.

Assailing the lost ship were not only the Russians, but nature as well. "A lot of deterioration is taking place," Haas said. The rusticles were not only thicker than he remembered but had been joined by other harbingers of decay. "Just after we collected that piece of *Mir* submersible, I happened to look down and I could see through the deck, through a rusty hole in the steel. It was the first time I saw something that would ultimately lead to the demise of the ship."

Clearly upset by attacks on the recovery work, Haas said that awareness of the shipwreck's deterioration was wholly at odds with ideas about the remains serving as a permanent memorial to the tragedy's victims.

"If those items are left down there, eventually they'll disappear," he said, mentioning the decay of the telegraph as an example. "Government funding of deep-sea archaeology is zero. It's essential that corporate funding of one form or another be found" to save what remains of the famous wreck.

On *Nadir*'s stern another dive was about to get underway. The French crewmen, muscular from submersible toil, were scrambling around in black wet suits and yellow jump suits. A nearby chalkboard said this was the expedition's tenth dive. High on the submersible's side was a black decal, "RMS Titanic, Inc.," one of a number of commercial endorsements, like those on a hot rod. With the exception of the submersible *Nautile,* which was shiny yellow and quite modern, the stern had the atmosphere of an old factory—crowded with people, winches, cranes, vises, cables, toolsheds, greasy decks, and piles of unused gear.

Today was dive day for David Livingstone, a high official from Harland and Wolff in Belfast, Northern Ireland. Ever since the sinking, the builders of *Titanic* had kept their distance from the most famous of all maritime disasters. Livingstone was the advance wave of historic change. An engineer and naval architect, he was now preparing to venture beneath the sea to understand the wreckage better and help separate facts from the many myths that had come to swirl around the disaster like ghosts.

Over a recent lunch, Livingstone had told me that the company had no official position on the recovery work.

"I think ship parts are okay," he said of his own views on the subject. "But I'm uncomfortable with personal things."

Livingstone spoke with an engaging Irish brogue and was clearly a star of Discovery's program. Cameras whirled as he donned extra socks and pulled corduroy pants over thermal underwear. He climbed down *Nautile*'s hatch and its silvery cover closed behind him. Today's dive was to include checking on a light tower deployed last night, setting up an experiment to test for microbes on the *Titanic*'s bow, and letting Livingstone's expert eye roam over as much of the sunken ship as possible.

Machinery rumbled to life and *Nautile* moved down the deck. The yellow sub hit the sea with a splash and, after a checkout, vanished beneath the waves.

Later, up on *Nadir*'s bridge, near the submersible control center, I found William H. Garzke, Jr., poring over plans for the *Titanic* that were said to have been kept secret ever since the sinking but had now been brought out of the closet by Harland and Wolff. Garzke is a naval architect who has written widely on the forensics of the disaster and was here to help sift through new evidence. Now, as temporary guardian of the plans, he was preparing to help

Livingstone more than two miles under, calling down detailed information over the undersea microphone.[28]

"All the expeditions to this site have been showmanship," Garzke told me. "This is the first one that is doing this for science and engineering."

Beside him was a stack of fifteen different plans. The one he had open, "Shell Plating Plan," seemed to be about sixteen feet long and was dense with design minutiae. The same was true of a smaller one he opened up, "No. 401. Details of Expansion Joints." The *401* was Harland and Wolff's yard number for the doomed ship.

Later, the yellow blur of *Nautile* was sighted at the surface. After some chatting and filming on the stern, we repaired to the mess, where Livingstone ate a late dinner and recounted his impressions.

In speech and manner, he was normally quite reserved, an engineer's engineer, always sticking to the facts, always careful to hedge. Now he was almost sputtering with excitement. On the bow's port side, he had seen a huge indentation in the hull plating, a sudden fold of about ninety degrees where the whole side of the lost ship had crumpled like a pound cake as the *Titanic* slammed into the bottom.

"It looks like it stopped in a hell of a hurry," he said between bites. "It really is unbelievable. You could spend one dive solely in that area, bringing back the evidence."

Most surprising of all, he said, was that the folded steel had no hint of cracking, despite the huge forces that must have been at work. No rivets were sprung. And for some reason the area had no rusticles at all.

"You can see the steel and it looks as if it's primed, as if it's out of the paint shop yesterday," he said in amazement, his eyes wide.

Asked if he saw anything that made him think of the *Titanic*'s crew and passengers, he said yes, his voice low.

"I saw shoes and I saw crockery.

"When I was . . ." He paused. "When you are on the bridge, perhaps you have a few thoughts back about the people, about the people who actually walked on that bridge—Captain Smith and Andrews."

Thomas Andrews was Harland and Wolff's managing director and the *Titanic*'s main builder, a man in his thirties. Captain Smith was in his sixties and was to retire after the *Titanic*'s inaugural voyage, the event capping his distinguished career. Both men went down with the ship.

"Yes," Livingstone continued, "you wonder what they were doing when you see the machinery still there."

The sunken bridge was famous among *Titanic* enthusiasts for the presence of a tall bronze telemotor that once held the ship's wheel, which had vanished. The telemotor was in mint condition and eerie because of its high state of preservation. It was also the site where the watch officer and helmsman that long-ago night had tried frantically to dodge the looming iceberg.

I asked if the return to the wreck after eighty-four years was perceived at Harland and Wolff as bringing a sense of emotional closure, of confronting the ghosts of the past.

"Yes," Livingston said, "for some people in the company. These are personal things," he added ever so gently. "It's not something that engineers discuss."

On Wednesday, *Nautile* prowled across the bottom of the North Atlantic to arrange lights for the illumination of the *Titanic*'s stern. It was also a moment of truth for Tulloch, the man responsible for putting together all the pieces of the expedition puzzle.

"Yes! Yes! Yes!" he cried as he ran down *Nadir*'s corridors, slapping walls, hugging crewmen, ecstatic with the news that the first tower was lit up, all five bulbs burning bright. "And they said it couldn't be done!"

Later, three other towers were successfully lit. It was a big win for Tulloch's agenda and augured well for the even bigger job of trying to raise the hull plate at the end of the expedition.

After lunch, I went up to *Nadir*'s bow deck, a quiet place on a crowded ship that I had begun to visit daily for stretching exercises and meditations on the surrounding sea. I was about to leave when two crewmen showed up. They activated a small crane and lowered into the sea off the starboard side a device about the size and shape of an American football, but with tiny wings. It was a special microphone, they said, one that could pick up bursts of undersea sound. Moored five meters below the surface, past the ship's keel, it was about to receive signals from *Nautile,* miles below.

We went to the submersible control area behind the bridge. There the crewmen switched on a big machine. It was to turn the sound transmissions not into human speech but into images from the submersible's cameras.

After the usual delays and false starts, the pictures began to appear on the computer screen. They were stunning. And spooky.

The images were of the *Titanic*'s reciprocating engines. These were the largest of their kind ever built, rising to a height of more than three stories. They worked on the same principle that a car engine does today, their huge pistons moving in enormous cylinders, up and down, up and down, the force turning crankshafts and propellers. Miles down at the bottom of the sea, the reciprocating engine visible to us was sitting upright within the ship's body, discernible despite its interior position because it was exactly at the seam where the liner had split in two, separating bow from stern. We were gazing at the lost ship's mechanical soul.

Eerily, the engine looked like a monster. The cylinder was its head, and the metal struts and supports its arms. Brightly lit but surrounded by gloom, it was one of the strangest things I had ever seen, very much an alien from another world.

Such images were to be sent to the cruise ships and were technically a product of Tulloch's new lighting system and *Nautile,* a cold-war device that debuted in 1985 amid a final international round of saber rattling and nuclear brinkmanship. Curious about its new role, I tracked down Federico Muñagorri, a neat, enthusiastic man who spoke good English and directed the commercial arm of IFREMER, which is headquartered in the ancient port city of Brest on the westernmost tip of France.

Budgets in Paris were shrinking, he said, and IFREMER was increasingly in search of commercial work. He added that the same was true of all the governmental gear around the world that was devoted to probing the deep, including American assets at Woods Hole. Naval and scientific work in the service of geopolitics, he said, no longer sustained such endeavors as it had in years and decades past.

"We must work harder because of the budget cuts," Muñagorri told me. "Everybody is working more on commercial projects. They have no choice. They must go to the private markets." He practiced what he preached. Around *Nadir* I had seen him wearing a "Voyage to the Titanic" T-shirt from Bass Ale, one of the expedition souvenirs.

Friday was dive day for D. Roy Cullimore, an elfin man from Canada with a disarming sense of humor. He was a microbiologist from the University of Regina in Saskatchewan and a respected expert on all kinds of bacterial growths. He was fascinated by the iron-loving microbes that cover the *Titanic.* He believed the ship probably carried more life now, in terms of crude biomass, than it did during its inaugural voyage with more than two thousand people. It was a remarkable thought.

Today Cullimore was diving down to do tests, to retrieve biological gear, and simply to eye the aging wreck, trying to gauge how long it might withstand the microbial onslaught.

Late in the day, he and his companions in *Nautile* released from the seabed a lifting basket that eventually bobbed to the surface. One of its buckets carried more blackish *Mir* debris. The other held a two-foot chunk of *Titanic* steel plate twisted into a snarl of metallic agony and stained a mélange of reds and browns, some areas thickly encrusted by layers of rusticles. The plate, about an inch thick, was violently bent. Rivets had popped loose in places. Thick steel reinforcing bars were straight where still riveted to the plate but where free had been bent into tortured angles.

Many of the expedition's advisers gathered around, as did some crew members, who up till now had seemed largely indifferent to the recovery work. The general opinion was that the twisted piece of metal was the first thing to come up from the depths that hinted at the raw forces of nature that had ripped apart the huge liner.

"For me?" a French worker asked Tulloch, holding up a rivet.

Tulloch shook his head.

The plate and its rivets were destined for structural analysis by Garzke, the naval architect, as well as metallurgical and other kinds of analysis at the University of Leeds and the University of Missouri.

Back from his *Nautile* dive, Cullimore interrupted a late dinner to examine the plate and a particularly large piece of rusticle.

"You can see it's already cracking away," he said. "As it dries, it will drop off. It's going to be interesting to see how much pitting is under there. If we knew the original thickness of the steel, then we could gauge how much metal has already come out" of the ship since it sank, suggesting how long it might withstand the sea's onslaught.

He touched the wet rusticle as a mother would touch a baby.

"I'm totally amazed at the structure," he said, happily gazing into the bucket in the fading light.

By Saturday, the early successes of the expedition had been replaced by a pattern of woe. Bulbs in light towers had blown. Lifting bags had drifted away. Uncertainties had arisen over the timing and feasibility of raising the large hull section.

Still, core activities had gone ahead. *Nautile* had succeeded in diving every day and had continued to bring up intriguing artifacts from the depths. Yesterday, it was an old chronometer.

The timepiece now sat in a special preserving bath, its hands askew and face clouded. A French conservator gently held the chronometer and its supporting parts, revealing that much of its square wooden frame was intact, somehow preserved from deep worms. Despite some decay, it was easy to see how the round chronometer could pivot on bearings inside the square wooden frame, allowing the timepiece to be turned in any direction. Clearly, it was an instrument for the ship's crew, not a decorative piece for the passengers. Perhaps it had sat on the bridge, ticking away the minutes and seconds before the disaster.

To me, the chronometer and its delicate handling seemed light years from the more commercial aspects of the expedition, from the overstated ads and the Bass Ale promotions.

Before the expedition, I had talked to Tulloch about his disparate agenda. He seemed comfortable with the Bass Ale boat and the cruise ships and other commercializations, saying they were responsible popularizations that helped raise money to deepen his goals, to make possible all of the investigations and lighting and filming in addition to the routine work of artifact recovery. He added that real profits had so far eluded him and his investors at RMS Titanic, despite nearly a decade of effort.[29]

"We're not complaining," he said. "It's just that a lot of capital investment has to be laid out over a decade before you're ready to give the public a proper presentation of the objects." He said that the public's paying to see such presentations, at a price of roughly ten dollars per person, was the only reliable way to turn a profit in the business of *Titanic* recoveries.

He implied, but did not say, that the investment phase of the project was ending and that the profit phase was about to get underway.

Now, after lunch, Tulloch and I talked about a number of residual issues, the picture on the wall behind him pivoting back and forth as the ship rolled in the waves. He wore sandals, shorts, and a yellow T-shirt emblazoned with "Lighting the Titanic," one of his promotions. His ever-present cigarette was there, too.

Smiling, Tulloch explained in some detail how a Virginia court's recent decision to forbid the Russians from getting near the *Titanic* carried international weight. Yet again, he damned the *Titanic* damage from the *Mir* submersibles.

Toward the end of our conversation, I asked Tulloch if he was glad he had gotten into this business.

"I'm not sure," he said, his face clouded. "That's a dangerous question. I haven't thought about it a lot. I know I don't like being called a grave robber. I never thought of grave robbing as a calling.

"On a personal level, while the criticism is humiliating, it also brings the satisfaction of knowing that what you've done will be appreciated some day. After we go, nobody will much care about what the critics said."

The people of the future were his reward and responsibility, Tulloch said. It was they, the museumgoers of the twenty-first century and beyond, who would appreciate what had gone on here, who would learn about the life and death of the great liner by gazing at its remains.

"They won't much care about me," Tulloch said. "They'll only care about the objects."

FEW PARTS OF THE GLOBAL SEA are as desolate as the wilds of the Atlantic just below the Tropic of Cancer at a longitude some fifteen hundred miles east of Barbados and fifteen hundred miles west of Dakar. It is right in the middle of nowhere. The waters are unusually deep, with no nearby land, no islands, no oceanic ridge, nothing to break up the monotony of the sea's surface. On the moonless night of June 23, 1944, under a clear sky, two submarines met there, one from imperial Japan, the other from Nazi Germany. The place was chosen for the rendezvous precisely because of its desolation and remoteness from land.

The Japanese sub was a giant, longer than a football field or, for that matter, any American submarine of the day. That made her ideal for a secretive exchange between Hitler and Hirohito. Early in the war, the Axis powers had conducted such trade on the Trans-Siberian Railway, but the Nazi invasion of Russia in 1941 ended that option. They tried surface ships in 1942 and 1943, but Allied attacks and blockades led to so many sinkings that they abandoned that approach. Finally, they resorted to furtive trade by submarines slipping halfway around the globe, avoiding coasts and the possibility of Allied planes, trying to hide whenever possible beneath a watery veil. By this point in the war, Japan was desperate for German technology—for optics, engines, and electronic gear. And Nazi Germany was desperate for raw materials.

The big Japanese submarine, the *I-52,* left Japan in March 1944, her ponderous bulk like that of an enormous whale 357 feet long. The warship carried 109 men, including fourteen experts from such places as the Mitsubishi Instrument Company, who were along to study and procure German technology. She also carried two metric tons of gold, 146 bars of it packed neatly in metal boxes.

The sub made one stop, in Singapore. There it picked up a massive load of raw materials, including 228 tons of tin, molybdenum, and tungsten, fifty-four tons of raw rubber, three tons of quinine, and three tons of opium. In late April, the sub set out again, sailing through the Indian Ocean and around the tip of Africa, bound for the seaport of Lorient in Nazi-held France. It traveled the usual way, submerged during the day and surfaced at night, charging its batteries with a diesel generator as it sped through the water. Back in Tokyo, a summary of the sub's route and manifest was transmitted over the airwaves to Berlin in secret code.[30]

Three months into her voyage, the *I-52* rendezvoused in the Atlantic with the German supply submarine. She took on food, fuel, and two German technicians. She also took on a radar detector, meant to warn of enemy planes as she neared Europe.

But that safeguard came too late. In the dead of night in the middle of nowhere, Allied radar was sweeping the Japanese sub and targeting it for destruction.

During the war, unbeknown to Tokyo and Berlin, the Allies had broken the main Axis codes and were extensively spying on Axis communications. From eavesdropping on naval transmissions, they knew all about the *I-52,* its cargo, and its route. Early on they laid plans for its ruin. Eventually, an American task force left Norfolk, Virginia, to track down the two Axis subs during their mid-Atlantic meeting.

Lieutenant Commander Jesse D. Taylor took off from the deck of the aircraft carrier *Bogue* in an American Avenger bomber and searched the dark Atlantic on the night of the planned rendezvous. Near midnight, his radar came alive with a glowing blip. It was the *I-52.* Zeroing in on his target, Taylor dropped parachute flares and two five-hundred-pound bombs and watched in dismay as the diving Japanese submarine kicked up white water and successfully evaded his attack.

Taylor had one more weapon. Working methodically but quickly across more than a mile of ocean, he laid six acoustic buoys in a circular pattern around his prey. Upon hitting the water, each buoy dropped a microphone on a thirty-foot cable, turned itself on, and started broadcasting undersea sounds over the airwaves, each buoy on a different frequency. Taylor and his crew listened in, switching between buoys to track the sub's direction of travel. The chug-chug-chug of its propellers was clearly audible. Taylor saw his chance. Swooping low, he dropped his only torpedo. The shaft of steel hit the water and sped below, tracking the sub's propeller noise. Listening through the acoustic buoys, Taylor and his crew heard a long silence, followed by the thud of a loud explosion.[31]

"We got that son of a bitch," Taylor swore.

As he listened, the sub's propeller noise faded, followed by what sounded like the groan of collapsing steel.

Far away, both the *Bogue* and the escaping Nazi submarine had seen the parachute flares above the distant battle, and both had noted the position of the blaze. The next day, two American destroyers that were escorting the *Bogue* found a variety of floating debris, including bales of rubber, fragments of silk, and a Japanese sandal.

Nearly a half century later, maritime researcher Paul Tidwell was digging through a pile of declassified documents in the National Archives, looking for evidence of wartime gold shipments. He stopped cold. The text he had stumbled on was a summary of German Enigma radio traffic that the Allies had intercepted and secretly decoded. (ULTRA—TOP SECRET read the warning atop the formerly classified page.) The old transmission told of a Japanese submarine that was preparing to go halfway around the globe carrying raw materials to the Nazi war machine. Remarkably, it said the sub was also to haul two tons of gold. Fascinated, Tidwell quickly searched out other documents and ascertained that the submarine, the *I-52,* did in fact leave on its voyage. But—what luck!—it never arrived, as was the case with so many Axis submarines. Attacked by the Allies, the sub had been lost to the Atlantic in more than three miles of water.[32]

There is nothing flashy about Tidwell. Short and portly, fond of white shirts and conservative ties, he can strike one as a quiet, unassuming type, happy to think his thoughts without trying to impose them on the world. That impression is dead wrong.

Tidwell grew up in New Orleans, the son of a retailer. In 1966, just out of high school, he enlisted in the Army for two tours of duty in Vietnam, serving in the infantry from 1967 to 1969, the hellish years that included the Tet offensive. He was wounded a couple of times and won a purple heart, reaching the rank of sergeant. After Vietnam he went to college at Southeastern Louisiana College, a big state school in Hammond, majoring in American history.

During this period he was bitten by the shipwreck bug, bitten hard. It happened on vacation in the Florida Panhandle when he stumbled on some old coins in the sand. They were Spanish. Timeworn and heavy, they struck him as far more interesting than the coins he had collected as a boy. Eventually he went back and looked for more, increasingly drawn to the undersea world. Tidwell learned to scuba dive and in time he opened a dive shop in Covington, Louisiana, near Lake Pontchartrain and the Gulf of Mexico. As a sideline he began to investigate the fates of modern shipwrecks, often for clients who would try to salvage them. In the early 1980s he made his own recovery debut. The effort centered on a freighter that had sunk off Florida in World War Two with a cargo of tungsten and tin. The ship lay only seventy feet down. Tidwell and his colleagues recovered much of the valuable metal, giving him confidence.

By the late 1980s he had quit the dive shop altogether and gone into the wreck business full time, specializing in World War Two cargo ships. Living in Louisiana, he would fly periodically to the nation's capital to dig through boxes of old documents at the National Archives and the Navy Historical Research Center. Mainly he hired himself out to clients, proud of his precision and thoroughness.

Two things changed during this period. First, Tidwell became convinced that he had learned the salvage business better than many of his clients, and this prompted him to focus on his own recovery projects. Second, he and many other experts in the field became convinced that it was poised for advance. The *Titanic* was found in 1985 and the *Central America* in 1988. If such legendary wrecks could come into play so long after vanishing into the sea's darkness, he reasoned, so could virtually anything else.[33]

It was at this point in his career that Tidwell came across the documentary evidence of the *I-52* and its tons of gold. The date was May 1990. Like a platoon leader who suddenly sees an opening on a battlefield, Tidwell proceeded to rally his troops and take action. Leaving New Orleans (while keeping an apartment in the French Quarter), he moved himself and his wife and two children to the Washington, D.C., area, throwing himself into finding out all he could about the lost submarine. This, he was sure, was his big opportunity. Tidwell was intrigued by the military history, by the amount of gold, and by the unusual depth of the wreck, which, significantly, put it in international waters and left it unbounded by the territorial jurisdiction of any state. He worked like a maniac. The locations where he did his research included not only the usual archives in the Washington area but overseas ones as well, places off the beaten track. Eventually, he pieced together a puzzle that revealed solid clues about the *I-52* as well as a number of frustrating ambiguities.[34]

One of the nightmares was the question of whether the submarine actually sank, despite all the evidence that it did. The deceptive use of debris was not unknown in submarine warfare. Such tactics were meant to distract an enemy from further attacks, especially if wounds had been inflicted and the victim was desperate to get away for repairs. Doubts increased as German records were declassified that supposedly had the sub requesting new routing directions a whole month after the midnight attack, and as Japanese records came to light that hinted at the sub's eventual loss in the Bay of Biscay. These records, too, might have been part of wartime disinformation campaigns meant to throw enemies off balance, a common enough practice. But questions about her fate were clearly on the rise. *The Atlas of Shipwrecks and Treasure* concluded in 1994 that the *I-52* "may have only been damaged but not sunk during the attack."[35]

And the ambiguity got worse, at least from Tidwell's point of view. Even if the Japanese submarine had in fact gone to its grave that evening, many puzzles lingered over who sank it and exactly where the sub lay. American forces took little for granted in the war. Thus, even as Commander Taylor's attack was underway, other torpedo bombers were being dispatched to the scene of the battle and were told to pick up the hunt as Taylor flew back to the *Bogue.* With their own acoustic buoys, they listened for any suspicious sounds—those of a miraculously intact *I-52* or the escaping Nazi sub.

About two hours after Taylor's encounter with the *I-52,* an Avenger piloted by Lieutenant William D. Gordon picked up strong sounds and fired a torpedo. Eighteen minutes later a violent explosion ripped the water. No one knew the target's identity or if the explosion represented anything at all, since the acoustic torpedoes had a reputation for exploding spontaneously as they sped deeper. The next day, the location of the debris field gave no unambiguous clues as to which action, if any, was responsible for the floating wreckage. Given the confusion, the War Department ended up awarding both pilots Distinguished Flying Crosses for the apparent destruction of the Japanese sub.[36]

Tidwell was nothing if not thorough in his *I-52* research, amassing a warehouse of facts and clues. As he sifted the evidence and weighed its nuances and contradictions over the years, he became convinced that the submarine had in fact been destroyed during the first attack. If true, that meant Tidwell had a reliable fix on the sub's grave, since he had tracked down field reports from the *Bogue* and the Nazi submarine that gave the headings of the distant parachute flares. The intersection of lines drawn from the two headings marked the spot.

Armed with such information, Tidwell set out to court investors and put together a recovery team. By this time, in the early 1990s, the undersea-equipment market was awash in gear and methods once reserved for the superpowers. Eager to exploit them, Tidwell made contact with Sound Ocean Systems, a Redmond, Washington, firm that built and ran the tethered robot that salvaged the *Central America*'s gold. SOS, as it is called, had diversified in the 1990s to include working as the American agent for a Russian oceanographic group that put up a fleet of seven ships for hire at fire-sale prices, typically renting them for about $10,000 a day instead of the usual $20,000 or even $30,000 for comparable Western ships.

The deal was too good to pass up. Tidwell chartered a Russian oceanographic vessel some 340 feet in length that was manned by a Russian crew of fifty. A big spool on board the ship carried more than four miles of coaxial cable for towing instruments in the deep. The cable, akin to that for televisions, could transmit less data than the more advanced kind of conduit made of glass fibers, but was nonetheless judged adequate for the job of relaying images from the ship's deep sonar and camera sled. To oversee the sonar work, Tidwell hired Meridian Sciences, Inc., of Columbia, Maryland, a Navy contractor skilled in teasing the most information out of such signals. Also from Meridian, Tidwell hired a seasoned expert to oversee the American operations, Tom Dettweiler, who early in his career had worked for the Navy and Woods Hole and had helped find the *Titanic*.[37]

Thus, the team that Tidwell put together was a mix of Eastern and Western strengths, and in some cases drew on military skills that civilians were just acquiring. He did all this with the financial aid of one main investor, a businessman, who put up nearly $1 million in a bet that two tons of gold lying on the Atlantic seabed for a half century could now be found with relative ease.

The team assembled in the Canary Islands off the coast of Africa in

February 1995, spirits high, ready to go. But the Russian charter ship was not. The *Gelendzhik* turned out to be a rusting monster. More troubling, Dettweiler discovered that the ship's deep camera sled was battery powered instead of getting its electricity through a long cable, meaning its bottom time would be severely limited. Dettweiler advised that the expedition be delayed until the right gear was in hand. Tidwell concurred, even though he was itching to put to sea after laboring on his project for so long.[38]

Part of the team reassembled in early April 1995 for the shakedown cruise of a different Russian ship that had docked at Long Beach, California, the port south of Los Angeles. The *Yuzhmorgeologiya* looked like a luxury liner compared to her predecessor. And to the team's relief, her electrical gear was excellent, both down below and up top. To track her position at sea, the ship had American receivers to pull in signals from the Global Positioning System of navigational satellites, giving the team a locational accuracy that was essential for the hunt.

Before the Soviet collapse, the ship had been used by Moscow for deep mining, mapping, and secret naval operations, including listening with deep microphones for the telltale sounds of American submarines. On board, Tidwell was told that the dreaded KGB, Moscow's top spies, at one time posted agents aboard who carried guns.[39]

It was during the April shakedown cruise that a bit of vital brainstorming occurred that in retrospect marked a turning point.

Tidwell had hired Meridian mainly for its sonar expertise, which it had developed for the Navy. Its specialty was making and using powerful computer programs that removed distortions from sonar images, transforming them from fuzzy impressions into sharp renderings that could be merged with existing seabed maps and geophysical data. The process was akin to removing wiggles from the reflections of a funhouse mirror. The firm called its computer program Orion, after the constellation. The company's other work for the Navy was more sensitive militarily but also involved the improvement of geographic accuracy, a theme of the firm's work (and thus its name, Meridian, a reference to the great circles geographers draw through the poles). Using secret Navy data on the undersea travels of nuclear submarines, Meridian would analyze the routes with its computers and remove navigational errors introduced by such things as currents, geophysical quirks, and recording delays and flaws, allowing Navy managers to better understand the stealthy routes taken by the submerged warships through the sea. The arcane process is known as renavigation.[40]

During the April 1995 cruise, the idea arose that Meridian might renavigate the half-century-old sighting data, in theory giving a better fix on the *I-52*'s grave. Tidwell liked the idea, but the job was given no special urgency.

After all, no one knew whether the analysis of information that was so old would substantially aid the hunt. Since the computer software for the renavigation process had been declassified only recently, Meridian had never before had an opportunity to do such work for civilians. It was a first.

An element of urgency and intrigue did soon arise as Tidwell learned that a rival British group had hired the Russian ship *Keldysh* to hunt for the *I-52.* Its twin submersibles could dive nearly four miles deep and had excellent claws for dissections and retrievals. Tidwell offered to join forces, but the rivals declined and set sail before Tidwell could.[41]

The race was remarkable, given the icy solitude that had enveloped the lost submarine for a half century. Now, quite suddenly in historical terms, rival groups were about to crisscross the sea to locate the old wreck, each side armed with different cold-war spinoffs. The competition was a measure of how fast the field was accelerating.

After passing through the Panama Canal, the *Yuzhmorgeologiya* stopped in Barbados to pick up Tidwell, supplies, and the rest of the team. On April 12, 1995, a Wednesday, the big ship pulled out to sea, sailing due east for the desolation of the mid-Atlantic.[42]

The search began six days out, some fifteen hundred miles east of Barbados. Cheerful and confident, the team expected to find the wreck in anywhere from a few hours to a few days. In an important bit of news, Tidwell learned from the Russians, who were in radio contact with their colleagues on the *Keldysh,* that the rival group had come up empty-handed, despite its head start. The lost treasure was still waiting to be found.

The upbeat hunters of the *Yuzhmorgeologiya* scanned the seabed more than three miles down, finding it to be in a transitional state from the relative smoothness of the Cape Verde abyssal plain to the rocky foothills of the mid-Atlantic ridge. Their key tool was the side scanning sonar, which looked like a fat torpedo in a boxy cage. It was tethered on a long cable, so that, as the ship inched forward at one knot per hour, the sonar was towed several hundred feet off the bottom, its pulses of sound sweeping outward horizontally through the darkness for a kilometer on each side. Characteristic of its class, the sonar had a blind spot where the lobes neared one another on the seabed but did not touch or overlap, the gap showing up as a band down the middle of a glowing computer screen.

Bats flying through the dark locate food by sending out high-pitched squeaks and listening for revealing echoes. In the darkness of the deep sea, sonars do the same kind of thing, emitting sound waves and measuring, with extreme precision, the time it takes echoes from distant objects to bounce back, as well as the strength of the returning signal. Intensity readings can reveal whether the bottom is soft sands and sediments or hard rocks and metallic objects. All echoes are arranged by a computer into a visual mosaic that resembles a photographic picture. The image, though fuzzy by nature, in some respects is more informative than a photograph.

Right from the start, the team saw tantalizing hints—long narrow images roughly the right size. The ship would then zero in on these targets for a closer look, its sonar towed closer to the bottom and switched into a higher frequency to see more detail.

Excitement ran high as the targets were investigated one by one, the sonar room packed with spectators. But each turned out to be nothing more than rocky outcroppings. The bottom was dotted with an amazing number of them, all roughly one hundred meters long.

For more than a week the team searched the seabed, moving back and forth in parallel rows, covering dozens of square miles, working around the clock in two shifts. But nothing of significance was found. Tensions rose. Some crew members knew of the location dilemma, and some began to fear that, after all, the submarine wasn't where Tidwell thought it was. Tidwell showed the most strain of all, losing sleep, looking haggard.

After two weeks at sea, on April 26, 1995, a Wednesday, Dettweiler and associates began to pore over Tidwell's raw research materials, hunting for overlooked clues, trying to do their own crude renavigation in order to narrow down the search area. The move was a measure of growing desperation. To make matters worse, the Russians announced that the fuel they had bought in Panama was bad. It was causing engine trouble. Supplies of good fuel, the Russians warned, were so low that the expedition soon would be forced to turn back.

A bit of hope materialized as the Meridian people in Maryland called in with the first tentative results of their computerized renavigation, talking to Tidwell over a satellite radio link. The hunt was off the mark and needed to move westward for a distance of many miles, they advised. On April 29, a Saturday, Meridian transmitted to the *Yuzhmorgeologiya* the exact predicted coordinates of the Japanese wreck, sending the data in an encrypted transmission to thwart any possible eavesdroppers. The data was accurate to the nearest arc minute, which at those latitudes translated into a distance of about a mile.

The news was bad. The spot predicted by Meridian lay outside the normal search plan, far westward. The team had already been moving in that general direction, based on its own crude renavigation. Now, a decision was made to increase the pace of westward advancement, rather than picking up the whole operation and moving it to the west, which would create a gap in the seafloor coverage. To date, the back-and-forth search pattern had been conducted with a fair amount of overlap, ensuring thorough coverage. Now it was changed so that the swaths abutted one another. The new approach was a gamble. It accelerated the pace of westward progress, but if the sunken submarine lay at the juncture between swaths, or in the shadow of a rocky ridge, it might escape detection.

The fuel got lower. The food got worse. Gloom spread among the crew like a creeping fog. To everyone's dismay, the accelerated search turned up nothing but the usual rocky outcroppings. The team had covered more than one hundred square miles of seabed, finding nothing at all. It was too much. Tidwell, who was accumulating bigger bills with each passing hour and was accountable for roughly $1 million of other people's money, announced that he wanted to call it quits and cut his losses. He would mount another expedition on another day, he reassured the team, even while unsure himself that he could muster investor support for a follow-up voyage. But this expedition was over.

With customary stubbornness, Dettweiler kept pressing for a little more time. Just one more run, he kept saying. After all, the hunt was only now getting to the heart of the area that Meridian had predicted as the *I-52*'s resting place. Tidwell was too tired to argue.

On the morning of May 2, 1995, a Tuesday, the sonar tracings of the seabed revealed a tiny but odd sliver of blackness. It was darker than the usual returns. And it was just the right size, about a hundred meters in length. Possibly, just possibly, it might represent metal, which tends to bounce echoes more forcibly through seawater than rock. Taking a closer look was no easy matter. Turning the *Yuzhmorgeologiya* around took many hours, a big investment for an expedition at its end, but Tidwell gave the go-ahead. This was it, the last chance, maybe the very last chance after five years of dreaming.

Tidwell and many members of the team crowded into the sonar room as the results of the second pass were sketched out on chart paper from a printer that showed the sonar's raw output. Nothing. No sub. The sonar was too close to the target, right over it, in fact, so that the hundred-meter object remained hidden in the gap between the lobes of sound that swept the sea's blackness. The gap appeared on the chart as a white stripe, a band of frustrating emptiness that marked the sonar's path across the bottom. Tidwell ordered another pass slightly to one side of the target, even though repositioning the ship would take hours. Killing time, he slumped over the ship's railing with a beer in his hand, overcome by fatigue and anxiety.

It was dark outside as curious team members reassembled in the bright lights of the sonar room at 3:30 on the morning of Wednesday, May 3. Five minutes later, the full image of the mysterious object spewed out of the printer. There it was. The thing was incontrovertibly a submarine. The fuzzy image gave just enough information to disclose intriguing details. The sub lay jet black against the grays of the seabed, sitting upright, its bow and stern clearly visible, its conning tower casting a shadow. The craft was apparently intact save for what appeared to be minor bow damage and a torpedo hole in its starboard side. Just discernible atop its sail was a railing and, nearby, some booms. In front of the submarine lay the clutter of what was clearly a debris field—material torn from the sub during the torpedo attack more than a half century earlier that had slowly drifted to the seafloor.

There were a few congratulatory handshakes but the physical pain and emotional exhaustion were too great for there to be any real celebration. Tidwell himself had been without sleep for two and a half days, brooding and fretting, trying to figure out where his logic had gone wrong, preparing to explain the failure to his investors, his wife, his family, and most especially to himself. His anxiety was so great that he had cramps in his sides.

Now, in a flash, a warship sunk long ago with two tons of gold in her belly had been discovered under more than three miles of water, deeper than the *Titanic* or the *Central America,* deeper than the average depth of the global sea. It was no small accomplishment. Moreover, despite the team's frustration over the duration of the hunt, it had in fact been relatively short. The *Titanic* had been found after a search of fifty-eight days. By contrast, the *I-52* had been found in seventeen days, a quickness that suggested the field was maturing.

Most remarkable of all, the sub had been discovered only a half mile from Meridian's calculated spot, which turned out to be tens of miles from the Navy's estimate of its position. For renavigation, it was a bull's-eye that marked the public debut of a powerful new oceanographic tool.

Despite the collective exhaustion of the crew, plans moved ahead quickly to document the find. Photographs would most aid that end, but such operations were tricky, since towed camera sleds could easily get hung up in the wreckage. So a decision was made to first investigate the debris field, a less perilous place. That job was also very important to Tidwell, since he needed to recover some debris if he was to claim ownership of the wreck under international salvage law.

The camera sled would be the Russian one, it was decided after some discussion. Also available to Tidwell was an American camera sled that SOS had built as a backup after the early troubles with the Russian gear. But fierce Russian pride won out. Known as *Neptun,* the sled was an old warhorse about the length of a man, covered by perhaps a dozen coats of paint. It had no thrusters or sonars or other advanced gear, but it did possess two thirty-five-millimeter still cameras and a small Sony video camcorder. By way of lights, it had floods for video and strobes for still photographs, all ready to illuminate the depths.

In the early daylight hours of May 3, 1995, *Neptun* was lowered on the long cable and towed across the seabed. No trace of the sub or the debris field could be found. The problem turned out to be navigational. The team, in a rush because the ship's fuel was so low, had tried to accomplish the pass without the aid of pingers. Now that was corrected. Four orange transponders were heaved overboard, sinking to form a diamond on the dark seabed, which helped the crew track the sled's location and progress. Over the next two days the debris field was successfully explored. It was found to be full of rubble and artifacts, but apparently no gold.

The team was attempting another run through the debris field on Friday, May 5, when the video camera revealed that the smooth bottom was suddenly giving way to rockiness, as if sedimentary muck had been scraped away to expose the rocky seabed. A cable flashed by. Dettweiler cried, "Bring it up! Bring it up!" fearful that the camera sled was about to hit the submarine's side. His worries were well taken. As *Yuzhmorgeologiya*'s winch turned and *Neptun* rose, a dark presence suddenly loomed in the video monitor. For too long it remained a blur as the sled's inexpensive camera struggled to regain focus. The strobes of the still cameras flashed away. Then the hulk suddenly came into view. It was the submarine. The video camera revealed a long glimpse of her rear deck. Then, as the ship and *Neptun* moved forward on their relentless path, the old submarine vanished as abruptly as it had appeared, its memory hanging in the air of the control room like a fading dream.

Tidwell waited impatiently as the still film was developed after the sled's return. Finally he gazed at the negatives and photos. They were perfect. Razor-sharp, they revealed a wreck that could only be a Japanese submarine from World War Two, his judgment based on years of studying photos and diagrams, anticipating this moment. The sub was in surprisingly good shape, structurally intact and bearing none of the rivers of rust that marred the *Titanic.* Eerily, many lines were strewn across her rear deck, some wound around two large bollards, as if the ghost ship was just finishing its transfer of men and material from the Nazi submarine when the operation was suddenly interrupted.

The sub was the *I-52.* It had to be, by logic if nothing else. The wreck was right on the predicted spot. And the Navy, which was more careful than even Tidwell about such things, knew of no other sunken submarines in a surrounding area that extended for hundreds of miles.

The *Yuzhmorgeologiya*'s dredge was lowered into the debris field the next day, Saturday, May 6. The process teased little from the seabed—bits of wood coated with red paint, fragments of steel, an electrical insulator, and two pieces of dark-blue wool cloth.

The team had run out of time. The ship had to head back for the Caribbean immediately or be stranded at sea for want of fuel. It was time to go home.

Tidwell called me a little more than a month after his return and before there had been any national publicity about the find. He was clearly excited yet managed to sound remarkably cool, making no sweeping claims. The rough coordinates of the find, he volunteered, were fifteen degrees north and forty degrees west.

"There was a lot of hardship getting to this point," he said of his successful hunt and the long history of gnawing doubts. "I was right. And that's a good feeling." We agreed to rendezvous at Meridian, where he would go over the voyage's details and the experts there would show me their methodologies.[43]

Curious about what a scholar unconnected to the find would say, I called Carl Boyd, a military historian at Old Dominion University in Norfolk, Virginia, and an expert on World War Two submarines. He told me that the discovery and further inquiry would be important because so little was known about Japan's undersea warships. The ones seized afterward had all been destroyed. "This is going to be hot stuff," he said of the *I-52.* "It's not the *Titanic* or the *Bismarck,* but in a more subtle way it's perhaps more interesting because they were pretty well known." Boyd said historians had no photographs of the *I-52,* one of Tokyo's mightiest subs. In contrast, the *Titanic* had been photographed hundreds and perhaps thousands of times, including every stage of its construction.[44]

The brick-and-glass building of Meridian Sciences was surrounded by a sea of green in the planned city of Columbia, Maryland, close to the Applied Physics Laboratory of Johns Hopkins University, a brain bank for naval research, where the company's founders had once worked. Early on Thursday, July 13, 1995, Meridian was all jeans and shorts, sneakers and sandals. The average age was thirtyish. In all, Meridian had fifteen employees—five computer programmers, five sonar and navigational analysts, and five administrators.

Its president, Dave Jourdan, a former nuclear submariner, was relaxed and friendly. Meridian was founded in 1986, he said, and for years did nothing but Navy work on renavigation and sonar processing. He added that the *I-52* success, spreading by word of mouth, had already prompted many international inquiries.

"I have a trip to London at the end of the month to talk to people looking for colonial kinds of shipwrecks," he said.[45]

We walked down a spacious wooden stairway toward the heart of Meridian's operations, a room where the paths of nuclear submarines were recalculated. CLOSED AREA, read a sign. Built into the door was a black combination lock like that on a safe. The door's handle bore a separate five-button lock. Jourdan punched in some numbers and the door swung open to reveal a small room filled with computers and filing cabinets and safes. One wall held a large map of the global seabed.

Jourdan slid up to a computer and, employing the company's renavigation software, prepared to redo the steps that disclosed the *I-52*'s resting place. He said the software was free of all government secrets, which in a stroke of luck had been carefully stripped away just in time for the *I-52* search. The goal had been to recalculate the route of the Japanese submarine with a precision much greater than even its officers knew.

"From course and speeds, we created a relative track," he said, "and then set up tables about the performance of the navigation." After that, "it's just crunching numbers." Jourdan tried to make it sound easy, but the procedure in its details seemed quite complex, especially as armies of numbers marched across the computer screen.

The finished product was a map that showed the renavigated route over which the American battle group and the German submarine had traveled on that June night more than a half century ago. Radiating out from the American task force was a line along the heading of the observed battle flares, 235 degrees, or to the southwest. A similar line radiated out from the German submarine at 0 degrees, or due north. Where these two lines crossed, Jourdan said, "was the best thing I had to go on," the best indication of the position of *I-52*'s grave. Then came the double checking, the testing of the prediction against such things as the renavigated position of the discovered debris. Finally, Jourdan factored in the distance he felt the sub had gone after diving beneath the battle flares to escape the attack and, after it was hit, how far the lifeless sub had drifted in ocean currents as it fell more than three miles to the seafloor. He concluded that the two movements more or less canceled one another out.

On the map, a small dot with a circle drawn precisely around it marked the calculated spot of the *I-52*'s resting place.

"The Navy's position was way out here," Jourdan added, pointing to an empty spot on the map in a sea of white, "tens of miles away."

We left the renavigation room and wandered into the work area of Dave Wyatt, a bearded computer expert, who led Meridian's sonar work aboard the *Yuzhmorgeologiya.* Tapping away on his computer, Wyatt zoomed in on a sonar image of the *I-52*'s debris field, showing how the processing could tease out information. "There's a lot of stuff here," he said, pointing to the luminous image of submarine rubble, "not just six or seven pieces but hundreds of different objects."[46]

Eventually Tidwell showed up. He was groggy again, still pushing himself hard, having flown in late from Texas and gotten just a few hours' sleep before driving to Meridian, which was more than an hour from his home. In his forties, he was a bulldog of a man, with gray hair and dark eyebrows, just over five feet tall. He spoke slowly and carefully, with a slight southern accent. Every once in a while his somewhat dour expression would break into a smile, a big one.

The analysis of the find was still underway, Tidwell said as he spread out a large number of deep-sea photographs on a conference table. The pictures were remarkably clear, showing not only overviews of the sub's deck but tiny details as well—bolts, railings, and a ribbed vent—all covered with a patina of fine sediment. Then he rolled out a photo mosaic that was still being pieced together. Already it revealed the sub's girth and how thickly its deck was covered with cables and ropes, evidently leftovers from the submarine rendezvous.

Tidwell said he planned to return in a few months to photograph the hulk and its debris field as thoroughly as possible. All kinds of scholars and experts would be along, including corrosion engineers to help judge the steel hull's integrity. By way of still cameras he was considering two cold-war spinoffs: a digital camera Woods Hole developed for the Navy, and the laser line scanners, whose thin beams of bright light penetrated far through water and whose declassification made such a splash at the 1993 conference I attended. More than three miles down, in pitch darkness, only the scanner could get a continuous image of the big submarine.

His goals were to get the gold, to assess the feasibility of raising the whole submarine, and, not inconsequentially, to learn more about the last hours of the *I-52*.

Even in his slightly dazed state, he fairly vibrated with excitement as he foresaw a flood of new information about the *I-52* that would flow from "photographing every square inch from every conceivable angle. And hopefully, from that, we'll be able to reconstruct the event in such a way, graphically, three dimensionally, that it will give us the truth—that it will allow us to say, 'This is what happened.' "[47]

The recovery expedition to gather up the gold would follow on the heels of the photographic voyage.

"We want to disturb the wreck as little as possible," he said, his voice low. "Something happened to me out there. It's hard to express. But I feel like I was offered a responsibility to make sure we treat it with respect. Those people were doing their jobs and died bravely, regardless of their nationality. We don't want to leave the thing scattered all over the place just because we were greedy to get the gold. It costs more—a lot more, actually—to be very careful. But we don't have any choice in it. I'm the leader of this program, and that's how it's going to be."

Tidwell said he had vowed to use no explosives or clamshells, dredging buckets with giant steel jaws that could bite through a wreck, smashing it to bits. Instead, he would work like a surgeon. There was a wealth of advanced gear available for such jobs, Tidwell said, including the *Keldysh* and its *Mir* submersibles, which would be ideal for photographing the sub as well as cutting it open.

Ultimately the sub might be raised. "If we do, we have offered to return it to Japan," he said.

At such a depth, Tidwell added, the pressure was too great to allow air to be pumped into the wreck for buoyancy. But a special marine foam existed that could be injected, displacing the water, strengthening the sub, lightening it enough so the mass of metal could be lifted to the surface. "We have discussed this with a lot of people," Tidwell said. "It's doable. It's just very expensive."

Far less costly was the job of simply recovering the two metric tons of gold, which at current prices was worth about $25 million, Tidwell said. The total bill for the *I-52*'s discovery and recovery might eventually reach $7 million or $8 million, at most.

"Yes, it will be profitable," he said. "But my main goal, besides finding it, is that it opens the door to a lot of other deepwater projects that are challenging and even more profitable."

What, I asked, could be more profitable than finding two tons of gold?

"A shipment of fifteen tons of gold," Tidwell replied. "I know of one. And there's another with seven tons. And there's several with five tons."

I wondered aloud if all kinds of ships lost in deep water, old and new ones, were suddenly coming into play.

"It's true," he said. "It really is happening. I know of three deepwater projects underway right this minute, no, four. Two recoveries are being made and two deepwater searches are being made, right now. Strangely enough, they're all modern ships."

I asked Tidwell what the future might hold for deep recovery. To me, he seemed like a logical person to ask. After all, he knew not only his own projects but, by virtue of having worked as a maritime researcher, knew of many other projects around the world, an unusual thing in so secretive a business.

Over the next fifty years, he said, salvors would find most of the big gold shipments that had been lost at sea. And by the end of the next century, he predicted, the number of all kinds of ships that had been investigated in the deep might reach into the low thousands.

I expressed surprise at the large number.

"We all read books about salvage," he replied. "But a lot more actually goes on that is never reported—deep salvage. There was a platinum recovery in the Mediterranean, five thousand feet down, that you never heard about. It was done three years ago, a private thing out of Europe. A lot of people don't realize that gold was taken off the *Lusitania.* And there are recoveries made off the Middle East that people don't talk about, cultural stuff, twelfth-century wrecks. A lot of it is done by people with money who, for whatever reasons, want to keep their finds discreet."

His hope, Tidwell said, was that some of the raw history, especially the human dramas aboard the lost vessels, would be captured along with the cargoes. Market forces and high technologies were now driving the hunt. But there was more to be gathered than gold, Tidwell suggested. This former Army sergeant who battled his way through the streets of Saigon during the Tet offensive, who became fascinated with the misfortunes of the Japanese military, was especially conscious of the fact that the sea's depths hid many untold chapters of war.

"You have to justify the cost of doing these recoveries," he said. But beyond that lay the prospect of human revelation.

"It can get very personal, the hardships and all. We have a general conception of some of it. But what really happened is often unknown."

FIVE

Canyon

MISTS COOL enough to warrant a sweater hid the coast road and the rocky shoreline around the edge of Monterey Bay, even though it was a midsummer morning in 1994 and, back in the Central Valley of California, the skies were sunny and the temperatures were headed for the nineties. But here on the coast, the sea was chilly, cooling the air and producing the fog that cloaked the day with a curious sense of expectation. By midafternoon, with luck, the grayness would be gone.

Part of the unpredictable Pacific, the deep blue bay nonetheless looks approachable when the fog has burned off and the water is calm, almost like a large lake cradled by the land's arms. The semicircular bay, some twenty miles from tip to tip, can seem smaller if you drive along the coastline, past gnarled evergreens bent from the wind, and enter the town of Monterey. Like many seaside resorts, its heart is a blur of bars, signs, souvenir shops, and fried food. It's hard to imagine John Steinbeck wandering its byways, as he did before the arrival of the tourists and the honky-tonk atmosphere. The bay can be glimpsed occasionally between some of the shops, or can be viewed directly from the waterside eateries. It is still lovely despite the chaos of commercialism and the din of nearby traffic.

But all this, even the panoramic views along the coast, turn out to be something of an illusion. Nothing the eye can see, nothing the camera can record, holds a hope of revealing the reality of the bay. It is a great wilderness, its dark waters running extraordinarily deep, teeming with unfamiliar forms of life. The depth is what makes the place so unusual.

Most of the earth's continents are girded by shallow shelves that slope gently away from the land for many miles, their depths seldom greater than a few hundred feet. On the East Coast of the United States, the shelf stretches for more than one hundred miles in many spots, before giving way to a steep slope that drops into the abyss. On the West Coast, the shelf is narrower but still a substantial barrier to the deep. Off Monterey, things are different. Here an enormous canyon cuts the continental shelf in two, bringing deep water remarkably close to land. The canyon is so close that erosional forces at its tip keep nibbling away at the old pier at Moss Landing, located at the bay's midpoint. The dilapidated pier used to be twice as long, but with troubling regularity its leading edge keeps tumbling into the head of the invisible canyon.

Studies over the decades have revealed the immensity of the gash that stretches out to sea. As you head outward, it quickly drops away beneath you into a maze of rugged tributaries. Go two miles and it's six hundred feet deep. Go ten miles and it's a half mile deep. Before long it's more than two miles deep, an invisible scar in the earth's crust grander than the Grand Canyon. In fact, the Monterey Canyon is the largest submarine fissure along the lengthy continental coastline of the United States. Its dissimilarity to the seafloor just up the coast is telling. Around San Francisco, some eighty miles northward, the waters are relatively shallow. Ten miles out from Golden Gate Bridge, the sea is only about sixty feet deep. Twenty-five miles out, it's still only about two hundred feet deep. In short, the ocean off San Francisco is a bathtub compared to the canyon's deep waters.

The contours of Monterey Canyon, as well as its location along the California coast, make it an intersection of complex winds, currents, and nutrient flows that feed a cornucopia of life. There are massive kelp forests, thick beds of mussels, droves of sharks, squids, salmon, rockfish, anchovies, and tuna, as well as commercial fisheries that tap this wealth. Decades ago, the bay was one of the world's great sardine fisheries, until that industry was killed by overfishing. Millions of seabirds still flock here to feed, as do countless marine mammals, including sea lions, harbor seals, porpoises, humpback whales, and blue whales. In all, the bay and its environs support more than three hundred species of fish, nearly one hundred kinds of birds, and two dozen kinds of marine mammals, including such endangered ones as the southern sea otter and the gray whale.[1]

For more than a hundred years, the bounty of the canyon has been a magnet for scientists intent on understanding the sea's richness. The Hopkins Marine Station, an arm of Stanford University and the oldest marine laboratory on the West Coast, was founded in 1892 at Pacific Grove as a stronghold of bay studies. One of the most famous investigators was Ed Ricketts, whose 1939 book *Between Pacific Tides* helped introduce marine biology to the wisdom of ecology—the then-radical idea that an organism can be understood only when its neighbors and neighborhood are taken into account. Ricketts was famous for other reasons as well, his eye for new creatures being as sharp as his eye for new women. His adventures were chronicled in Steinbeck's 1945 book, *Cannery Row,* as well as local folklore and literature. Today there are nearly a dozen bay institutions that pursue oceanography or explore the canyon, often with the same kinds of nets and traps used by Ricketts and scores of other scientists over the decades. A trend over the years has been to investigate the bay's deep waters with increasing thoroughness. Their proximity cuts transit time for such research and aids repetitive studies, casting light on seasonal changes and the shifting character of deep life. The closeness also means oceanographers can go home every night rather than bedding down in cramped bunks aboard ship. It's like having the abyss in your backyard.[2]

One institution has gone further than any other in disclosing the canyon's deep secrets and has done more than any other exploratory group in the world to reveal the overall richness of midwater life. It is known as the Monterey Bay Aquarium Research Institute or, for short, MBARI (pronounced em-BAR-ee). With great regularity, it plies the waters of the bay with an advanced robot and other high-technology gear, illuminating the depths, finding new creatures at a rate of about a dozen species a year, discovering how hundreds of little-known animals behave and hunt, live and die. The key to the accomplishments is a robot of great agility. Its name is *Ventana,* Spanish for *window,* a reference to its being a window on a dark world. No net or diver or submersible can move as fast to examine or capture sea creatures, especially in the remoteness of the deep. *Alvin* is a snail. In comparison, *Ventana* is a dolphin, its masters having it speed through the deep in search of the unusual. The institution's logo is a gulper eel. The emblem shows the gulper's enormous mouth opening wide and its sinuous tail trailing off into the canyon's darkness. And the gulper, an animal rarely seen by scientists, turns out to be the least of it.[3]

THE MAN BEHIND MBARI was David Packard, the billionaire cofounder of the Hewlett-Packard Company and a father of Silicon Valley. He was its patron and visionary. Among other things, Packard made MBARI a leader of deep privatization—of moving cold-war ideas and equipment into the private sector to advance deep exploration. As a top Pentagon official in the late sixties and early seventies, Packard ran Washington's deep war against Moscow. Years later, knowledge of those feats helped inspire MBARI's founding, gave the institution much of its high-tech flavor, and provided a rough blueprint for its exploratory work. Packard set up MBARI in 1987 and proceeded to spend more than $60 million of his own money to probe the mysteries of the watery unknown. With the cold war's end, the civil institution acquired all kinds of military gear and personnel and became a prime example of the synergy that is fast illuminating the sea's darkness. For Packard, the canyon was a test. Eventually, the devices and methods perfected here are to tour the world to further man's understanding of the deep, particularly its enigmatic middle waters, the vast darkness located between the surface and the seabed.[4]

MBARI CAN BE HARD to find. You pass from Monterey into the calmer streets of Pacific Grove, driving past rows of tidy shops. Having missed it, you enter a neighborhood of old Victorian homes and turn back. Finally, you locate a modern, airy building of sloping roofs secluded behind a profusion of greenery. It is very new, very discreet. Two blocks from the waterfront and its sister institution, the Monterey Bay Aquarium, the building is the antithesis of tourist-strip garishness. It is Monday, July 11, 1994, and you are awaited inside.

MBARI's president is Peter G. Brewer. British born and a geochemist for two decades at Woods Hole, he gave me an institute overview and tour, starting with a videotape of robot discoveries. The film is unabashedly known as the Greatest-Hits tape.

"We're out on the water four days a week, leading the world in terms of dive experience," Brewer said as he narrated the silent video. The images of the undersea domain were arresting, some of the best and clearest I had ever seen. The first showed *Ventana* plunging beneath the water, its lights on, looking like something menacing from outer space. The robot's lights and camera revealed creatures that were both strikingly odd and beautiful, full of shimmers and unearthly body parts. In one scene, a semicircular jellyfish fled the pursuing robot and, unable to escape, dropped all of its long, luminous tentacles in a burst of light, apparently a tactic meant to bewilder a pursuer. A long marine worm corkscrewed in endless circles, its motive a mystery. A blue shark sped by, as did the shadowy blur of a great white shark, one of the sea's fiercest predators. The largest creatures were also the strangest: gelatinous siphonophores, snakelike animals that MBARI had discovered growing to lengths of 130 feet, longer than the blue whale. Each siphonophore had a cluster of pulsating bells at its head for propulsion and, hanging down from long bodies, willowy tentacles dense with stingers for fishing and drawing prey up to dozens of waiting stomachs. They were like nothing I had ever seen or imagined.

The tape also featured rare footage of a gulper eel, its sinuous body expanding into a huge head and hideously large mouth. It turned slowly to face the camera, its jaws opening and closing in eerie silence. A final haunting image was that of the wispy and mysterious ribbon fish, a smaller cousin of the monsterlike oarfish. MBARI had managed to photograph the fish for the first time in its habitat, capturing its slow, elegant movements as well as a series of hypnotic wavelike motions along its dorsal fin. Curiously, even as the fish hung motionless in the water, opening and closing its mouth and gills, the long fin continued its pulsations, beating in a baffling rhythm. "Seeing these kinds of animals in their natural habitat and how they feed and behave is very unusual," Brewer said. "It's giving us insights into a very different world."[5]

Brewer said the secret of MBARI's success lay in its ability to plunge advanced gear into deep water day after day, year after year. The general goal was to take the robot out Monday and Tuesday, Thursday and Friday. Wednesday was usually a down day for maintenance. At this rate, Brewer said, deep waters were explored on average about 155 days a year, far more frequently than anywhere else in the world. "This is the only place in the country where you can leave the dock and, before you've finished your cup of coffee, you're at true oceanic depths."

Ventana can dive more than a kilometer deep. A new robot being built at MBARI, Brewer said, will go to depths of up to four kilometers, or two and a half miles, opening to study the hidden recesses of the canyon and the majority of the world's deep waters as well. The new machine is to be unusually smart, swift, and silent, thus its name, *Tiburon,* Spanish for *shark.* "It's probably the most intelligent and sophisticated underwater vehicle in the world," Brewer said. "It will do amazing things." He showed me a section of its tether, made by the Rochester Corporation, an old-line Navy contractor. It was heavy. Five-eighths of an inch in diameter, it was made of three copper lines and three fiber-optic lines surrounded by layers of steel-mesh jacket. The tether was meant to supply the robot with physical support, electrical power, and directions from human controllers. More important, it was to send back pictures from down below.

The birth of the new robot at MBARI was aided by one of Packard's guiding principles—the close collaboration of scientists, engineers, and operators, a secret of Hewlett-Packard's success. Brewer noted that the rule was hard to follow at MBARI because the ninety-person staff was split between two sites, the headquarters in Pacific Grove, where most scientists and some engineers worked, and the operations center up the coast at Moss Landing, where mechanics and engineers labored and gear was kept for exploring the bay, including the research vessel *Point Lobos* and the robot *Ventana.* Pack-

ard's solution was to house everybody under one roof. The walls of Brewer's office were covered with colorful plans for the new complex under construction at Moss Landing. By virtue of its location at the bay's midpoint, the complex would be at the very tip of the invisible canyon.

Packard's commitment to MBARI and its explorations was suggested by the scale of his funding. The new laboratories and offices at Moss Landing will cost $21 million. The new robot will be about $7 million. A new ship being built to guide the robot will be about $14 million. In addition, Packard expanded MBARI's operating budget so new workers can be hired. In the next few years the number of personnel is to grow from 90 to about 250 people, nearly tripling MBARI's staff.

On our way to lunch, Brewer took me by MBARI's video library, the heart of the discovery zone. The room was packed with row upon row of videotapes—5,300 of them from 772 dives, according to a sign. At the center of the room, two women were viewing tapes, adding electronic notes to them so countless findings and observations could be easily relocated, cataloged, and digested for the writing-up of scientific studies.

The lode included data gathered by the itinerant Russian *Mir* minisubs. In a cold-war spinoff, they had descended more than two miles to the far bottom of the canyon, deeper than *Ventana* can go. During the 1990 dives, a camera was set up to monitor the dense beds of clams thriving at cold seeps. Cold-seep communities are akin biologically to volcanic hot vents but in some respects are more surprising, since the waters from the seabed that feed them have temperatures virtually the same as the surrounding sea. Discovered in 1984, the communities are believed to be based almost entirely on symbiotic microbes that eat either hydrogen sulfide or methane, a common by-product of oil deposits and decaying organic matter. Cold-seep communities have been found to include clams, crabs, mussels, fish, shrimp, snails, limpets, anemones, sea stars, brittle stars, sea cucumbers, and tube worms, often being rich in unfamiliar species. The methane seep communities of Monterey Canyon, first discovered in 1988, are populated by dense colonies of giant clams, such as *Calyptogena,* which harbor chemosynthetic bacteria in their gills. The microbes feed on methane, and the clams in turn feed on the microbes. One of the mysteries of the canyon is why the communities form where they do, a topic of inquiry to which MBARI is devoting much time and energy.[6]

As we walked around after lunch, Brewer talked of another benefit from

FISHES OF PACIFIC ABYSS. Big fish never before seen in their deep habitats include the gulper eel, the great white shark, and the ribbon fish, its dorsal fin alive with hypnotic undulations.

the end of the cold war—the new ship under construction, a remarkable type originally developed by the Navy. Known as SWATH, for Small Waterplane Area Twin Hull, it is extremely stable, even in treacherous seas. The secret is the hull. Instead of a single one that rides the waves, a SWATH ship has two hulls, similar to a catamaran, except that each is located deep underwater and is basically a submarine speeding beneath the waves. Subs, of course, are largely immune to surface pandemonium. So, too, the submarine hulls of a SWATH ship dramatically reduce the vehicle's roll, pitch, and heave. The twin hulls hold engines and propellers. Running upward from the twin hulls are thin, strong struts that join together to form the ship's superstructure. Seen head on, the ship's overall shape is like the letter *n.* The Navy perfected SWATH ships at the end of the cold war and used them mainly for a secretive class of surveillance ships that lowered microphones to spy on enemy submarines. But SWATH ships turned out to be ideal for oceanography because of their stability, spacious decks, and ease of ocean access. The one being built for MBARI, more than one hundred feet long and fifty feet wide, a vast breadth, is to have a central well through which the robot will be lowered into the sea below, rather than over the side. As important, it is to have twenty-four bunks, allowing scientists to roam the world for months at a time. MBARI's current ship, *Point Lobos,* has no bunks and by definition is a day sailer. By contrast, the new ship's range is to be more than two thousand miles.[7]

"The run to Hawaii is within our reach," Brewer said. And from that refueling point, the ship will be able to reach anywhere in the world. With long range and great stability, the ship will allow MBARI to compare Monterey Canyon to the rest of the global sea.

Back at Brewer's office, I leafed through a book describing other MBARI projects, including an Ocean Acoustic Observatory. Its purpose is to listen in on the calls of whales and other sea life, initially with some of the military's SOSUS microphones on the seabed and eventually with new ones of MBARI's own design. Quite close by the institute is a Navy microphone array moored in the depths off Sur Ridge, an undersea promontory on the canyon's south rim. The microphones there are anchored at a depth of more than a mile, far from the noises of ships, waves, and rain and close to those of deep animals, which initial studies suggested are more vocal and mysterious than expected. The plan is eventually for MBARI to listen to the deep sounds as they happen, live, as a way of discovering where to look for new creatures. "When we hear something interesting," said Brewer, "we can go out quickly and investigate it."[8]

Brewer had to go, but the main person I had come to see would not arrive back in the office until tomorrow, Tuesday, July 12. He was Bruce H. Robison, a biologist in charge of MBARI's midwater research and an important figure in the founding of the institute. I had first met Robison and learned about MBARI during his talk at a New Orleans conference in 1993. Robison caught my attention by saying that old sampling methods had overlooked a sizable part of the deep's creatures, and that MBARI's robot was discovering them in droves. My notebook where I jotted down his remarks was cluttered with underlinings.

In the meantime, I decided to go to the aquarium. Near MBARI's front door, I noticed a crate of apricots. People at the front desk said they were from Packard's ranch and were there for the taking. The fruit was good and sweet, a tingle of California sunshine.

The Monterey Bay Aquarium is another part of the Packard vision, a gift to his sea-minded children. It is a $60 million architectural gem set on the bay, full of sealife exhibits and panoramic views. Even the restaurant has great scenery out its cathedral-like windows. As I sipped coffee, harbor seals frolicked around a nearby rock, diving and sunning themselves. The aquarium has no tropical fishes or exotic fauna. Its focus is regional, an extended meditation on the bay. Its centerpiece is a huge three-story tank featuring the lush complexities of a kelp forest. The only exhibit of its kind in the world, the tank gives a unique glimpse into a shallow ecosystem that teems with hundreds of plants and animals. Leopard sharks cruise by constantly, back and forth, back and forth. Fish move in close synchrony or hang in the kelp. Rocks are covered by fields of multicolored sea stars and anemones. Jungles of kelp stretch from the sandy floor up to the tank's heights, their leaves a medley of greens as rays of natural light filter down through the water. Most intriguing of all, this parade of life throbs back and forth as water sloshes though the tank's upper reaches, mimicking the action of waves. The sloshing is a prerequisite for kelp survival and growth, I was told. It is also hypnotic. All the creatures and kelp fronds move in rhythm with the waves.[9]

BRUCE H. ROBISON is known universally around MBARI as Robey, pronounced ROW-bee. He is informal and gregarious. Visually, he is a man of rugged good looks, deeply tanned, broad-shouldered, his hair longish, his beard gray. He could easily win a Hemingway look-alike contest and probably break a few hearts along the way.

Sitting in his cluttered office on Tuesday morning, leaning back, his hands clasped behind his head, the scientist exuded an air of relaxed vitality. He had just returned from a trip down the Salmon River in Idaho, which is undammed for its length, the West's last wild river. The walls of his office held a picture of a big sailboat as well as a large, detailed topographic map of the Monterey Canyon. His life seemed to be filled with water, whether at work or play.

We talked of MBARI and its robot discoveries and made plans for a voyage I was to accompany him on later in the week. I expressed wonder at the Greatest-Hits video, noting the eeriness of the gulper eel.

"Nobody else has seen them alive in their habitat, except Beebe back in the thirties," he said, referring to the midwater pioneer. "We were able to confirm some of his observations. Some of the bioluminescence he saw was indeed there. It's kind of nice to wait sixty years and find out that old Bill Beebe was right. Usually the only way we see gulpers is all scrunched up in bottles, wrinkled, all pasty and lifeless and gray and gooey. They're such nifty animals. They live quite deep. I think we're just skimming the top of their range."

He described the exploratory setup for stalking creatures aboard the research vessel *Point Lobos,* saying the pilot of the robot *Ventana* and the mission scientist sat side by side, the pilot controlling the robot's motions, the scientist controlling the lights and cameras, both of them viewing monitors to track the deep action. "The scientists and ops people work together," he said. "There's no question that's been a key to our operational success. You'll see."[10]

Robey proceeded to tell me of the long professional journey that led him to probe Monterey Canyon with an advanced robot. It was a tale of nonstop inquisitiveness. Born in Los Angeles, he wandered from college to college, going to Purdue in Indiana for an undergraduate degree, to William & Mary in Virginia for a master's, and to Stanford in California for a doctorate in biological oceanography. As a graduate student at the Hopkins Marine Station, he studied the bay's deep fishes, working with nets and traps, curious about the mangled bits of gelatinous material often hauled to the surface. Doctorate in hand, he went to Woods Hole on Cape Cod, diving repeatedly on *Alvin.* He was amazed by the deep life, especially the flashes and shimmers of the bioluminescence. But he was frustrated at the long distances the *Alvin* group had to travel to get to deep water and at the difficulty of making midwater observations. *Alvin,* like most deep-diving submersibles, was built to investigate the bottom. It repeatedly passed through the middle kingdom but seldom paused to explore its mysteries. The reasons were twofold. First and perhaps most important, scientists of that era viewed the bottom as the most interesting and challenging area of exploration. The deepest areas of the earth's surface were a frontier, just as the highest mountains had been decades earlier. Second, stopping in midwaters was hard to do. Typically, deep-diving submersibles carry two sets of weights that make them heavier than seawater. The first set is dropped upon reaching the ocean floor, at which point the submersible becomes neutrally buoyant, so it can conduct investigations. The second set is dropped to make it lighter still for the ascent. The first set of weights can be dropped in midwater, but that step is usually taken reluctantly, since it inhibits further descent.

After a postdoctoral fellowship at Woods Hole, Robey landed a job at the University of California at Santa Barbara. There he worked for twelve years, growing increasingly disenchanted with the old investigative methods and constantly trying new ones. "As I told my students, 'There are only so many things you can learn from dead fish,' " he said. "There I was calling myself a deep-sea ecologist. I was. But I had never really seen the habitat I was trying to describe. What self-respecting desert ecologist or other kind of ecologist would describe his or her chosen area without visiting it?" So in the early 1980s he experimented with small, one-man submarines that had been developed for the offshore oil industry, using them to study sea life in the relatively shallow waters of the Santa Barbara Channel.

The turning point came in 1985. Graham Hawkes of Deep Ocean Engineering, based in the San Francisco Bay area, had extended some of the Navy's pioneering work to build a new submersible known as *Deep Rover.* In essence it was a plastic bubble that could descend to a depth of one kilometer, or six-tenths of a mile. Instead of heavy weights, it used electric motors and ballast tanks to descend, making it ideal for midwater work. And its field of view was enormous. Unlike the tiny portholes of *Alvin,* its viewing area through the plastic bubble was essentially unlimited, giving the operator a breathtaking view. As *Deep Rover* underwent sea trials, Robey and some of his Santa Barbara colleagues were able to lease it to explore the Monterey Canyon.[11]

"We had it for a month, fifty-one dives in thirty days," Robey recalled. "It was pretty remarkable. Every dive was a revelation in one way or another. We kept having our socks knocked off."

Years earlier Robey had gotten skeptical reactions at scientific meetings when he told of uncommon observations he was able to make off Santa Barbara in the small oil-industry submersibles. His peers were understandably wary of things for which there was no evidence other than Robey's descriptions. Now he had hard data. *Deep Rover* was roomy enough to hold a large video camera in addition to a human operator. In a first, it repeatedly carried a low-light camera into the depths of the canyon. Lights extinguished, hanging motionless, surrounded by nothing but a blackness more intense than any terrestrial night, the sub would begin to move slowly through the deep water, stimulating waves of bioluminescent fireworks. In all, Robey and his colleagues captured on film the pyrotechnics of thousands of deep creatures of at least thirty different species. Most were gelatinous, their delicate bodies perfectly adapted to an alien world of no surfaces, no waves, and no sunlight. The more fragile the creature, the brighter the bioluminescence.

One of the most dazzling displays came from a siphonophore about twenty feet long, its tentacles rich in stingers. As the sub brushed against one end of its long body, the whole animal lit up and glowed with an unearthly light for nearly a minute. The illumination was so bright that, even when the siphonophore was ten feet away, the camera's lens aperture had to be reduced two stops, from f/1.8 to f/4, to avoid overexposure. In the long paper that Robey and his colleagues eventually wrote about the series of dives, the scientists said the purposes of such illuminations were a riddle but might be a defensive mechanism. Animals poised to prey on a gelatinous creature, they speculated, might be blinded, distracted, warned of nearby stingers, or their presence revealed to other nearby predators who would then turn on them.[12]

Those videos and revelations were shared with the people who had paid for and supported the unusual dives, including the newly established Monterey Bay Aquarium, headed by Julie Packard, one of David Packard's children. Robey had known Julie's sister, Nancy, ever since she was in graduate school studying the bay's biology. He was also good friends with Robin Burnett, the man Nancy was then dating and eventually married. The Packards and their circle of friends and spouses grew increasingly fascinated by the deep discoveries.

"Folks would look and get more and more excited," Robey recalled. "The aquarium was thrilled." Eventually, the videos helped inspire Packard, his marine-biologist children, and the Monterey Bay Aquarium to set up a research institute devoted to exploring the canyon. Robey was asked to aid the effort and soon presented a plan that envisioned using two vehicles of the *Deep Rover* type. David Packard—uncomfortable with the inherent dangers of manned exploration, knowledgeable about what unmanned government gear had done in the deep during the cold war, and sure that undersea robots were poised to make prodigious strides—insisted that the effort focus on unmanned gear.

"That was very appealing to him," Robey recalled. "He's a technologist. He saw there was a lot more progress to be made. I had strong feelings about the goodness of manned systems. After all, nobody had ever done decent science with a robot. That was the challenge. And it turned out that one of our principal achievements was taking a system that had no track record and turning it into a first-rate scientific tool."

THE CAR THAT GLIDED up in front of MBARI on time at six-fifteen in the morning of Thursday, July 14, was a black Corvette. I don't know what I expected Robey to be driving—maybe something bland with a save-the-whales sticker on the bumper. Not a power machine. Then again, on reflection it seemed like the car probably fit the man. It was jet black, like the midnight world he loved to study. It was bold and racy, the sort of thing that might appeal to an explorer. And it was small, not unlike those cramped one-man submersibles he had learned to pilot through the deep, whizzing through a dark wilderness. I squeezed into the passenger seat. We were on our way to Moss Landing. It was dive day.

We sped past houses and then dunes, hugging the coast as we drove northward. The dawn was gray. Mist hid the water. Robey said the bay was usually calm this time of year but in winter could be fearsome. The new facilities Packard was building at Moss Landing were designed to withstand not only earthquakes but waves that once every hundred years or so grew to gigantic proportions, thundering down with devastating force. Robey told of the care and attention to detail in all Packard did, especially engineering. "It's not flashy or ostentatious," Robey said. "It's good."[13]

Moss Landing is no tourist mecca but a rough-and-ready fisherman's village cluttered with parking lots, docks, boats, warehouses, and a few restaurants that probably serve good fish on paper plates. The cry of gulls filled the air as we parked the Corvette and got out. Nearby, behind a long chain-link fence, diggers and dump trucks stood mute among big piles of rock and dirt at the breakwater site undergoing development for MBARI. Adjacent to the construction site, one MBARI building was already completed and occupied, its grayish front cut by two large blue garage doors. Next to it was Phil's Fish Market, its weathered wood exterior surrounded by plastic tables and chairs. FRESH FISH, HERE, cried a sign. A dock across from the MBARI building was home to the gleaming white *Point Lobos,* 110 feet long. The boat was a swirl of activity as seamen and scientists carried supplies aboard. At surrounding docks, boats were tied up in long lines, quiet and still in the water. Some were sailboats with tall masts. The majority appeared to be work boats or fishing boats of one sort or another. A bluff overlooking the landing held a Pacific Gas & Electric plant, its big smokestacks vanishing into low-lying clouds. The morning was damp and chilly, the temperature perhaps in the low sixties. Robey wore a down vest over a denim shirt.

We boarded ship. Soon, the rumble of engines rose from a dull throb to a sustained roar. It was just after seven in the morning as we moved from Moss Landing into a bay whose waters were relatively calm, preparing to search for deep life.

The robot *Ventana* dominated the rear deck. It was the size of a small car and nearly as tall as a man, its welter of cameras and parts overlaid by a top of white syntactic foam that, once in the water, would act as a float to counteract its two-ton weight. Emblazoned on the robot's side was the blue MBARI logo, a gulper eel. *Ventana* was tied down to the deck by three or four cables to keep it steady. A retinue of technicians came and went with wrenches and other tools, attaching parts and preparing the robot for the dive. Robey was a picture of casual authority. Smiling and telling jokes, sipping a cup of coffee, he unobtrusively played a central role in directing the serious business of readying the robot for its plunge into the salty bay.

Robey explained the robot's history and operation. The basic robot had been built for MBARI at a cost of $600,000 by International Submarine Engineering of British Columbia, a company with ties to the Canadian Navy. The general model, often used by oil companies to erect and inspect offshore platforms and pipelines, had been heavily customized by MBARI. In addition to the usual lights and mechanical arm, it bore special sensors, sampling devices, and cameras. Most prominent was the main camera, a Sony DXC-3000, its three chips generating broadcast-quality color. Bulging outward from its front was a Fujinon lens that could zoom from 48 millimeters down to 5.5 millimeters, giving tight closeups. Higher up on the robot's front end was a smaller camera for overviews. To the right, instead of a second mechanical arm, the robot bore a suction tube used to capture undersea creatures. Working like a vacuum cleaner, the tube snaked through the robot to the rear, where it deposited its catches in a clear rotating drum, which was internally subdivided. Next to the rotating drum was a tiny light and camera so operators aboard ship could inspect the catch. "Depending on how we rig it, we can go down with as many as eight cameras," Robey said.

Mounted high on the robot's front was a long metal bar with a tiny sonar transmitter on one end and at the other a small device that looked a bit like a tin can, but fluted so it would spin on its axis as the robot moved forward. "It's a low-speed flow meter," Robey explained. "What I asked for initially was an odometer. It's easy to calculate the area that you're looking at from the videos. But to know the linear distance that you're driving through is very difficult, particularly at low speed. It took us a long, long while to come up with a reliable, low-speed odometer. It's been a watershed. Now we know how far we're going so we can make volumetric measurements."

As we talked, a technician installed gallon-size clear plastic jars high along the robot's front bar. They were open top and bottom, their motorized lids able to shut rapidly to trap creatures too big for the suction arm. Robey said four of the samplers would be installed.

Ventana was powered by the ship and its generators, the electricity flowing through a long orange tether that entered the robot at its top. Robey stooped down to point out one of the prime destinations for the electrical current, a large electric motor and attached pump at the robot's core. This hydraulic system, he said, pushed fluids through a maze of tiny pipes and valves to power the robot's six thrusters: two fore and aft, two lateral, and two vertical. The system was very efficient but noisy, Robey said, and would be replaced on the new robot with quiet electric motors that would be less acoustically intrusive.

As our boat sped outward, another passed us coming in. Greg Maudlin, our bearded chief mate, said it was a squid boat, heavy in the water, filled with a night's catch. Its stern was dominated by a big reel and net. The fishermen, Greg said, lit the dark waters with floodlights to lure squid to the surface and then scooped them up. The squid were taken from relatively shallow waters around the upper fringes of the canyon and were the common market variety, growing up to lengths of about a foot. Many became fishing bait and calamari. The squid boat soon disappeared from sight into the gray mist. "On a clear day you can see Monterey," Greg said as it vanished.

I went below. A mess area with two roomy booths was busy with people eating breakfast and reading newspapers. Next to it was a small but well-stocked galley containing a stove, microwave, refrigerator, freezer, and all kinds of storage cabinets packed with food. It was self-service. I grabbed a granola bar and had breakfast on the run.

Through a nearby door was a brightly lit control room. It was known as The Hole, a name that belied its importance as the mission's nerve center. One wall was covered floor to ceiling with consoles, keyboards, computer screens, and television monitors, and in front of them was a row of comfortable chairs. The computers in the consoles were made by Hewlett-Packard, no surprise considering who paid the bill. On the screens, one image came from the camera eyes of *Ventana* on the rear deck. It showed a technician's lower legs and shoes, all in vivid color. Another showed overall deck activity, taken by a camera mounted high overhead on the ship. The view made it easy to see the comings and goings of technicians, a handy thing as the robot's various systems were checked out.

The pilot in the control room was running the checkout and wore a headset with a microphone so he could talk to colleagues on the rear deck. Buried inconspicuously in one console was a key item of the whole setup, a videotape recorder that would capture all the robot saw, creating a scientific record of the dive. Overall, the room was very high tech. The banks of advanced gear and monitors made it look like a miniature version of the space agency's control center in Houston. Moreover, it was pleasingly spacious, with comfy seats and room to stretch. The area clearly had many advantages over the cramped quarters of a manned submersible, not the least of which were a toilet, mess, and well-stocked galley right next door.

In the mess, Robey paused to tell me about the day's plan and to link it to past research. We were to revisit an area he had studied before, seeking both new creatures and a better understanding of known ones. He had first observed the area with his own eyes nearly a decade ago, when he went down in *Deep Rover.* Today, Robey said, the cutting edge of midwater ecology was discovering the subtle relationships between this dark habitat and its odd fauna. "In the past, we'd go to sea three times a year if we were lucky. But at this site alone we now average three dives a month, so we have a unique capability to examine patterns and processes in fine-scale, temporal resolution. As you'd expect, there are significant ecological processes that take place on that kind of time scale—molting or the coming and going of certain kinds of animals. What we're trying to do is to develop the big picture of how midwater communities are hooked together. It's never been done in any comprehensive way. Nobody's had the opportunity. But it's also such a diverse and confusing fauna. It's organized along such apparently conflicting lines that acquiring the big picture has eluded us to date, except in the sketchiest details. That's what's driving us—to understand it in one gulp, in one comprehensive picture."

The canyon was the perfect place to develop a standard by which the rest of the world's oceans could be measured, he said. Its biological richness was such that dives tended to reveal much. But species diversity was right in the middle of the global range—lower than tropical seas, higher than polar ones. "You want to be sure you have enough diversity so that the basic elements of ecological structure are all represented," but not so many species that the patterns are obscured, he said. "That's why this place is ideal. The target is a universal ecological model." In MBARI's next stage, he said, after the new ship and robot debut, the insights gleaned here will be important to understanding the ecology of the rest of the global deeps.

A crew member interrupted to say dolphins were off the bow. I ran up to the bridge, eager to see the mammals and their graceful movements. But by the time I got there they were gone. It seemed like a lesson in field ecology. As Robey had pointed out, the observations had to be repeated if there was any hope of becoming familiar with the local fauna.

In the mess, Robey described the day's activity in detail, saying we would first slowly send the robot straight down to a depth of one kilometer and then punctuate its slow return with a few crosswise sweeps through the deep. The normal exploratory plan was to zero in on any animal that was unusual or was part of ongoing studies, including deep-sea squids.

A main goal of the initial descent was to locate the oxygen-minimum layer, an enigmatic zone where oxygen concentrations fall to the sea's lowest levels. In theory, oxygen should be plentiful everywhere. Surface waters are full of it, owing to atmospheric mixing and plant respiration. These saturated waters, traveling in currents, descend to refresh the depths whenever they encounter waters that are warmer or lighter. Such mixing occurs especially near the Earth's poles as cold and warm waters meet, these areas being described as the ocean's lungs.

Despite such mixing, studies over the decades have shown that an exception to the oxygenation rule occurs in the upper kilometer of the ocean, where oxygen levels quickly drop with depth by as much as 90 percent. Why it should be so is largely a mystery. The area of greatest deficit usually lies somewhere between six hundred and eight hundred meters down, after which oxygen levels increase steadily all the way to the bottom, no matter how deep. Understandably, the minimum is hard on animals, most of which, after all, need oxygen to survive. "It's an area where it's difficult to breathe and that has a profound influence on the structure of the community," Robey said.[14]

Surprisingly, he added, the zone was far from lifeless. "It has a resident fauna of some of the most bizarre animals we see," including predators that hunt creatures in search of more agreeable climes. One of its most interesting residents, Robey said, was *Vampyroteuthis infernalis,* a living fossil from a stage before cephalopods evolved into squids and octopi. "It's this big"—his hands outlining something about the size of a loaf of bread—"and dark brown. It's got beautiful blue eyes. If we're lucky, we may see one." I learned later that the creature's name arose from its alleged resemblance to a vampire. The webbing over its eight arms was supposed to be the cape, and the two sensory tendrils that extended from beneath the cape were fancifully taken to be the fangs. *Vampyroteuthis,* first described in 1903, was one of the missing links found in the wake of the Darwinian revolution as scientists probed the environmentally stable sea in search of ancient species.

Robey said that in six years of robotic research he and his colleagues had found perhaps one hundred new species of animal life, but that the number might be higher. The team was unsure since it focused more on ecology than taxonomy—the science of identifying, naming, and classifying organisms. Echoing his talk that initially fired my interest, Robey said oceanographers in their rush to the bottom had overlooked perhaps one-third of the sea's large fauna. "On a global scale," he said, "we may find that's conservative. It could be half." The idea struck me as extraordinary. For every kind of large creature that we know—for every angler and jellyfish, for every squid and shark—another kind of animal might be lurking in the sunless depths, unfamiliar and unexamined.

"If an alien civilization came to look at the dominant life form on the planet, they'd be out looking at midwater creatures," Robey said, his hands splayed open as if to embrace the unknown. "In terms of biomass, numbers of individuals, geographical extent—any way you want to slice it—these are the biggest ecological entities on earth. But we know virtually nothing about them."

The throb of the ship's engines dropped.

We were at the dive site, some twenty miles out to sea. Our trip had taken a little more than an hour and a half. We were above the axis of the canyon, idling over waters nearly a mile deep.

On deck, no land was visible. The overall grayness had lightened up enough so that a clear horizon was visible. The day was still cool, the sky overcast. The sea was almost calm.

Two or three technicians gathered around the robot, completing last-minute checkouts. One had on a headset and microphone so he could talk with controllers in The Hole. Adjustments were made. Lenses were cleaned. A small strobe light was attached to the robot's top—a beacon if the vehicle should be lost. Then the technicians stepped back as *Ventana* came alive. Cameras tilted up and down, back and forth. Lids on the large sample jars swung in and out of place.

Suddenly, the robot began to emit a loud whine, sounding like the world's biggest dental drill. Bill Wardle, the ship's captain, a large man with a bushy gray beard, said it was the activation of the robot's hydraulics. The pitch got louder as thrusters were engaged one after another.

"It's just like an airplane," Bill said, his head bundled inside a hooded sweatshirt. "The robot's going into an environment that's probably the harshest you'll find anywhere on this man's Earth. So every day they check it. Every night they check it."

Bill watched the action from a perch behind the bridge from where he could steer the ship and aid the checkout, communicating with the team over a headset and microphone. Among the instrument displays at his side were latitude and longitude, courtesy of the military's Global Positioning Satellites. Bill said the gear pinpointed the ship's location on the globe with an accuracy of about one hundred feet, less than the length of the ship.

Techs unhooked the robot's hold-down ropes. Then a large crane worked with surprising grace and quickness to lift up the two-ton vehicle and move it over the side of the ship. The machine entered the water with barely a ripple. Its thrusters were immediately fired up for more checkouts, spewing white froth into the dark water. In its element, rising and falling with the swell, its thrusters boiling, the robot looked like a mechanical monster ready to pounce on some victim below. With surprising speed, it moved away from the ship and dove just out of sight for what Captain Bill called "a bounce check." This was to make sure its buoyancy was slightly positive so it would float back to the surface if there was a major breakdown. The test was performed at the start of each session because the robot's weight changed

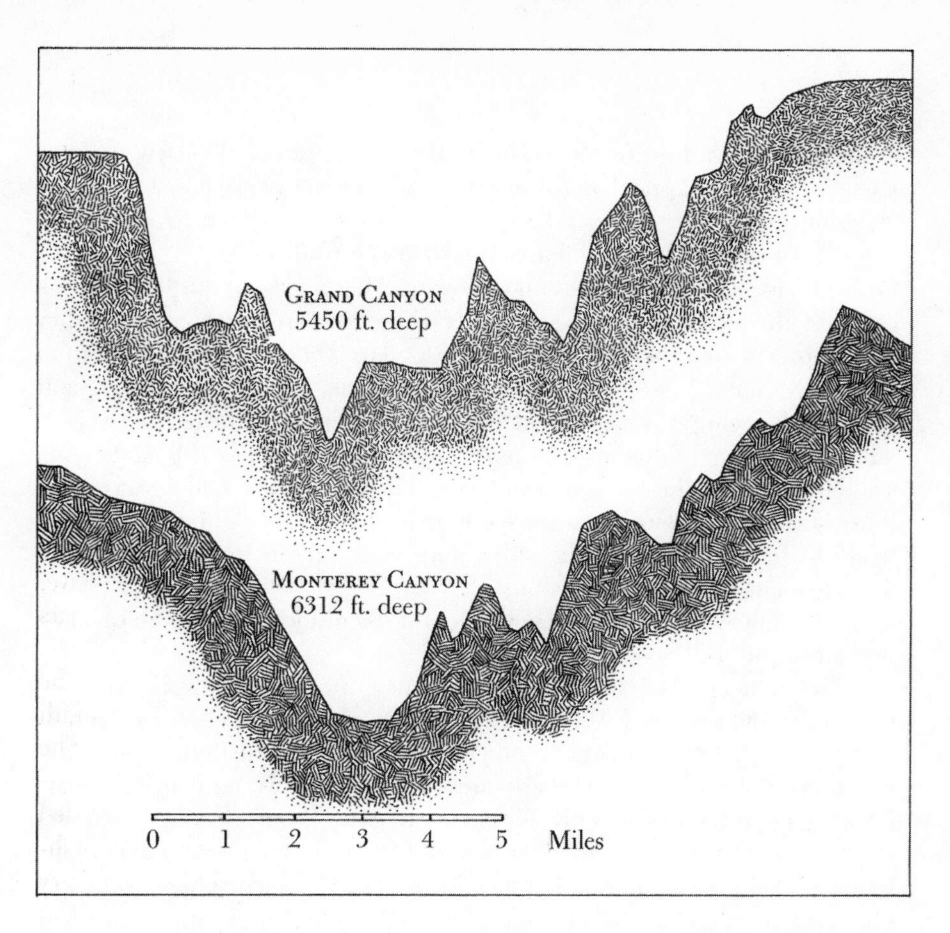

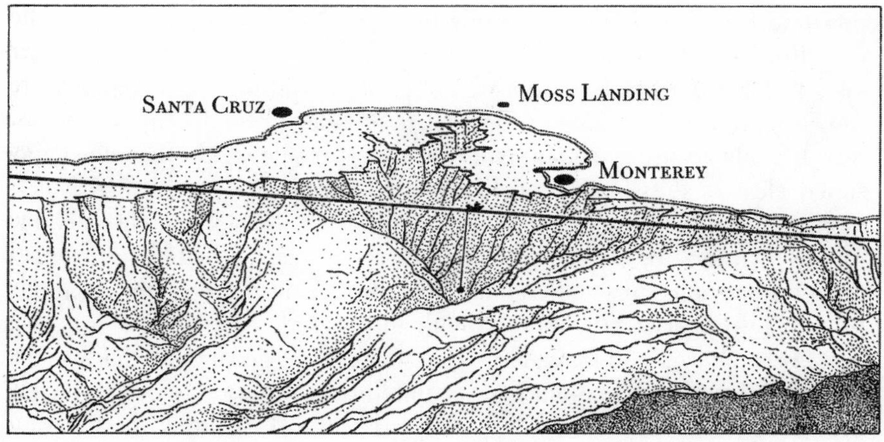

GRAND AND MONTEREY CANYONS. The largest submarine fissure along the lengthy coastline of the continental United States, Monterey Canyon, is deeper than the Grand Canyon. A tethered robot is regularly lowered into its depths to seek out unfamiliar forms of life in the darkness below.

slightly from dive to dive, depending on the gear it carried. A few minutes passed. Nothing happened. If the robot was floating back, it was doing so very slowly.

"It's heavy," Captain Bill announced after a while.

The mechanical dip was run in reverse. After the robot was hauled back on deck, the techs removed ten pounds of small, strap-on weights as Robey peered over their shoulders. Back in the water, *Ventana* sped far away from the ship, trailing froth. The technicians paid out bright orange tether, and every so often clipped on a small float, which counteracted the tether's weight. The robot moved about two hundred meters away, which Bill said was a standard distance for collision avoidance. Then it dove. The time was nine-thirty, some forty-five minutes since our arrival at the site. As *Ventana* slipped beneath the waves, trailing the tether and floats, an albatross sat nearby on the water, curious at the goings-on.

"Usually when he sees these kinds of floats in the water they're carrying nets and fish," Bill observed.

Below deck, The Hole had been transformed. Rock music thumped from a pair of speakers. Four people sat in front of consoles, tweaking controls and readying various systems, seemingly oblivious of the rhythm. Most different of all, the room was extraordinarily dim, all the overhead lights having been extinguished. The only illumination came from glowing computer screens and television images. The brightest was from the main camera of the robot, its lights probing the darkness below us. The picture was a pleasing blue and revealed a constant rain of marine snow that, as the robot dove downward, appeared to be flying upward.

Robey wore a headset with a microphone, as did the person sitting on his left, Chris Grech, the robot's chief pilot, a balding, clean-shaven man. Also tied into the communications net were aides in the room and, on the rear deck, the techs paying out tether and Captain Bill, making sure the tether stayed clear of the ship's propeller. Chris kept his eyes glued to the large monitor in front of him as his right hand worked a joystick that governed the robot's thrusters. To his left was a computer touch screen that let him control the robot's various subsystems and motors. Among other things, the computer prominently displayed the robot's depth and bearing.

A monitor up and to the left was lit with a large glowing orb that seemed to gaze outward at the room like an unblinking eye. It was data from the robot's sonar, whose sound waves radiated outward to probe the darkness beyond the range of lights and cameras. The orb was evenly illuminated, save for a bright spot at its center that represented the robot. In truth, the system was a monster scope. Its acoustic searches were too crude to track single fishes or gelatinous creatures whose lengths were measured in inches and feet. But if something really big and solid showed up in the robot's vicinity, the screen would come alive, perhaps outlining a big shark or the kind of dark shapes that swept past Beebe during his Bermuda dives. Perhaps it would be something like the giant squid Robey once snagged with a deep trawl off the California coast, his only trophy a piece of tentacle fourteen feet long. Perhaps it would be something like the giant siphonophore known as *Praya,* the one discovered growing up to lengths of 130 feet. On the Greatest-Hits video, I had seen readings from *Ventana*'s sonar that showed a number of *Praya* coiling through the water, long and sinuous and very big.

Sitting at the sonar console was James Hunt, known as Jay, a doctoral candidate from the University of California at Los Angeles, who worked at MBARI to study squids. He wore wire rims, a mustache, and long red hair pulled back into a ponytail. Jay flipped some of the console's switches to show me the sonar's various ranges, saying its waves could penetrate through the water for distances up to 50 meters, or about 160 feet.

Suddenly, the blur of marine snow on the main television screen slowed and then halted altogether.

"All stopped at one hundred meters," Chris said, eying something small and indistinct on his screen. The robot was now 325 feet down—already, at the start of its dive, well beyond the range of all but the most adept humans equipped with scuba gear.

Robey, at the console controlling the robot's lights and cameras, gazed at his own large monitor as he zoomed in on the object, whose image grew to fill the screen. The thing looked like a crumpled balloon that was both gooey and gossamer.

"It's a discarded larvacean house," Robey announced.

I knew from earlier reading that larvaceans are gelatinous creatures that surround themselves with fans of fragile mucus up to a meter in length. The animals were so named because their bodies retain larval traits, looking like tadpoles with long tails. The mucus houses, which are some of the most complex external structures built by any organism, act as filters to remove coarse bits of marine snow from the creature's domain and allow the penetration only of finer bits, which are then ingested. The houses are abandoned when they become clogged with too much detritus or when the animal is disturbed, after which it flits away to start life anew. This particular discard was slowly sinking into the canyon's depths, where it would undoubtedly become food for a variety of bottom-dwelling animals.[15]

As the robot renewed its downward trek, Kim R. Reisenbichler, the senior aide of the midwater operation, asked Chris, the pilot, to turn on the robot's suite of nonvisual sensors.

"Sure," Chris replied as he touched the control screen, sending a flow of electricity to a small system somewhere in the robot's heart. "Power on," Chris said.

Kim sat facing a large console to the right. He had worked with Robey since his Santa Barbara days, and the two had coauthored scientific papers together. Quiet and clean-shaven, Kim booted up a computer at his console and began loading programs that would turn the signals flashing up the robot's fiber-optic line into precise readings of water salinity, temperature, and oxygen concentration. The last would be the key measurement to finding the oxygen minimum, the zone of the unusual.

High up on a console, a small monitor showed three technicians on the rear deck, two of them putting another float on the tether. One gazed at the water. All were bundled up. We in The Hole were working in comfort. Robey and Chris both had drinks they occasionally sipped.

"Ah, there's a *Bathochordaeus*," Robey said as he eyed his screen, zooming in on a larvacean until it filled the screen with an array of translucent curves.

"All stopped at one thirty-eight," said Chris, referring to the depth in meters. The robot was now more than 450 feet down—below where submarines of the Second World War routinely used to roam, windowless, ignorant of the life around them.

The outer detritus filter of the animal was indistinct and poorly developed. But its fist-sized inner sanctum was sharply defined, looking like fine crystal, delicate and luminous. Its overall shape very much resembled that of a human brain, only the lobes running down either side of its hemispheres were very evenly spaced. The creature itself dwelled somewhere inside this gelatinous inner house, largely out of sight. It seemed that *Bathochordaeus* was the kind of fragile animal that, if caught in a net and hauled to the surface, would seem to be a blob of jelly, its subtleties gone, a puzzlement to anyone trying to understand its workings and way of life.

"It's not pumping right now," Robey said of the larvacean. Then, as we watched, a delicate flap over its home suddenly started to beat back and forth, moving particles into the creature's mouth. Fine marine snow swirled about, but it was hard to tell what part of the circulating fluid was being eaten.

The image of the larvacean was remarkably steady on the screen, moving ever so slightly up and down, probably owing to wave action transmitted along the robot's tether. Chris closely watched the creature and gently tweaked the joystick, keeping the robot stable. His stream of little adjustments seemed similar to those the driver of a car makes when headed down a straight road. Of course, his job was harder. Chris was driving in three dimensions, not two. And his goal in tracking an animal was an accuracy measured in fractions of an inch.

The robot left the larvacean behind and started moving rapidly downward, causing an upward rush of marine snow. Although the robot in theory could descend vertically, it never did so. Chris always kept its forward thrusters on slightly so its path had a forward slope. That way the camera never saw water or animals that the robot had just thrust through. The scene was always fresh and undisturbed.

"*Nanomia, Nanomia,*" Robey said as the robot sped past two small stringlike siphonophores. He called out the names of the creatures as part of the scientific inventory. Back at MBARI, the tape of the dive would be analyzed by technicians, who would compile a statistical portrait of the area under investigation, which would then be compared to the growing collection of previous ones.

We kept passing more and more of the siphonophores as the robot sped downward. All siphonophores, Robey said, were colonial animals whose different members worked in unison. The most famous of the breed was the Portuguese man-of-war, which had adapted itself to surface life and was famous for its danger to humans. Most deep-sea siphonophores had a cluster of swimming bells at the front and long elastic tentacles for fishing and drawing prey to waiting stomachs. Robey ordered the robot stopped at a depth of about 200 meters, or 656 feet. Four or five *Nanomia* were visible on the screen, each looking like a short segment of fuzzy rope, some moving slowly through the water.

"They're strong predators," said Robey. He touched the screen lightly, highlighting some nearby insectlike blurs. "And these guys zipping around in here are *Euphausia pacifica,* the krill that they feed on."

One *Nanomia* grew until it filled the screen, its image larger than its real-life length of about a foot. Its head was a delicate aggregation of translucent bells, which were pumping and moving the forward part of its body upward while the rear trailed horizontally. Robey said its body was a hollow stem through which digested food was passed. Attached firmly to the ropy body were five or six fingerlike protrusions—empty stomachs. Tentacles dangled from the body along its length. They were transparent and almost invisible except where they were highlighted by what appeared to be beads or bits of marine snow. The tentacles were covered with hundreds of stinging cells, which could quickly disable prey.

The tiny shrimplike crustaceans known as krill danced and darted nearby, unaware of the danger. The siphonophore's empty stomachs swayed gently back and forth in the currents, awaiting a feast.

If a tentacle snagged a krill, the siphonophore would shrink its afterbody and swim rapidly away as the prey-bearing tentacle trailed behind. Playing the krill like a fisherman, reeling it in relentlessly, the siphonophore would slowly contract the tentacle until the krill was close to the body, where other tentacles and fingerlike palpons would then grasp the captive and cram it into a waiting stomach. Obviously, such intricate behavior implied that this animal—this slender rope of gelatin—harbored a relatively well-developed nervous system.

Robey's studies with the robot over the years had not only uncovered the subtleties of such feeding behavior but had revealed that the hunt could be remarkably bold and rapid. An individual *Nanomia* could hunt, move on, and redeploy its tentacles at a rate of fifty or sixty times an hour—about once a minute. Robey's studies had also shown that the small siphonophores were so abundant and aggressive that they ate about a quarter of the canyon's krill population, competing successfully against such big predators as squids, albacore tuna, and blue whales.

"They're fast little swimmers," Robey noted.

The scene around the siphonophore pulsated with life. Small fish swam by. Krill wiggled. Robey zoomed in on a bell-shaped jellyfish with short tentacles that was rhythmically throbbing. It had captured a smaller jelly it was trying to eat.

"He's chowing down," Robey observed.

The robot continued its dive, passing great numbers of *Nanomia* and another larvacean house.

Pausing from his driving work, Chris, the pilot, explained to me the workings of his glowing touch screen, which not only controlled the robot's various subsystems but displayed an amazing amount of data. A schematic diagram of the robot showed which lights and thrusters were operating. A compass gave the robot's heading. A bar graph showed the pressurization of the hydraulic system. A row of other bars indicated possible warnings of system failures, one of them glowing. "That shows we have a logic ground fault right now," Chris said. "It's bad but not catastrophic to where you'd abort the dive."

A graph to the left had a wobbly trace that showed the robot's depth over time. It leveled out horizontally as it reached the robot's current position, which was motionless. At its fastest, the robot could descend about one hundred feet per minute, or a little more than one mile per hour. As Chris touched a bar at the bottom of the screen, a series of menus flashed on the monitor that allowed the powering up and down of all the robot's various subsystems.

Munching on a snack, Robey said the high degree of automation in the control room allowed a tight focus on research and was envied by other scientific groups that probed the deep with robots. "We had a pilot from another system out with us and he saw that touch screen and nearly came unglued. He said they still had to type in commands on a keyboard. He was almost weeping."

The robot dove deeper. At a depth of about 800 feet, or 246 meters, a beautiful fist-sized comb jelly came into view, speeding along like an alien spaceship. Rows of fused cilia along the rear of its transparent body moved in waves, shimmering and iridescent, pushing the animal forward to sweep up prey and particulate matter. The lobes of its mouth were open wide, ready to apprehend any small animal foolish enough to be in its way.

"This is the kind of critter that's so fragile you can't catch it in a net," Robey said.

Such difficulties had misled generations of biologists into thinking the creatures were only modestly abundant. Not so. As Robey and other marine ecologists discovered in their deep studies, comb jellies, technically known as ctenophores, could be found almost everywhere in the sea's upper mile or so. Indeed, sometimes they were the dominant form of life, their abundance occasionally making them infamous. In the early 1980s, a ctenophore known as *Mnemiopis leidyi* invaded the Azov and Black seas, probably carried there in a ship's ballast water. The animal was soon eating huge quantities of zooplankton, crustaceans, fish eggs, and larvae, destroying fisheries directly and by depriving fish of food. The cost to the Black Sea fisheries was hundreds of millions of dollars. The Azov fisheries simply shut down. Not just numerous and voracious, ctenophores could be large, one type growing to a length of more than three feet.[16]

A glance at the monster scope showed no unusual activity in the robot's vicinity, just the flicker of the tether every so often. If any behemoths were lurking out there in the darkness, they were beyond the sonar's range.

Nine hundred and fifty feet down, we came upon a short, squat siphonophore with a tassel of tentacles adorned with row upon row of yellow dots, as if strung loosely with pearls. "It's an *Erenna*," said Robey. "We think all those little yellow things are luminous." Couldn't the robot's lights be switched off to find out? "This camera is not sensitive enough to see bioluminescence," Robey replied.

The siphonophore was meatier and larger than *Nanomia,* its body about two feet long, its tentacles rippling down for another meter or so. Dozens of swimming bells were packed together at its head like leaves on a bushy plant, their numbers much larger than on *Nanomia,* the bells pulsating in time to some arcane pattern that involved alternating rhythms. As Robey zoomed in close, we could see the animal's tentacles contracting effortlessly with much greater elasticity than rubber bands, responding to some internal command or invisible stimulation.

The downward-speeding robot passed two large umbrella-shaped jellyfish known as *Solmissus,* which were blobs of translucent pulsation. Then Robey had the robot stop to view a large *Nanomia* that had just eaten. The ropy body of the siphonophore was sharply kinked at a stomach, which hung low and swollen, apparently filled with a krill. The animal's overall shape was almost like the letter *V*, with the heavy stomach dangling down. Robey got excited and pointed on the screen to the bottom of the stomach. "See the eyes? That's the giveaway. He's fed on the Euphausiid."

Two little dots, apparently the krill's stalked eyes, hung out of the bottom of the stomach. Though a bit unsavory, this scene lay at the heart of the kind of analysis that seeks to understand the feeding relationships among animals and to learn all the links in the great chain of being.

Just behind the siphonophore, a huge shrimp swam by. It was headed downward, its antennae extraordinarily long, twice the length of its body.

A few meters below the siphonophore we found another iridescent comb jelly, its mouth closed and cilia pumping away. "He's in moving mode rather than fishing mode," said Robey.

Past it swam another comb jelly moving even faster. This one, about the size of a plum, trailed two tentacles that were more than a meter long. With no stinging cells, the tentacles used sheer stickiness to capture such prey as eggs, larvae, and tiny worms. Robey said they were highly elastic. To illustrate, he had the robot creep up and bump the comb jelly with the suction arm. In a flash, the long tentacles contracted into its body.

"See?" Robey said.

Robey called out the depth to make an audible note on the dive tape. The robot was 335 meters down—about 1,100 feet, a fifth of a mile. Our trek into the depths of the canyon had barely begun.

In the old days people thought that life got increasingly sluggish as depth and temperature dropped. But now the robot spied a small siphonophore that was zooming around like a rocket ship, much nimbler and faster than *Nanomia* or *Erenna.* Its aim was a mystery. Robey said it was a calycophoran and that it roamed beneath *Nanomia*'s range but had a similar diet. Whenever the robot stopped to examine one creature, it inevitably saw others. This area seemed to have a dozen critters moving around, including a pair of large umbrella-shaped jellyfish that throbbed their way through the blizzard of marine snow.

I went up on deck to look around. Unexpectedly, the experience was a jolt that left me wobbly. My head was filled with images of unfamiliar life that had been hidden from man for millions of years. But here I was, standing on deck, gazing at the commonplace. I felt schizophrenic, torn between unrelated worlds. Squinting, I saw that the sky had lightened up and that uniform grayness was giving way to patchiness. The sea was broken only by gentle swells. Surprisingly, nobody else seemed to be on deck. For the moment, the tether was managing itself. I was alone. The whole experience was eerie and disorienting. I quickly retreated to the reassuring darkness of The Hole, picking up a couple of granola bars on the way. It was ten-thirty in the morning.

The music was milder, featuring an acoustic guitar and gentle rhythms. To my surprise, the images on the big screens had changed completely since my departure. The robot, now more than a quarter mile down, was closely tracking a delicate squid, its eyes big and its ink sack dark and visible through its transparent body. The animal was probably a juvenile and perhaps six or eight inches long. Its eyes were large for its body and looked quite intelligent, squid eyes in general being said to rival human ones in complexity. The animal was moving downward at a leisurely pace, its cream-colored arms folded tightly to the rear, unopened. Our squid, if it was like all the rest, had ten arms, two of them longer and thinner than the others and known as tentacles. The animal's downward movement was driven by its clear mantle, which contracted every so often like a bellows to push water out its siphon—a form of jet propulsion. At the front of its body, protruding from a thin stalk, were paper-thin fins that undulated back and forth in a fascinating, corkscrewlike rhythm, aiding the downward journey. With oscillating fins at the front and a jet engine at the rear, the creature was obviously an adept swimmer. Right now it was just idling along and performing no evasive maneuvers, despite the robot's bright lights. The camera slowly panned its body, revealing minute details.

"It looks different," said Jay, the doctoral candidate, who had been studying the squid of the canyon for two years.

"It probably will turn out to be *Galiteuthis*," he said, referring to the genus name of a fairly common deep-sea squid. "It's just that, because these things are so rarely seen, especially in the younger stages, we don't have a good record of what they look like. Most of the information we have historically on younger squids was taken from trawling samples, when the animals were chewed up a bit."

Silvery lantern fish with big eyes kept darting through the dim background behind the squid, the reflections of their bodies flashing every so often like distant lights. A chain of salps floated by. These small creatures with cylindrical bodies, though gelatinous and transparent, like much else in the deep, belong to the phylum Chordata, and therefore, remarkably, are related to vertebrates and man. All chordates at some point in their development have a notochord, a rodlike structure that is the animal's main source of axial support. Salps, too, have one at some point. But to the eye they looked like part of the gelatinous goo that seems so widespread in the deep.

Chris, carefully working the joystick, slowly positioned one of the sampling jars over the descending squid and then sped up the robot's rate of descent. The jar encircled the animal. Then a lid swung shut, capturing it for later inspection.

"Nice job," Robey said.

"You can't imagine how difficult that is," Robey said, turning to me. "The skill that these guys have had to develop to judge distance when you have only a two-dimensional representation is really something. The fact that they've become so skilled at it is a real accomplishment."

The robot was 465 meters down, edging toward the area of minimal oxygenation, the zone where some of the strangest creatures live and hunt.

Suddenly, something long and black sped by, very close to the robot's camera.

"All stop!" Robey cried. We halted and pulled the robot back and scanned the region to try to relocate the dark shadow. But it had vanished from sight.

"It looked like a *Vampyroteuthis*," said Robey, stroking his beard, referring to the living fossil that is half-squid, half-octopus. "I wonder what that was."

Soon, the robot's eye fell on a pinkish squid that started moving away, rapidly flapping its fins like bird wings, trying to escape the onrushing machine. We chased it and zoomed in on its image with our camera lens. It performed an abrupt left turn. We followed and nearly hit a lantern fish. It sped up. We struggled to keep pace. Swimming relentlessly, it pulled ahead.

"He's going up," said Robey. "They usually head down."

Abruptly, the vanishing squid inked and inked again, leaving tubular patches of inky material floating motionless in the water. We could see no trace of the animal itself. It had escaped the planet's most deadly predator, man.

"The ink is a pseudomorph," Robey said. "It's suppose to look like the body of a squid," to fool a predator into attacking the ink rather than the escaping animal.

"A long, thin squid makes long, thin ink," Jay, the squid expert, added. "A short, round squid makes short, round ink." With no robot to disrupt the darkness, he added, shimmers of bioluminescence from microscopic fauna in the agitated water would surround the dark ink, highlighting the ruse.

At 470 meters, or about 1,500 feet, the robot spied a small undistinguished fish swimming vigorously back and forth. As we closed in, Robey said it appeared to be a juvenile of some sort and then proceeded to get very excited.

"Oh, my God, it's a *Scopelarchid*," he said. "This is the baby of a deep-sea fish, with really nifty telescopic eyes." It looked rather mundane now, but as the fish grew, its extremely sensitive eyes would become long and tubular, pointing straight up like binoculars, to gather in fleeting shimmers of light from above, perhaps powerful enough to see at a distance the bioluminescent flash of a siphonophore, which it could then seize. Anatomical studies had shown that such eyes had very large lenses as well as twin retinas, one close to the lens and one far away, suggesting two distinct levels of viewing and magnification. Strangely, at least nine different classes of midwater fish had evolved such telescopic eyes over the eons, implying that this type of organ was very good at doing whatever it did.[17]

Robey asked Kim to keep an eye on the fish as he prepared the suction sampler, flicking through the menus of his computer console and making various adjustments.

"Okay," Robey said, "pump's on. Chris, he probably won't be a fast swimmer but he may be able to dart away. I don't think he'll be able to go very far."

Conveniently, the little fish began to hang motionless in the water, its head up. The main camera was zoomed all the way back so we could see the suction arm as the robot slowly advanced on its quarry. The arm bumped the fish, causing it to dart away. But then the small creature again hung motionless. This time the arm was right on target, near but not touching the fish. Suddenly, the whoosh of the arm's suction engulfed the fish and whisked it tail first into the robot's interior.

All attention in The Hole turned to a small monitor that showed the lighted sampling drum at the robot's rear, which was the receiving end of the suction arm.

"Come on, little fish," Robey said impatiently as he waited.

Finally, the *Scopelarchid* completed its trek through the robot and plopped into sampling jar number two.

"Pump off," Robey sang out, adding a satisfied "okaaay."

The robot dove relentlessly downward. There seemed to be a slight lessening in the density of life. Then it picked up again. Around 520 meters, or 1,700 feet, the robot came upon a pulsating jellyfish known as *Colobonema.* It was bell-shaped and bore long tentacles. As the robot drew close, the jelly suddenly dropped all its tentacles and sped away, pumping its bell frantically. The forsaken tentacles slowly turned in the currents set up by the fleeing jelly. It was like the episode in the Greatest-Hits video.

Robey was delighted. "How nice." He laughed. "It's a kind of antipredation behavior. So these," he tapped the image of the tentacles turning in the water, "are glowing right now but we can't see it." Apparently, the acoustic vibrations of the robot had triggered the flight of the jelly, which had no eyes. In real life, in inky darkness, a predator distracted by the light show might have failed to see the animal's escape.

I asked how long it would take the *Colobonema* to regenerate its tentacles. Robey said he had no idea. "That's something we'd like to find out."

Close to the slow gyrations of the abandoned tentacles was a long, thin siphonophore that was less fuzzy than the others, its body dappled with dozens of bright dots that appeared to be stomachs. Robey said it was an undescribed species of *Praya,* the genus of the siphonophore that Robey had found could grow to exceptional lengths. This particular species was obviously small. And a glance at the monster scope showed no signs in the vicinity of its big cousin.

The robot continued its downward dive, passing another big shrimp with long antennae. A small fish darted by, Robey saying it was a paralepidid, a cousin of the lantern fish, with a ducklike bill.

We came upon what looked like a bulky siphonophore, with a thick head and long tail. But it was not. Robey said it was a protochordate or a doliolid, one of the odd animals of the phylum Chordata that, like the salp, was distantly related to man and all vertebrates. Its gelatinous, barrel-shaped head was pumping vigorously away, jetting the animal through the water. Trailing behind, four or five times the length of the head, was a taillike structure that, again, turned out to be very different from what it seemed.

"That's not a tail," said Robey. "It's babies."

The thick tail was subdivided into dozens of tiny compartments, each a developing larva. The animal engages in no sex but reproduces by budding, simply growing new copies of itself and then shedding them into the water.

We passed another salp as the robot dived.

I glanced at my watch. It was 11:08 in the morning. Though the day was still young, we were light-years from the precincts of man, heading deeper and deeper into the unfamiliar.

The density of life seemed to lessen as we sped into the heart of the low-oxygen zone. The robot was now at 650 meters, or four-tenths of a mile down. No behemoths showed on the monster scope.

We spied a translucent squid at a depth of 682 meters. It hung motionless in the water. As we approached, the reason for its immobility became clear. It was eating a small fish.

"Nifty, nifty," said Robey, ever the ecologist, ever interested in who ate whom. "This is the sort of thing you really hope to find," he told me. "When you bring up a net and a squid is eating something, they say, 'Oh, that's just net feeding.' But there's no way this can be an artifact."

The suction sampler was revved up. And the squid headed down, taking the fish with it. We followed it down a few meters. Finally it paused and we proceeded to stalk it, closing in for the kill. With a whoosh, the squid was scooped into the robot, dinner and all.

"Got 'em," Robey cried. "Got 'em both. Good job, Chris."

Jay eyed the squid in the rotating drum. It was changing its colors, displaying a vivid range of hues. "I think he was about to ink. He looks really mad."

"Can you imagine dining in Carmel and getting sucked up by aliens?" Kim joked from his console.

Chris deadpanned that he recalled reading something about that in the pages of the *National Enquirer.*

"Yea," Kim answered, his voice changing slightly, taking on an extraterrestrial tone. "We'll put him in the lab and probe him and he won't remember a thing."

Robey laughed. "It must have been something in the special sauce on the burger." Then, more seriously, he remarked, "All in all, it's been a pretty good squid day."

Musing on the big picture, Robey said making direct observations in midwater was the only way to discover many food chains and fill in key parts of the deep puzzle. Indeed, his own research had shown the shortcomings of an old method. Marine ecologists had often analyzed the stomach contents of creatures caught in nets, thinking the half-eaten remnants had been ingested in the wild. But Robey and his colleagues reasoned that the nets themselves, stews of immobilized sea life, were the sites of feeding frenzies in which squids and jellies and fishes would try to eat one another in ways that were quite unnatural. To test the hypothesis, they peppered some filled nets with chopped-up rubber bands and foam cups. Sure enough, the artificial debris showed up in animal stomachs frequently enough to let the scientists calculate a general contamination factor. "That did not make us popular with our buddies," Robey recalled. "It was the late seventies. By then we were becoming disenchanted with nets as sampling tools."

Jay echoed the analysis. "Last year we found a squid from net trawls that was holding seven shrimp and two larval fishes. But it was all bogus because it was taken in the net."

We were now in the core of the oxygen minimum. The robot was down more than seven hundred meters, nearly a half mile. Conversation was picking up because the density of animal life was plainly going down. The marine snow was just as thick as before. But the twitchings and gyrations and flashes and shimmers and throbs of life that had been so plentiful during most of the dive were now diminished. The unstated question of the moment was whether we might stumble upon one of the peculiar inhabitants of the zone, perhaps a *Vampyroteuthis,* with its blue eyes, cape, and fangs.

About 780 meters down, we came upon a small owl fish, a resident of the minimum. It skittered quickly out of sight. "They're so spooky we have a hard time catching them," said Robey. Owl fish grow up to a foot in length and have conspicuous eyes, thus their name. "He's living at a depth where there is no sunlight left, but he's got these enormous eyes," Robey remarked, alluding to one of the riddles of the domain's higher animals. It is well known that eyes of midwater fish can be up to one hundred times more sensitive than those of humans. What look like shimmers of bioluminescence to man might look like dazzling fireworks to creatures of the endless night. In all, it is estimated that about 80 percent of the inhabitants of the realm produce some type of bioluminescence, the cool luminosity of their light similar to that of fireflies.[18]

Oxygen levels began to rise by the time we dipped to a depth of nine hundred meters. Disappointingly, the part of the minimum that we crossed turned out to be largely a biological desert, bearing only a few wisps of life and no signs of bizarre inhabitants. I took comfort in the knowledge that we would pass through the zone again on our way back to the surface, perhaps then making some unusual encounters.

A giant grayish, bell-shaped jellyfish loomed up out of the dim background 946 meters down, its tentacles long and thin and menacing. As with all such creatures, the pulsations of its bell forced water out its rear, propelling it forward. But its hugeness made the pulsations somehow more ominous. Around the bell's periphery were fingerlike swellings, where the tentacles attached, and the skin around the fingers was infused with a faint yellow glow, giving it an alien air. A peek beneath its bell revealed the presence of thick feeding arms, apparently strong and meaty. They looked like those of an octopus, able to pull a victim inexorably toward its mouth. The thing was formidable. Throbbing with slow pulsations, contracting and expanding its leathery bell as it moved through the water, dangling its giant tentacles, the creature looked ugly enough and powerful enough and strange enough to be an extraterrestrial from the Planet of the Jellies, ready to ooze down on the unsuspecting for acts of unspeakable horror.

Robey said it was all an illusion. "It looks tough but actually it is extremely fragile. When we've tried to catch them, they just come to pieces with any contact. We've never been able to bring one back in any kind of shape."

We moved on. A half mile down, we spied a small squid that went into a strange pose as the robot approached. Head down, it put its tentacles together, as if praying. It splayed its shorter arms outward toward the robot. All the while its tail fins beat hard. Jay said the posture was defensive, readying the squid to do battle with an unknown threat, in this case a monster machine with bright lights. The locked tentacles, Jay said, apparently acted in an aerodynamic way to stabilize the squid and keep it from tumbling as it drifted slowly downward. As the squid girded for battle, its skin went through a subtle shift of hues, tiny dots on its body going from ocher to salmon pink.

The squid did a flip and was instantly gone.

Six-tenths of a mile down, at 960 meters, we came upon another small squid, this one relaxing in a vertical pose, its arms pointed down, its fins barely moving. The camera zoomed in. Just visible along either side of its arms were faint translucent lines.

"You know how some aircraft have fins up by the nose?" Robey asked. "Those are canard fins. We've discovered that a number of squids have little flared, flangelike additions to their arms or tentacles that they use like canards. We think it allows them to steer better."

The robot's suction arm was quickly revved up and put into action. With a whoosh, the squid was sucked tail first into the robot's core.

The rear camera showed an empty bottle on the rotating drum and, next to it, a full one that held the earlier squid, still going about the business of eating fish. Suddenly, the new arrival plopped into the empty bottle. He took stock of his situation and then started to whirl violently on his axis, apparently using a jet of water from his siphon to spin his body like a pinwheel, his arms flailing about, driving himself rapidly downward. He paused at the bottom of the bottle, looked around, turned around, and began the rapid whirling motion again, spinning to the bottle's top, obviously trying to escape his cell.

"Wow!" said Jay, nearly as excited as the squid. "That's unbelievable. Nobody ever thought that a squid could do that!"

"Maybe this is the Nureyev of the squid world," Kim remarked.

The squid settled down. So did Jay. The robot dove deeper, its lights revealing no lessening in the swirl of marine snow.

We spotted a little bell-shaped medusa hiccuping along at a depth of 982 meters, its tentacles short and knotted.

"This is a new one," said Robey. "I don't think we've seen him before. He's a pretty little thing." Visible through its translucent, pulsating bell were a series of thick reddish bumps that ran top to bottom in knotty vertical lines. Robey said they were probably sex organs. If the animal was like other medusae, the lines were either ovaries or testes, and the animal was either a female or a male. Once every so often, eggs or sperm would break through the inner surface of the sex organs and be shed through the medusa's mouth into the water. The random meeting of the two in the sea would result in fertilization, at which point a larva would begin to form.[19]

Medusae got their name because their tentacles were likened to the snaky locks of Medusa, one of the monsters of classical mythology that had snakes for hair and turned anyone looking at them into stone. Modern biology holds that all hydrozoan medusae, like the reddish one we were tracking, were among the oldest large, multicellular forms of life on the planet. Their lineage dates back to the Precambrian, an ancient time that antedates the explosion of life on Earth. Fossil medusae had been found in south Australia that were seven hundred million years old. Such animals were part of the first wave of complex life on the planet, the forerunners of all else to come. Unlike the life that came to dwell on land, and that evolved quickly with changing habitats, the medusae lived in a relatively stable environment that allowed them to go on much as they had for eons, perfectly adapted to their dark, watery world. It was conceivable that our reddish medusa might represent a type of life that man was seeing for the first time since its predecessors beat their way through the primeval seas a billion or more years ago.[20]

We left the jelly hanging motionless in the water, its erratic pumping momentarily suspended.

We headed deeper. A kilometer down, we came upon a small, sinuous animal that seemed more snake than fish. Robey said it was an eelpout, a midwater version of a family of fishes that live on the bottom. It curled its rubbery body around in one tortured shape after another, first like the letter *U,* then like the letter *S.* Finally, it straightened out to swim violently away, faster and faster, its body snaking back and forth. We pursued. The suction arm closed in, and soon, with a visual whoosh, the animal was sucked up and dropped in a sample bottle.

After that, we dove no further.

"We're bottomed out," Robey announced.

Our downward journey of nearly three hours was at an end. The robot was 1,012 meters down, its tether at its limit. My watch said it was 12:18. Robey, glancing at Kim's console, observed that the oxygen levels were still a bit depressed. Normally at this depth they could have been rising more vigorously. The temperature around the robot was frigid, about forty degrees Fahrenheit.

With that, we started the trek back to the surface, heading again for the oxygen minimum. The marine snow now began to fall downward across the glowing monitors, the way it would if it were real snow. Surprisingly, I found myself heartened by the change. I hadn't realized that the unnaturalness of the upward-falling snow had somehow made me uneasy.

Immediately we spied another eelpout going through its elastic dance of a million shapes. It swam vigorously away and then became listless. The robot's suction arm crept forward and accidentally hit its tail, sending the animal racing away. It then stopped and looped itself head-to-tail into a perfect doughnut. Robey said it was apparently a type of mimicry in which the fish was trying to look like a circular jellyfish to throw off a predator that preferred flesh to jelly. Undaunted, the robot rushed forward. The fish sped away, but the suction arm won the race.

After the capture, the robot sped upward into the core of the oxygen minimum, the twilight zone of the midwater world.

It was at a depth of 788 meters, or about a half mile down, that we came upon what was, far and away, the strangest creature I had ever seen.

At first, the blur of marine snow on the monitor was cut by a thin line that looked quite insignificant, like a shoelace abandoned in somebody's bathtub or swimming pool. The line kept growing thicker and larger as the robot moved closer and closer, ultimately revealing an unconventional blur of life that spilled off the screen.

"Ahhh," Robey hummed contentedly, "an *Apolemia.* A big one."

Previously, the long, thin siphonophores we had seen during the dive were either a few inches or a few feet in length. But this one was in another league altogether. Head to tail, it probably consisted of some twenty feet of gelatinous goo. Thousands of willowy tentacles hung from its slender body, stingers ready to immobilize prey. Along its length were dozens of stomachs, perhaps a hundred or more, some fat with digesting kills.

"Look, he's got a little *Aegina* in his tentacles," said Robey, referring to a small medusa that the siphonophore was slowly drawing up for ingestion.

As we watched, that catch was joined by another, as a small crustacean that was flitting through the water suddenly came to a halt in the siphonophore's tentacles.

The pilot maneuvered the robot close to give us an intimate tour of the siphonophore's body. A dozen or so translucent swimming bells at its head pumped away, keeping the creature in a slight state of tension, moving it slowly forward in a long arc. The bells were clear and worked in alternating rhythms of contracting pulses. They seemed to harbor some kind of gas, for when the bells occasionally stopped pumping, the head area slowly began to rise. The ropy body of the siphonophore was cream-colored, and its tentacles hung down in wispy profusion, so thick in places they looked like cobwebs. While the tentacles were now arrayed in curtains, Robey said they could also be dispersed radially around the siphonophore's body, forming a kind of translucent cloud that enlarged the animal's kill zone. The longest tentacles were a meter or more in length, each apparently covered top to bottom with stinging cells.

"I've felt them," Robey noted wryly.

The animal's many stomachs were strung along its body like beads on a necklace, each one associated with a group of fingerlike projections. Such fingers helped move captured prey into a stomach, Robey said. The animal had no single mouth. Instead, as with all siphonophores, each stomach had its own mouth. Some of the stomachs on the living chain were dark and swollen with captured animals, which Robey said might include jellies,

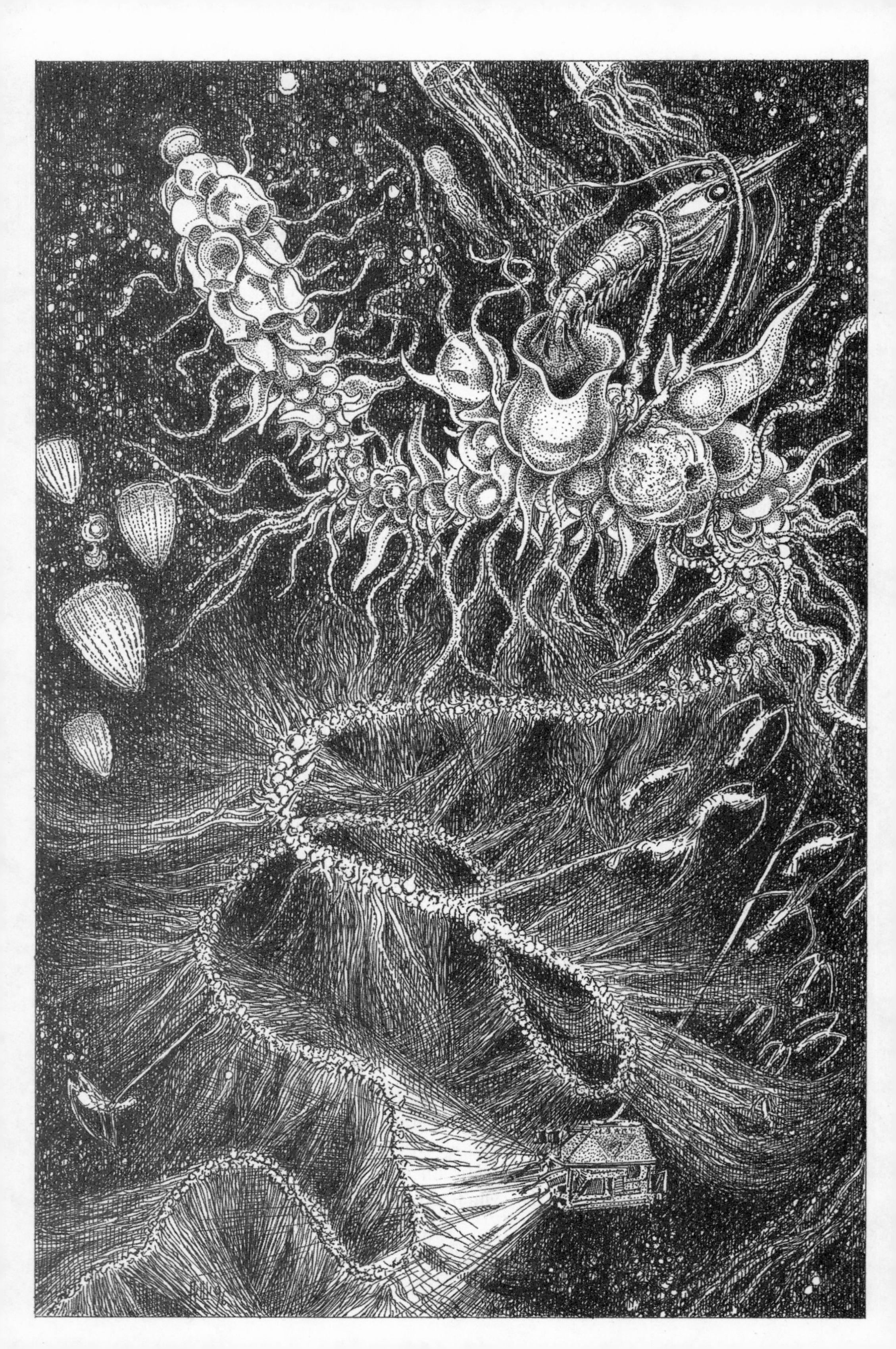

shrimps, and small fishes. One dark stomach was banana-shaped, perhaps holding a marine worm.

"That one has food in it," Robey said of a stomach as we moved down the chain. "That one has food in it. That one has food in it. . . ."

Some were empty. But many if not most of the stomachs on this long siphonophore were swollen, the animal clearly making a sizable dent in the fauna of the oxygen minimum. How was it, I wondered, that the biggest animal we'd encountered during the whole dive inhabited a region that was the most desolate, an area shunned by most creatures? How did it eat enough to sustain its large body? Its bigness seemed like an ecological paradox.

Robey said he pictured the big siphonophore working like a living drift net, hanging passively in the oxygen-poor zone to capture animals that were actively migrating through. At times, he added, the traffic could get quite heavy.

For decades marine biologists have known that the upper kilometer of the sea possesses a thick stratum of deep life that migrates upward at night, presumably to feed in rich upper waters that quite suddenly become comfortably dark after the sun sets. It is known as the deep-scattering layer, since the upward migration is dense enough to scatter sound impulses from sonars. Studies had shown that the layer is composed of siphonophores, copepods, euphausids, cephalopods, and small fish.[21]

A riddle of the canyon was the intensity and composition of this nightly migration and what role it plays in the sustenance of large siphonophores, such as *Apolemia.* To find out, Robey was periodically taking out *Ventana* at night, probing the depths to explore daily changes in faunal stratification. Even the bright orb of a full moon, he was discovering, could influence the behavior of some of the canyon's deep residents.

However it got its food, *Apolemia* clearly got enough to increase its body size to impressive dimensions, even though it was unclear whether such growth occurred over the course of years, decades, or perhaps centuries. Robey had discovered that the animal could grow to lengths of nearly one hundred feet. Nothing like that had ever been pulled to the surface in a scientist's net. The creature was too delicate. It simply broke apart or was shredded into unrecognizable goo. *Apolemia* was the kind of creature that had to be studied on its own turf. When so examined, it was easy to see that its body, as with so many gelatinous animals, was a remarkable adaptation to a place that had no boundaries, no turbulence, and no firm surfaces, a place

SIPHONOPHORE. Thousands of willowy tentacles radiated from the gelatinous body of the siphonophore Apolemia. *Stingers along the tentacles were ready to immobilize prey. Scores of stomachs dotted the creature's body, some fat with digesting kills.*

where animals never encounter waves or the bottom, a place where the only run-ins are with other animals. The stability of the dominion was driven home by the amazing lengths to which *Apolemia* grew despite its extreme delicacy.[22]

In general, Robey said, the gelatinous animals of the deep are a mystery just starting to be addressed, thanks to advanced technologies, such as deep-diving robots.

"Their abundance and obvious importance to the ecology down here has never been assessed," he said, his face lit by the monitor's glow.

The robot left the *Apolemia* and continued its upward trek.

I kept glancing at the sonar screen, hoping to catch a glimpse of *Praya.* But the scope revealed nothing that big or that bizarre. *Apolemia* was too scrawny to show up on the monster scope. But *Praya* was meaty. Its body on the Greatest-Hits video looked to be about the thickness of a human leg.

At a depth of 750 meters, a little less than half a mile down, we began a series of crosswise sweeps, the aim of which was to assess the fauna over a horizontal distance of a half kilometer. The sweeps were quite rapid, since the goal was simply to record the presence of animals and their numbers, rather than to make detailed observations of them and their behavior.

"Full wide," said Robey as the sweep began.

A blizzard of marine snow rushed toward us from a point in the middle of the screen, as if we were a spaceship going through a time warp. Amid the whoosh was a blur of animals that zoomed by in surprising abundance, considering that we were still within the oxygen minimum. Robey called out the names of the familiar ones.

"*Apolemia.* Calycophoran. Owl fish. Squid at left. *Aegina.* Mysid. *Aegina.* Mysid. Sinking larvacean. Calycophoran. Mysid. *Aegina. Aegina. Aegina.* Lots of it. Calycophoran. Squid. Red cydippid. Fish, looks like owl. Calycophoran. *Aegina. Apolemia,* smashed through it. *Aegina.*"

The ship shook as we raced forward, apparently taxed by the tether's resistance to rapid movement through the water and by a struggle to keep up with the speeding robot far below.

The sweep took about ten minutes. Afterward we slowly ascended to five hundred meters and made another sweep that featured a cast of new animals, all interesting. After that, we headed for the surface.

"It was about an average day," Robey said as the robot ascended. "We didn't run into any of the super exotics, like *Vampyroteuthis* or one of the really bizarre siphonophores."

He described a kind of siphonophore he had found a few months earlier that was unlike anything he had ever seen before. It was bunched in a spherical mass, its stomachs pointing outward and writhing like snakes. "We caught it and brought it back," he said. The animal lasted a few months in the laboratory, slowly shrinking in size. "A lot of these gelatinous animals cast off body parts with little regard," Robey remarked. "Tentacles and body parts are cheap and can be cast off readily for whatever reason. Similarly, the animals seem to survive periods of no feeding by getting smaller. We were trying every trick in the book to feed this critter in the lab. Over time, it diminished in size, still sending out tentacles."

Beginning at a depth of about forty meters, the water visible through the camera began to get bluer and bluer.

At around three in the afternoon, some five and a half hours after slipping beneath the waves, *Ventana* gently broke the surface of Monterey Bay. Before I knew it, the robot had been hoisted out of the water and was back on the rear deck, transmitting pictures of a technician's legs and feet.

I followed Jay up to inspect the catches. He and several technicians quickly detached the jars and bottles and took them into a laboratory van next to the robot staging area. The delicacy and beauty of the creatures was easy to admire and hard to comprehend, especially the squids. Their transparency was so absolute that they almost seemed to defy existence, their body parts trembling on the verge of disappearance, like bubbles in a glass of seltzer, their presence only faintly suggested by the interplay of light and motion. An exception was their dark eyes and ink sacks, which made them all the more mysterious. The transparent fins of one squid moved back and forth in a delicate wave motion that seemed far too ethereal for any kind of propulsion through the water.

A riddle of science is why some inhabitants of the sea's depths should be so filmy and transparent. One theory is that it is a defensive mechanism aimed at concealment from predators. Another, increasingly favored of late, is that pigmentation is unnecessary in the deep because there is no penetration of damaging ultraviolet rays from the sun. Our own skins are opaque in order to give that kind of protection, and darker human skin usually has its evolutionary origins in regions of intense sunlight. Perhaps deep-living jellies and squids represent the antithesis of such protection. None was needed in the womb of the deep sea.[23]

As Jay and I watched, a squid began to turn milky white, its chromatophores appearing as a pattern of exquisitely fine dots.

Though the subtleties of such behaviors are just being uncovered for the first time, scientists have long known that squids are very smart and that the roots of their intelligence lie partly in an evolutionary gamble hundreds of millions of years ago in which ancestral squids abandoned the safety of external shells, the kind still employed by their cousins the nautili. A shell gave good protection from predators. But it was heavy. Freed of that burden,

the ancestor of the squids evolved into a quick hunter who sought protection for his exposed fleshy parts with camouflage and ink production, with great mobility and the kind of complex behavior that favors the evolution of elaborate brains.

Jay looked at another of the captured squids. It was a different genus and species but still starkly transparent and bewitching.

"Nobody really knows anything about it," he said, eyeball to eyeball with the creature.

Later, as the ship headed back to port, Jay played a videotape (a "squid-eotape," he called it with a chuckle) of earlier discoveries and observations that he and Robey had made. Crew members who were off duty gathered around the monitors in The Hole, interested to see the result of some of their labors over the years. The tape was a revelation. It quickly became clear that squids, some of the smartest and most elusive of the sea's creatures, were rapidly giving up many of their secrets.

The parade of images included squids that looked like peacocks, ones brightly colored in oranges and reds, ones with striped tentacles, and a really odd one that somehow managed to look like a ripe strawberry. One squid's whole body kept changing color as it swam, pulsating between light orange and translucent white as its fins flashed from light red to white.

The behaviors were subtle and disparate. One was obviously ready for combat, confronting the robot, its tentacles locked together, its other eight arms splayed outward in semicircles like a Hindu god. Another was frozen motionless amid a school of hake that darted to and fro, its body cavity filled with ink, bulging. Eventually, wisps of ink began to leak out its siphon like smoke from a cigarette smoker's nose.

"The fish don't touch it," Jay said of the squid. "But the water's disturbed. So it inks, retaining the ink inside its body. That has never been described before. To do that kind of thing, you have to pinch your collar closed, your vent closed, your siphon closed. You can't breathe when you do that, because water can't be brought in over your gills. This animal held its breath essentially for forty-nine seconds before starting to release that ink." Jay said the finding was important because it challenged beliefs that the ink of deep-sea squids was meant to repulse predators with chemical or olfactory cues. "Here, inking inside your body, holding it in for forty-nine seconds, no chemical or olfactory cues are being emitted, so it's got to be a visual response

SECRET LIFE OF SQUIDS. Smart and elusive, squids are giving up behavioral secrets as robots spy them fishing, fighting, hiding, and resting. Vampyroteuthis, *top, a living fossil with a milky blue eye, turns out to be far more energetic than scientists imagined, at times racing through its dark habitat.*

of some sort." In dark, disturbed water, he said, the bioluminescent shimmers of microscopic plankton would highlight the puffed-up squid, tricking an eyed predator into believing it was something else.

One of the oddest-looking squids bore no resemblance whatsoever to a squid. Instead, it looked more like a siphonophore, long and thin with knotty parts that could pass for stomachs. Its barely perceptible fins were in the center of its long body. Jay said the animal was a juvenile squid that had an extra-long tail section that disappeared as it grew. The belief, Jay said, was that it was a case of protective mimicry.

"In a postlarval stage, it loses that tail and goes through a whole metamorphosis. Its neck shrinks, the eyes become thick, and the bottom set of arms swell to three times the size of what they are right now. Robey wrote a paper that showed there are lots of different tail morphologies for these deep-sea squids. Before, they always broke off in nets. This is the first time anybody has seen those kinds of anatomies intact."

I had assumed that squids use their long tentacles simply for grabbing prey. But as the tape revealed, they can deploy them with great artistry to fish, turning them into lines and lures. One sequence showed a squid with its short arms held upward in gentle arcs, like petals of a flower, the pose apparently stabilizing it in the water. Meanwhile, its long tentacles dangled straight down, reaching far below its body, perhaps a few feet. In a way, this too seemed to mimic the siphonophore, the tentacle deployed passively to catch anything that might wander by. Some species, I was told, flash a light at the tentacle's end to attract prey. Another scene showed a squid lying in a horizontal pose, its fins moving gently, one of its arms holding a long, thin tentacle that ran over the arm's tip and dangled far below its body. The squid trolled the tentacle lazily through the water like a contented fisherman.

The most arresting image of all showed *Vampyroteuthis,* the living fossil. It was the first known filming of the half-squid, half-octopus in its dark habitat, ending a solitude of perhaps hundreds of millions of years. Surprisingly, given its primitive nervous system, the brownish-red creature swam with great dexterity, flapping large muscular fins that looked like those of a whale. It sped around, doing flips and turns, trailing a long filament whose function was unknown but probably sensory.

"Everything that was written about this animal was that it was a slow, sluggish sort of cephalopod," said Jay. "But when you actually see it alive, doing flips and cartwheels and trailing that thing and moving all over, it's clear all that has to be rewritten. We have to rethink the animal."

In general, Jay added, deep-sea squids were turning out to show surprising alertness and alacrity compared to the previous view of their being primitive and lethargic.

A close-up showed the animal floating in the water, slowly opening and closing its large arms, revealing them to be joined by thick webbing. The arms were extraordinarily flexible, like those of an octopus, curling in and out of one another like a swarm of restless snakes. Visible at the animal's head was a milky blue eye, the lack of a pupil somehow making it quite ghostly. In the final sequence, the animal's arms and webbing opened wide to reveal its underside and, at the center, a beady mouth, held motionless. It was easy to see how a victim caught among the meaty arms and thick webbing would have no hope of escape.

As the tape ended, Jay described how he had recently gone to Naples, Italy, for an international meeting of squid experts, or teuthologists (a word derived from the scientific Latin for squid), and there shown the tape. The reaction was one of amazement, especially for the images of the living fossil.

"I didn't have to worry about what I was saying," Jay recalled, "because nobody was paying any attention."

Eager for bigger things, Jay said he and several other MBARI personnel were preparing to travel afield with *Ventana* and *Point Lobos* in a few weeks to hunt for one of the greatest and most elusive of the sea's creatures—*Architeuthis,* the giant squid. They planned on searching for it around the fringes of the Pioneer Seamount, a mass of rock that rises from the seafloor some fifty or sixty miles from Monterey. The area was well known for the presence of sperm whales. Jay hoped the whales were diving into the deep to feast upon the big squids and that the robot might join the big mammals to track down one of the squid behemoths, filming it in its dark lair.

"I would do nearly anything to see one of them alive," Jay said. "That's every teuthologist's dream."[24]

I went on deck. In time, the light gray mist along the horizon eventually darkened in spots to reveal hints of the bay shoreline and the twin smokestacks of the Pacific Gas & Electric plant, which stood over Moss Landing like an old friend welcoming us home. We pulled into our slip at four-thirty in the afternoon, nine and a half hours after we left. It was still cool outside. And my head was still spinning with visions of the deep world, so different from our own.

Jay and the crew quickly loaded up the animals that we had caught and transferred them to the wet lab in the MBARI operations building. The room was dark and cold and filled with tanks. Robey checked for messages and made a phone call. Then we piled into his Corvette.

As we sped down the coast past the dunes, I mentioned Jay's hunt for the giant squid and asked about the likelihood that undiscovered monsters lived out there in the deep, thriving in happy isolation from man. Remembering John Delaney's sidestepping of such questions, I expected a gentle rebuff and was surprised to find Robey addressing them directly and going on at some length about animals that he had never seen and that existed only as statistical possibilities. Even Monterey Canyon, he said, undoubtedly concealed much.

"They're down there," he said. "I'm certain there are other large animals—cephalopods and marine mammals and other things—tapping the nutrient energy resources down there. There's a lot more biomass in the water column than we previously knew about. While much is cycled in ways we're familiar with from past research, I'm sure there are other branches of the food web and other critters that we haven't been able to resolve. Remember, we're working only down to one thousand meters, and the average depth of the ocean is four times that. I'm sure there's lots more out there."

Ventana, for all its strengths and capabilities, he added, was noisy and probably scary to much marine life—sort of like a helicopter landing on a quiet meadow in the middle of the night, its roar deafening and its lights blinding. The hydraulic system that powered the robot's thrusters created a constant din that carried quite far through the water and undoubtedly frightened away many creatures, including such exotic fish as anglers and sea dragons.

In comparison, Robey said, the new robot under construction would be virtually silent and better able to sneak up on its prey. Moreover, its more sensitive cameras would be able to operate without lights when so desired to enhance its stealthiness and reveal the world of natural luminescence.

Robey said that in the past when he personally went down in one-man submersibles such as *Deep Rover,* when he hung motionless in the water like Beebe in his Bathysphere, sometimes with the lights switched off, it became clear to him that the resident fauna of the deep included creatures that were much bigger and stranger than he and his colleagues were routinely able to observe.

"I've been aware of large animals—dark moving shapes," he said, his voice low. "You see them on the sonar, too."

Monster was the wrong word for undiscovered animals that live in the depths of the sea, Robey said. "They could be as unceremonious as ctenophores. They may be strange or bizarre. But *monster* has negative overtones. I don't think of them as monstrous or negative but as wonderful. They're so marvelously different from what we're used to, often in pleasing ways—the colors, the structures, the fluid ways that they move—all those are really beautiful adaptions to the unique habitat in which they live."

With a laugh, he added that there was only one creature out there lurking in the depths that he was the slightest bit worried about. It was a giant squid with nine arms.

"It may hold a grudge," he said.

• •

I WENT TO SEE the half-built robot and talk to its makers and architects, interested to learn more about a successor to *Ventana* that would go four times as deep and do so with greater finesse. Steve Etchemendy, head of MBARI's marine operations, gave me a lift to Moss Landing. The day was as gray as all the others.

Bearded and strong, Steve said he had started in the field as a diver and over the years had worked for many deep innovators, including Oceaneering, Deep Ocean Engineering, and Woods Hole, where he had been an *Alvin* pilot. He was now in his forties and a walking encyclopedia on the topic of undersea gear. As we drove up the coast in his pickup truck, Steve mused on the field's technological trends and said robots were destined to prevail, if for no reason other than their economy. *Alvin* and its support ship, he said, cost up to $30,000 a day to operate. In comparison, *Ventana* and *Point Lobos* cost $8,000 to $9,000 a day. Such savings, he said, were becoming more significant as robots began to do as much as submersibles. Steve named six major robots doing science around the globe and noted that all were first-generation machines that had suffered the usual kind of glitches associated with pioneering. The next generation, led by *Tiburon,* would have many of the bugs removed and would have many more features. "Over time," he said, "we'll do anything a submarine can do."[25]

The team designing the new robot was composed mainly of exiles from the end of the cold war, I learned. With Packard's eye for talent, MBARI had hired highly skilled individuals as Federal money for deep military work began to dry up in the early 1990s. Jim Newman, the chief designer, came from Woods Hole, where he worked on *Jason* and *Jason Junior,* both Navy projects. Ed Mellinger, an electrical engineer, was also from Woods Hole. Bill Kirkwood, a mechanical engineer, came from Lockheed, where he had designed Navy systems for undersea surveillance and for mine hunting and neutralization.

Bill's office at the MBARI operations building overlooked the bay and was dominated by a large Hewlett-Packard computer and a big table covered with blueprints. He showed me the plan for *Tiburon,* which looked a bit like a Volkswagen Beetle, compact and slightly rounded.

"We've taken most of the frontiers and pushed the technology," he said, rattling off a list of firsts. The robot would have side-by-side color cameras, giving it stereo vision and the basis by which a scientist could carefully calculate animal size. Its lights would be fined-tuned for truer color. The robot's thrusters would be powered by quiet electric motors instead of a noisy hydraulic system. Its titanium casings would hold a wealth of electronic controllers, including a central computerized brain. Anything that could conceivably make a noise, even electrical transformers, would be mounted on rubbery grommets to dampen vibrations and keep sound from radiating outward through the water. In miniature, the robot seemed to embody the kind of silencing philosophy that the *Thresher* had tried to pioneer for Navy submarines more than three decades earlier. Only in this case its stealthiness aimed at taking animal life, rather than Soviet submariners, by surprise.

Bill pulled out a large plastic jar full of tiny glass spheres. It was the stuff that gave syntactic foam its buoyancy, he said, and would eventually do so for the robot and its float, which was under construction. The jar, though full of glass, was light as a feather. The spheres were much smaller than grains of salt, looking like a fine white powder. I picked up a pinch and rolled it between my fingers, amazed that something so seemingly insignificant had changed the face of deep exploration.

We left Bill's office and walked to a laboratory where *Tiburon* sat half built. At this point it was mainly a tubular frame with housings for equipment and jet thrusters, yet enough parts were visible to give a feeling for its overall heft.

The most impressive thing about the emerging robot was its main computer. It was housed in a titanium cylinder about three feet long. For ease of inspection and repair, the computer slid out of the cylinder like a long, thin drawer from a dresser. As Bill pulled it out, the computer was revealed to be a mosaic of chips and colorful wires. Bill noted a cooling fan at its front end. The fan would draw warm air from the computer's core and circulate it over the inner wall of the titanium cylinder, where the chill from icy water during a dive would quickly soak up the heat. Even more impressive was the acoustic silencing. Bill pointed to the support legs of the cooling fan, where tiny rubber grommets had been installed to deaden vibrations. It seemed like extraordinary attention to detail.

"We're doing that everywhere," he said matter-of-factly.

THE ROAD TO THE HOME of David Packard rose slowly through wooded hills. The homes and estates along the way were the kind that people who are very rich tend to be very discreet about. White fences and horses were occasionally visible. Bicyclists sped by, slim and colorful in their tight-fitting pants and shirts, out for exercise on a sunny Sunday. Eventually, after a long climb, the road flattened out. At the Packard residence, the gate opened automatically and the long drive wound through a large orchard of apricot trees spread out over many acres, the land rising gently to the top of a hill where David Packard lived alone.

The entrance walk was shaded by two enormous pine trees. The style of his home was low ranch, built to conform to the hillside, surrounded by bushes and trees. A pair of wooden doors opened inward. Inside, an interior courtyard was green with plants and trees. Fresh flowers were about. No computers or electronic devices were evident, nor any aquariums or pictures of sea life. Windows and skylights were plentiful, giving the place a natural airiness. It was the abode of a billionaire, but it didn't feel like one. It was too comfortable.

Packard rose from his armchair. The picture window behind him overlooked his groves and, farther away in the distance, the industrial colossus he helped create, Silicon Valley, a global engine of high technology in the late twentieth century. He smiled. Tall and beefy, in his prime an athlete and outdoorsman, he was wearing a string tie and casual pants. A pen was in his pocket.

At eighty-one, he was old but hale. His handshake was firm and his eyes lit up as he spoke of the things that mattered to him now, as he neared the end of a long, productive life.

For his first seven decades or so, David Packard had showed little or no interest in the sea and its creatures. Born in 1912 in Pueblo, Colorado, the son of a lawyer, he grew up with the Rocky Mountains as his playground and early on learned to love the outdoors, beginning to fish in the wilderness as a teenager. He also discovered a knack for things mechanical and electrical and vowed in grade school to become an engineer. He soon got his radio license and became a ham operator, tapping out messages in Morse code around the world.

At the age of seventeen, in 1929, he drove his mother and sister to California to visit friends. They saw Monterey Bay and visited Pacific Grove. What impressed him most was not the bay or its deep canyon ("I don't think I knew about that") but an amiable radio amateur he visited in Pacific Grove, who had previously chatted with him over the airwaves. Also on that trip, Packard became interested in Stanford University and ended up getting a degree there in electrical engineering. He married a Stanford graduate, Lucile Salter.

The year after this wedding, in 1939, he teamed up with his college buddy William R. Hewlett to tinker on projects in the Packard garage—a small step that turned into an icon for generations of Silicon Valley entrepreneurs. Their first sale was eight audio oscillators to Walt Disney studios, which used them for the sound track of *Fantasia.* Profits were plowed back into the business, which grew steadily into one of the world's great manufacturers of measurement and computational gear. Hewlett was considered the engineering ace and Packard the deft manager. They kept the business close to Stanford, eager to hire the brightest graduates and to keep up with the latest research. The proven depth of his business skills caused Packard to be named to the boards of many large corporations, including General Dynamics, the maker of Navy submarines. His own company did work for the United States government, including the Defense Department and the National Aeronautics and Space Administration. Its atomic clocks for the Apollo craft synchronized the action in space during the voyage to the Moon and helped land men safely on its surface.[26]

His life changed dramatically with the rise of Richard Nixon to the White House. In January 1969, he took up duties in Washington as Deputy Secretary of Defense, serving under Melvin R. Laird. His job was basically to manage the far-flung operations of the United States military, including its deep forces. In his position he oversaw *Halibut* and similar Navy vehicles as they engaged in sensitive operations, especially those that probed the sunken gear and warships of foreign powers. "I was in charge of all those clandestine programs," he said, his head high, obviously proud. "That was something where we had a tremendous advantage over the Soviets. I had to approve all the missions and get approval from the White House and always debriefed the guys when they came in. It was really a great experience."

Packard also signed the orders that initiated the creation of the *Glomar Explorer,* the Nixon Administration ship that tried to haul up the sunken Soviet submarine from Pacific depths. He said the majority of the wreck did in fact drop back into the deep as the ship's claw broke. But he added that it was still a worthwhile venture, implying that much of value was retrieved. "The cold war probably would have turned out as it did without that endeavor," Packard said, peering into his memories. "But you never know."

America's military achievements in probing the deep were an inspiration to Packard later on as he, one of the richest men in America and a major philanthropist eager to foster science and environmental conservation, toyed with the idea of getting into the business of undersea exploration. Packard himself had no great love of the sea, though he was a dedicated naturalist and financially backed such organizations as the Nature Conservancy and the National Fish and Wildlife Foundation. But his daughters, Nancy and Julie, both marine biologists, were different. The sea was in their blood. The seashore was their playground. "We supported programs our children were interested in, even when I was not quite sure they made sense," he said, half serious. The $60 million aquarium opened in 1984 and, at Packard's urging, soon sought to expand into deep exploration. "My wife and children wanted to concentrate their efforts on education, so I decided we'd set up a separate program to do research with underwater vehicles."

Several factors influenced his decision. In 1985, Robey had his revelations in *Deep Rover.* Also that year, Ballard, working with the Navy-funded robot he had built, discovered the shattered hulk of the *Titanic* in waters more than two miles deep. Undersea prowess was increasingly on public display in May 1986 as Packard called together a group of senior oceanographers to meditate on the merits of a private institution dedicated to probing the deep. The group unhesitatingly gave its assent, adding that, in light of such surprises as the undersea hot vents, it was wise to expect the unexpected. In usual fashion, Packard went for it in a big way. Loving technology and knowing the extent of the Navy's undersea feats, he decided to bypass manned submersibles, judging them antiquated and deciding the new institute "could do just as good a job, maybe a better job, with unmanned vehicles." While small manned submersibles have to come up every day, he told me, robots can remain on the job indefinitely. "You can stay down for two weeks if you want to."[27]

The death of his wife, Lucile, in 1987 coincided with the birth of MBARI, to which he devoted himself. No detail of its development was too small for his attention, no idea too large. "Don't be afraid to fail," he kept telling his troops, adding that absence of occasional failure meant they were not trying hard enough. And he pushed himself. If the aquarium was for his children, the research institute was for him, an old man who had found a new love. "There's a tremendous volume of the ocean that has not been sensibly explored, that nobody knows anything about," Packard said, sweeping his long arm through the air.

Despite his dedication to MBARI, Packard had no favorite discovery or creature among the swarms of fishes, squids, and jellies that *Ventana* had viewed and discovered in the deep. The canyon seemed to hold no particular fascination for him. It was just a convenient starting point for global exploration. One got the impression that what excited Packard was the intersection of a great unknown, the deep, with a new way of doing great science.

"All this technology," he told me, "gives us an opportunity to see things and learn about things we would not have known otherwise." Moreover, it was very clear that Packard saw this explosion of knowledge as important not just academically but practically. A man of the world, an adviser of Presidents, a lover of nature, he saw MBARI and deep exploration as playing important roles in harmonizing relations between man and the sea. For instance, he saw them as throwing light on the mystery of the missing carbon.

It has long been known that levels of carbon dioxide are steadily rising in the atmosphere because of increases in automobile exhaust, industrial combustion, and the burning of oil, coal, and wood around the globe. Levels might double by late next century. Many climatologists believe this rise will hasten the trapping of sunlight in the atmosphere—a process known as *greenhouse warming* that threatens to heat the earth and disrupt its climate, intensifying storms and raising the oceans. But the exact dimensions of the danger are quite uncertain. About 20 percent of the carbon dioxide emitted by man is missing. Scientists want to trace the elusive gas, since knowing how it disappears might alter predictions of climate change or suggest a solution to the whole problem.

Enter MBARI and the deep blue sea. Just how the ocean absorbs carbon dioxide and what happens to its constituent parts after the molecule is broken down are poorly understood at best. Many experts suspect the sea soaks up more than is generally believed, acting as a giant sponge. Clearly, it is a stupendous reservoir for the element, holding at least fifty times as much carbon as the atmosphere. A main pathway is the dissolving of carbon dioxide in surface waters and its subsequent uptake in the photosynthetic reactions of plants, algae, and phytoplankton, which in turn are eaten by aquatic creatures in the great circle of life. Some of the carbon ends up in the calcium carbonate shells of zooplankton. Some ends up in larger animals and their living mosaics of fats, proteins, and carbohydrates, which to a large extent are just long chains of carbon and hydrogen atoms. But how much carbon? What percentage goes into plankton and what percentage into large animals? What is its fate? Is there a thick chain of life stretching all the way to the bottom? How much carbon reaches the ocean sediments? How long does it reside there? Do such processes change over time?[28]

Such questions are a central part of MBARI's agenda. One aim of Robey's census of deep creatures and food chains is to shed light on the sea's carbon cycle. The sinking larvacean houses, the armadas of squids, the strings of siphonophores, so big and so ubiquitous, all are part of the investigative puzzle. All are complex admixtures of carbon, and as such their fate was of great interest to Packard.

"It's an important storage mechanism we don't know very much about," he told me. "Bruce Robison can measure the amount of material that's transported down with all these organisms he sees, and ultimately that will give us a figure on how much is transported over time."

Another riddle that intrigued Packard is the extent to which man will be able to benignly tap the deep's vast productivity, especially in the next century, when the world's population is expected to peak at ten or eleven billion people, up from the current total of nearly six billion. The sea is already overexploited, with the global fish catch recently seeming to level off or decline after decades of steady rise. If the annual take stays at about one hundred million tons a year, the current level, then the catch per person could fall by about half as early as 2030. Some experts have argued that the moribund state of the world's fisheries is one of the first signs of a tightening food supply and impending global turmoil.[29]

For its part, MBARI is working closely with Federal and state officials to assess the possibility of harvesting some of the bay's deep species and protecting others. And it is gearing up to address the same issues globally. Packard saw the development of all kinds of fisheries as playing an important role in defusing the population bomb. But he also saw the dangers of unbridled exploitation and was eager to help forge a strong scientific understanding of deep life, lest it be hauled away haphazardly in nets and trawls. The best solution would be an increase in the global catch that did no damage to the subtleties of deep ecology that are just starting to come to light. The key, Packard said, is judicious action.

"It will be possible to increase substantially the food resources from the ocean if it is managed properly," the long-time manager said. "If we do things right," like putting a lid on population growth and slowly increasing the bounty of the sea available to man, "that means everybody in the world will have the same subsistence level we have in the United States. That's obviously a very important objective."

The interview over, I said my thank-yous and headed for the door. Waiting for me on the way out was a bag filled with fruit. Packard said it had been picked that morning. He waved goodbye. Outside his home, I paused near his orchards, enjoying the sun and the view of Silicon Valley, savoring one of the ripe apricots he had given me.

Packard, it seemed, was a rich man interested in opening a new frontier and doing some good in the process. Part of him clearly wanted to end world hunger. A technophile who found a new love late in life, he would leave a legacy not only of enriching science and of exploring an unfamiliar part of the planet but of helping people and the environment.[30]

In time, I learned that many experts other than Packard had gazed into the deep over the decades and beheld in it the promise of great natural resources that might benefit man. But as it turned out, these visionaries on the whole tended to be less altruistic.

SIX

Fields of Gold

AFTER WORLD WAR ONE was over and nearly 10 million men lay dead, the Allies demanded huge reparations in the form of gold—50,000 tons of it, amounting to the tidy sum of 132 billion deutsche marks. Germany had no gold to speak of. But it did have one of the world's great chemists, Fritz Haber, who had recently saved the nation by extracting nitrogen from the atmosphere. During the war, the Haber process allowed Germany to continue making explosives despite an Allied blockade that cut off all imports of nitrates. Had no such method been discovered, Germany would have simply run out of bombs and bullets and been forced to surrender. So afterward it turned again to its famous son. This time Haber's secret assignment was to extract gold from the sea—a dream of generations of scientists.

Every cubic mile of seawater contains about forty pounds of gold, and Haber, relying on faulty measurements, believed there was even more. In theory, the extraction of that dissolved gold, tons of it, would pay off the war reparations and leave lots left over for rebuilding Germany. In 1925, the German ship *Meteor* sailed for the South Atlantic, ostensibly to do basic oceanographic research. In great secrecy for two years, the Germans worked and assayed and tinkered, examining huge volumes of seawater from the surface as well as many parts of the deep. But try as they might to extract gold, the only thing Haber and his team won in abundance was frustration. In Germany after the voyage, the chemist broadened his search to examine samples from all the world's oceans—to no avail. Gold seemed to be everywhere but somehow managed to elude man's best efforts to mine it.[1]

For decades, the ghost of Haber has haunted most dreams of extracting riches from the deep. The field, regularly hailed as a new commercial frontier, just as regularly has gone through boom-bust cycles. And it may be headed for another.

The technological speedup in the wake of the cold war is stirring all kinds of dreams and actions aimed at tapping the nearly limitless wealth that lies beneath the waves. But just how far the field will progress this time around remains to be seen.

At some point in the future, despite the great uncertainties and risks, wide development seems likely to take hold. After all, the deep's reservoirs of copper, nickel, cobalt, manganese, zinc, silver, and gold are thought to frequently dwarf deposits on land. In a world that is increasingly crowded, the pressures to mine the deep seem likely to only increase. The ultimate question, and perhaps dilemma, will be how to reconcile such pressures with the responsibilities of environmental stewardship.

HABER AND HIS KIND hunted for gold in liquid seawater. But as the technologies of the twentieth century advanced, the hunt eventually shifted to the rocky seabed, a wilderness that businessmen hoped would yield fortunes more readily. No private patrons or naval organizations paid for the initial move. Surprisingly, the pioneer investor was the American government, in particular, its agencies that fund pure science. They created a remarkable tool that entrepreneurs quickly seized upon and deployed aggressively in pursuit of deep wealth: the drill ship.

The tool was perfected in the late 1950s and early 1960s as American geologists struggled to keep their field from being overshadowed by the space race. Their goal was to pierce the Earth's crust and sample the hot mantle below, drilling many miles down. The logical site for the venture was the rocky domain far beneath the sea, since that was where the Earth's crust was known to be thinnest. The question was whether the feat could be achieved. On land, long pipes and diamond bits had routinely bored through miles of dense rock, usually in search of oil. So, too, oil companies had probed shallow waters to some extent. But no one knew if drilling equipment could be developed that would operate at sea atop a mile or two of water.[2]

Things moved expeditiously as the government threw its backing behind the project. By early 1961, a converted oil-company drilling ship known as *CUSS I* (after the companies that owned it—Continental, Union, Shell, Superior) sailed to a predetermined spot off the northwestern coast of Mexico, lowered a long pipe through more than two miles of water, and began to drill into the seabed. As on land, a tall derrick aided the handling of pipe. But the operation also had many innovations. To avoid damage as the ship moved up and down in the waves, the drill pipes were specially built to slide over one another like sections of a collapsible hand-held telescope. Perhaps most important, the ship was equipped with a new method of side-to-side stabilization that kept it almost motionless relative to the faraway drilling spot on the seabed. Anchors would never hold in heavy seas. So four giant outboard motors were used, each with a shaft some sixteen feet long and a propeller four feet wide. The method, now known as dynamic positioning, eventually became nearly universal in such work.

That wintry day in 1961, the 27th of March, as winds blew to thirty knots and waves crested at fourteen feet, the innovations proved their worth. The drill pipe successfully dug into the dark seabed more than two miles down. The nature of the operation was such that no one aboard ship could actually see the progress at the bottom. At the control panel, the only signs of headway were wavering needles that showed such things as pressure, torque, and speed of rotation. After drilling several test holes, the scientists pressed the apparatus home, slicing through hundreds of feet of soft sediments until the bit encountered hard rock, as indicated by an abrupt slowing in the rate of descent. Eventually, a core of the rocky material was pulled up through the long pipe and laid out on deck for the excited scientists to examine. For decades, debate had raged over the composition of the ocean crust, the candidates including limestone, compacted sediment, and volcanic basalt. Now the verdict came in. It was basalt—the first of many secrets to be surrendered to drill ships.[3]

The goal of penetrating far deeper to actually pierce the Earth's crust required the construction of stronger gear and a much bigger ship. But Federal funds for that project never materialized. The civilian side of the government had nowhere near the political muscle of the Navy and the nation's intelligence agencies.

Even so, a new chapter had opened in man's relationship with the sea. The technologies of deep drilling were now understood in outline and were recognized as having the potential to haul up from great depths literally tons of rock and seabed materials, in contrast to the far lesser amounts that had been collected by dredges and would eventually be gathered by robots and manned submersibles. Drill ships could dig for deep oil. They could tap mineral deposits. When in motion, they could tow mineral collectors across the seabed with a kind of mobility undreamed of by drill operators on land. In time, experience would show that drill ships also had plenty of shortcomings and were far less sophisticated than the secretive probes that the Navy developed. The ships saw nothing of the seabed and were useless for wide exploration. Yet their brute strength quickly caught the attention of the business world.

As early as 1963, Willard Bascom, a leader of the scientific effort to pierce the Earth's crust, was applying the field's innovations to the undersea mining of diamonds, by weight one of the world's most valuable commodities. Hired by DeBeers, the international cartel, Bascom created a ship with a derrick that sailed off the coast of Africa and lowered a pipe more than two feet in diameter down to the ocean floor, where it sucked up diamondiferous gravels. By late 1964, the ship, *The Rockeater,* had recovered 129 diamonds. Based on the survey, Bascom's team estimated that the concession held more than ten million carats, or more than two tons of diamonds, an immense treasure.

Impressed by such numbers, DeBeers teamed up with Bascom to create a new company, Ocean Mining A.G., of Zug, Switzerland, to pursue such work around the world. While seeking the glitter of diamonds off Africa and other places, the company also broadened its agenda to prospect for tin off Thailand, zirconium off Australia, and gold off Alaska. Quite suddenly in historical terms, the bottom of the sea was perceived as a vast repository of mineral riches.[4]

The undersea mining of the early 1960s centered on the relatively shallow waters of the continental shelves. For the most part, the tin and gold and diamonds were extensions of land deposits, a fact that limited the extent and profitability of the harvesting. But the deep had its own glitter, in particular the irregular balls of manganese on the seabed that lay packed so densely in some places they looked like cobblestones.

The *Challenger* scientists were the first to discover the ubiquity of the nodules. Confirming the suspicions of the deep's first global investigators, later scientists correctly judged that the metallic balls formed like pearls as dissolved metals in seawater precipitated with great slowness over the ages. Exactly how the nodules formed and why they remained on the surface of the seabed, rather than being buried by sediments, were mysteries. Microbes were suspected of being important players. But it was indisputable that the nodules existed in numbers that were quite accurately described as gargantuan. Trillions of the potatolike lumps lay hidden in the icy darkness.

The potential windfall was scrutinized slowly and painstakingly in the 1950s and '60s as ships of many nations lowered old-fashioned dredges into the depths. The blackish lumps hauled into the light of day were ugly, looking not a little like slimy meatballs. But analysis showed them to be treasures of rare metals—up to 25 percent manganese, 2 percent cobalt, 1.9 percent copper, 1.6 percent nickel, and lesser amounts of dozens of other uncommon elements. Manganese is important for making all kinds of steel, especially the super-hard kind used in armor plating and bulldozer teeth. Cobalt is used in alloys for the high-temperature parts of jet engines and industrial gas turbines. Copper is the heart of all electric wires. Nickel is vital for making stainless steel as well as coins, plating, and electronic circuits. In short, the ugly lumps were industrial gold, and they were calculated to constitute a reservoir that dwarfed known land deposits. All told, the sea's inventory was estimated at trillions of tons, an astronomical sum that would take millennia of mining to start to consume. Some of the richest sites lay just off the United States. The Blake Plateau, a deep terrace adjacent to the continental shelf off Florida, was found to be covered with hundreds of square miles of nodules ranging in size from golf balls to big potatoes. The greatest concentrations with the highest metal contents were discovered in the Pacific southeast of Hawaii and just north of the equatorial zone. There, in waters up to three or more miles deep, the metallic nodules lay scattered densely over many thousands of square miles of seafloor. In theory, the manganese nodules of the sea were an unclaimed natural resource worth untold trillions of dollars. Sometimes they were referred to as black pearls.[5]

By the late 1960s, as dredges at sea were joined by waves of new gear and drill ships and whole fleets that daily proved their mettle in shallow waters, the deep prize began to look increasingly within reach. The development of big ships with long pipes and powerful pumps meant that, in theory, the "pearls" could be sucked up from the deep as if by giant vacuum cleaners.

What developed in the 1970s was a rush of preparations for deep mining, both technical and political. Some of the ado was defensive. For instance, the International Nickel Company of Canada, known as Inco, had huge land mines in Sudbury, Ontario. It needed to know whether these mines were fated to become uncompetitive, and, if so, it wanted to be ready to preempt competitors and dominate the new wave of mineral extraction at sea or perhaps even to feign interest in the deep in order to put off potential rivals who might otherwise see an easy opening. The same was true of Noranda Mines, Ltd., another Canadian giant. So these and other companies probed the deep with increasing vigor, gathering and smelting and analyzing.[6]

Very quickly this work got caught up in a noisy political clash at the United Nations as poor countries assailed rich ones. Seabed resources, the have-nots proclaimed in a phrase that would echo over the decades, were the "common heritage of mankind." The aggrieved pushed for a Law of the Sea Treaty, which in theory would create an international legal framework for divvying up the deep riches. The work was to benefit all countries, in particular developing ones.[7]

Thrown onto this smoldering fire was a good deal of gasoline in the form of the ultimate ship ever built for the extraction of deep wealth—the

Glomar Explorer, its 618-foot length packed with the latest gear for raising manganese nodules. Today we know that the secretive ship was actually made for recovering a sunken Soviet sub. But at the time its cover story was both plausible and widely believed. Longer than two football fields, its drilling derrick twenty-six stories tall, its long pipe able to suck up tons of seabed materials, the ship looked like a giant extrapolation of the existing art, boosting the field's legitimacy. So did its putative owner, the reclusive industrialist Howard R. Hughes, who was widely seen as just the type to try something bold in pursuit of new wealth. Scores of wide-eyed stories about the mining ship appeared in magazines and newspapers around the world, to the delight of disinformation specialists at the Central Intelligence Agency. The *Wall Street Journal* in 1973 said Hughes was "leading the pack" in the race for seabed riches. *Science* magazine in 1974 said the enigmatic Hughes had "bought the best expertise available and now is well in front." Details were usually sketchy, but experts said the nodules were to be gathered by a seafloor collector and then pumped up the ship's long pipe to the surface, the standard approach.[8]

This apparent escalation of seabed action set off alarm bells among poor nations, who saw the Hughes ship as a capitalist assault on their socialist dream. In 1974, as *Glomar Explorer* began its secretive maiden voyage in the Pacific, some five thousand delegates and observers from forty-eight nations converged on Caracas, Venezuela, for another round of United Nations deliberations to try to hammer out a sweeping treaty on the Law of the Sea. Understandably, much talk focused on the giant ship. In a mood of righteousness and outrage, the delegates worked on treaty language that, if it ever became international law, would put a heavy financial burden on entrepreneurs and in theory create a mechanism whereby poor nations had a hope of suddenly becoming rich.[9]

Remarkably, the field kept up its momentum even after the public disclosure in 1975 that the *Glomar Explorer*'s mineral work was a fiction. In a way, the continuation of the exploratory and political rush was a credit to the adroitness of the CIA's cover story.

Between 1974 and 1977, four American-based consortia formed to explore seabed mining. They were composed of a number of international companies, eventually including Lockheed, Sun, and Cyprus Minerals from the United States, Union Minière from Belgium, Inco and Noranda from Canada, ENI from Italy, Mitsubishi from Japan, Consolidated Gold Fields from the United Kingdom, and Preussag and Metallgesellschaft from West Germany. Scores of American patents for deep-mining equipment were issued, cresting in 1976 at a rate of nearly sixty a year. Expenditures for exploration work reached their peak in 1978 at nearly $100 million a year. A total of more than $650 million was spent as fleets of ships sailed the seas to probe its recesses for mineral wealth, especially in the Pacific, sometimes lowering nothing more advanced than deep dredges, sometimes miles of pipe. Research was also done on a variety of nodule collectors, including towed sleds, self-propelled miners, and lines of buckets that were to sweep across the deep in a continuous arc.[10]

With visions of great wealth dancing in their heads, the mining consortia laid claim (often illegally in the eyes of the United Nations) to enormous tracts of the Pacific in the Clarion-Clipperton fracture zone, an area southeast of Hawaii rich in nodules laden with nickel, which was judged to be the most valuable mineral. The four American claims were big enough to support mining for at least twenty years and in total covered an area bigger than Spain. They were soon joined by claims from such countries as Germany, France, Japan, China, India, the United Kingdom, the Soviet Union, and South Korea. The crowding forced new entrants to expand across the Pacific. China's claim lay due south of Hawaii. The United Kingdom's lay just off the coast of Mexico. Germany's lay off the coast of Peru some seven hundred miles south of the Galápagos Islands. In many cases, the states negotiated bilateral treaties with one another to try to legitimize their claims.[11]

In short, the world's dominant states quietly carved up enormous parts of the sea, often defying the UN and wagering that a deep boom would eventually take place. The action bore some resemblance to the colonial partitioning of Africa and the New World centuries earlier.

In parallel to this high-seas grab, many coastal states made big territorial claims, which the United Nations sanctioned and even encouraged if the state also subscribed to the emerging international sea law. By UN rules, a state could legally acquire a large offshore area known as an Exclusive Economic Zone, or EEZ, which extended out from a coast for up to two hundred nautical miles (equal to 230 statute miles or 370 kilometers). Between 1975 and 1980, the number of coastal states making such claims jumped from under thirty to more than one hundred. The action often represented a burst of nationalism from former colonies. Not that it was insignificant. The laws of geometry meant the two-hundred-mile area often translated into vast territories. Consider the Cook Islands, a tropic paradise of a dozen or so inhabited isles in the South Pacific with a total area of some ninety square miles. The tiny nation in 1977 claimed an EEZ and overnight gained an exclusive marine area of one million square miles, a domain nearly the size of Argentina. Much of that deep territory turned out to be densely paved with manganese nodules.[12]

Despite many visions of instant riches, and despite much work and even larger amounts of positioning, the activity largely dissipated by the end of the

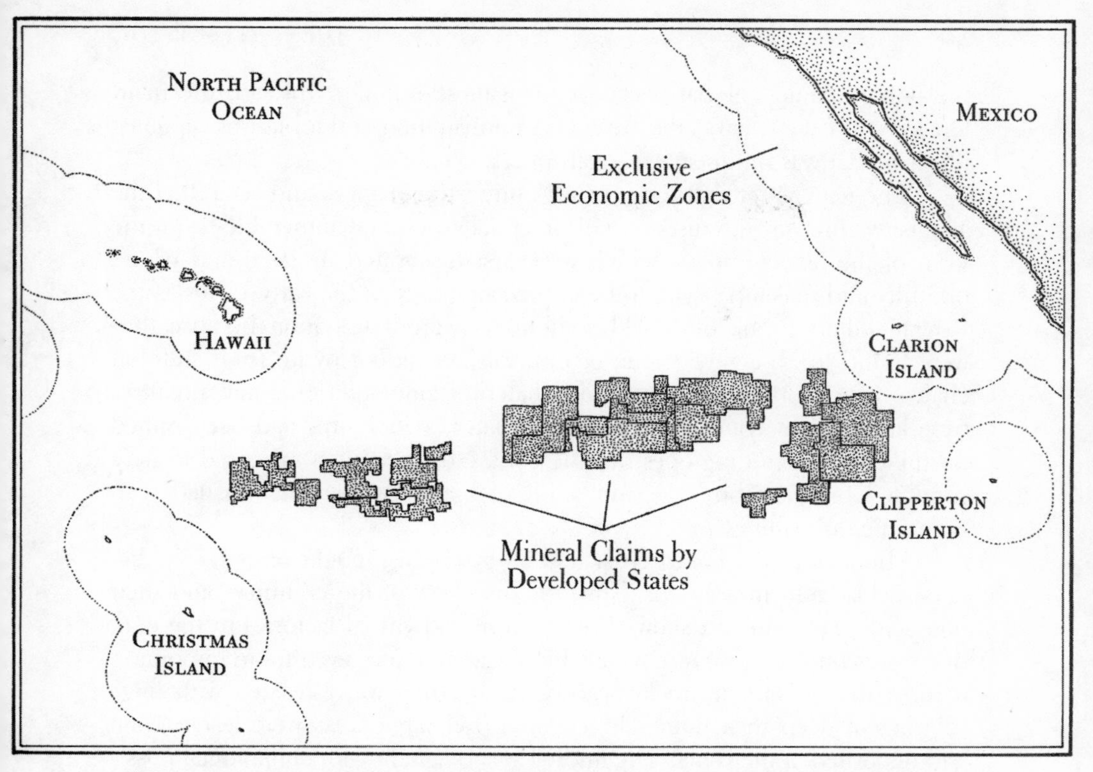

PACIFIC TERRITORIAL CLAIMS. Developed states, foreseeing new kinds of mining in the decades and centuries ahead, have staked out claims to large parts of the Pacific seabed laden with irregular balls of manganese. The nodules are so dense in places that they look like cobblestones.

decade, a victim of a global recession and an appreciation that the mining job was far more difficult than first envisioned. Hauling up small samples with a dredge or drill ship turned out to be a very different economic proposition from gathering up nodules scattered over hundreds of miles of seabed. The nearly universal conclusion was that manganese nodules *were* a potential treasure, but one that would be won only when and if metal prices and economic conditions warranted the great expense and difficulty of laboring in the sea's depths amid icy temperatures and crushing pressures. A seabed mine was estimated to have capital startup costs of up to $1.8 billion, and annual operating expenses of up to $440 million. The price of entry was seen as too high, and the frenetic activity slowly fell off to a relatively low level of exploration.[13]

A SECOND ROUND of deep commotion soon materialized that was quite unlike the first. New technology played no role. Nor, for that matter, did private companies, global fleets, or manganese nodules. Instead, the main force behind the stir was the Reagan Administration and a blast of capitalist ideology that was uniquely American in character.

The new glitter that caught the Administration's eye and rekindled the acquisitive fire was the discovery of potentially rich new mineral lodes in the form of the hot chimneys, which were first discovered in 1979 and whose ubiquity and makeup began to be appreciated only in the early 1980s as the Reagan officials took office. The chimneys were a sensation because they were found to be a potent mix of minerals, including pyrite (iron sulfide), chalcopyrite (copper sulfide), and sphalerite (zinc sulfide). Such mixtures were known as polymetallic sulfides. On land, sulfide ores had been mined for millennia, including ones bearing gold. Now, scientists appeared to have stumbled on their birthplace in a bit of happenstance that bristled with economic possibilities.

Minerals in the ocean crust had always been thought of as thinly dispersed, like gold in seawater. Now, the discovery of the chimneys and their hot, acidic environment showed that nature had virtual factories in the dark that were concentrating rare metals from seawater and, even more important, it turned out, from the rocky seabed. As scalding water reacted with huge volumes of deep rock, minerals scattered over large areas were leached out and deposited at the vents. The finding was pregnant with commercial possibility. In essence, nature for ages had been doing in secret the very thing that Haber and his German team had failed to achieve. The key questions of the early 1980s centered on whether the hot chimneys and their kindred formations were regular features of the deep, whether their composition was often rich, and whether the metallic deposits were superficial or extended deep into the Earth's crust. And beyond such resource issues was the question of whether mining could be economic, given the severe challenges posed by the deep's extraordinary cold, dark, and pressure.

The stir began in 1980 and 1981 as a team of researchers from the National Oceanic and Atmospheric Administration, led by Alexander Malahoff, a venturesome geologist, explored the Galápagos rift, the area of seafloor spreading off the coast of South America where the deep hot springs were first discovered. Using *Alvin* to probe the inky darkness, the team stumbled on the largest known mass of fossil metallic sulfides, topped by more than twenty extinct chimneys. The ore body beneath them was estimated to be 130 feet thick, 650 feet wide, and 3,280 feet long. If its several million tons of imagined ore could be mined efficiently and brought to market, Malahoff reasoned, it would be worth a fortune. The copper alone was valued at perhaps $2 billion.

The stir soon deepened as scientists in 1982 discovered that the sides of volcanic isles and seamounts in the Pacific were often covered with a thin crust of copper, manganese, cobalt, and nickel in an oxide slurry, with each site expected to yield several million tons of ore. In scientific shorthand, these deposits came to be known as cobalt-rich crusts, cobalt being one of the main metals needed for the Reagan Administration's military buildup.[14]

Upbeat estimates began to swirl around Washington amid talk of a new gold rush. It happened in the fall of 1982, when the Reagan Administration was little more than a year old. Giving the stir a sense of certitude were two dozen experts who hailed the findings in a special issue of the *Marine Technology Society Journal,* the field's top publication, based in Washington. "It is now clear," Malahoff enthused, that the hot vents had quietly concentrated metals "for several billion years and have led to the generation of ore bodies now mined on land." The deep harbored the birthplace of the Earth's greatest riches. Malahoff said a chimney sample from the Galápagos site was composed of sulfur, iron, copper, silica, zinc, manganese, aluminum, selenium, cobalt, magnesium, molybdenum, lead, arsenic, barium, cadmium, chromium, phosphorus, mercury, nickel, tin, vanadium, uranium, tungsten, and silver—the metallic underpinnings of modern industrial society.

Joining the chorus in the special issue was Conrad G. Welling, senior vice president of Ocean Minerals Company, one of the four seabed mining consortia based in the United States. He said the ores of the hot vents were quite potent compared to those of the icy manganese nodules, being "approximately one thousand times" as concentrated. One thousand times! It was like fistfuls of gold versus tiny flakes. Welling said it was possible that the seafloor was littered with "many thousands" of deposits similar to the Galápagos site.

The final article was by David B. Duane, a NOAA official. Driving the logic home, he noted that the United States by a quirk of nature was blessed with a likely mining site right off its shores—the Gorda Ridge, a volcanic spreading center that could be claimed as part of America's territory. Most volcanic ridges lie far out to sea, such as the Mid-Atlantic Ridge. But the Gorda was just off Oregon and northern California, much of it a mere seventy or so miles offshore. The hope, not entirely naive, was that America's backyard hid a new kind of treasure. Duane proposed a five-year research program to assess Gorda, whose mineralogical features at that point were still a mystery. Aware of the political winds blowing through Washington, he sought to tie these esoteric issues to perceptions of Western advantage in the cold war against Moscow. The possibility of rich polymetallic deposits lying so close to American shores, he said early in his article, was "timely in light of growing U.S. concern over supplies of strategic metals."[15]

It was martial music to the Reagan Administration's cold warriors, who at the time were strengthening their anti-Communist agenda in ways that were all but inconceivable a few years earlier. One of the Administration's goals was material self-sufficiency for the nation that bordered on physical isolationism. The Reagan people especially wanted to develop reliable (preferably domestic) sources of so-called strategic metals like aluminum, chromium, and cobalt, which were considered key materials for making arms and fighting wars. Right or wrong, the Administration worried that dependency on foreign sources, sometimes in unfriendly or unstable states, was an unacceptable risk that could weaken Washington in its war with Moscow. Navy Secretary John Lehman, speaking in May 1982, typified the inclination to look to the deep sea for solutions. "Of the nearly two dozen strategic minerals on which America is now virtually wholly dependent on overseas sources," he said, nearly all "could be provided from the seabottom." The discovery of polymetallic sulfides, Lehman added, suggested "prospects vastly more exciting than anything dreamed of just a few short years ago."[16]

The principal stumbling block for the Reagan warriors was the United Nations and its dream of a Law of the Sea, the framework of which took shape during the 1970s and early 1980s. As finally written in 1982, the proposed treaty said that all mineral resources on or below the ocean's floor belonged to the people of the world and proposed that the United Nations control their development. Among the law's provisions was a $500,000 fee to be paid by prospective miners, and a mechanism whereby half of any claim areas would be turned over to the Enterprise, a United Nations concern that would mine the sea on behalf of the world's people.

In December 1982, Fiji became the first nation to ratify the treaty, followed by such states as Angola, Belize, Cuba, Djibouti, Iraq, Mali, Yemen, and Zaire. But that same month the Reagan Administration denounced the process as socialistic nonsense and encouraged American companies to mine the seas freely in accordance with United States law. It also opened negotiations with other industrial nations to achieve bilateral agreements that would recognize ensuing claims on the high seas. In a further dig at the sea-law process, the Reagan people unilaterally seized one of the inducements treaty framers had offered coastal states—an Exclusive Economic Zone, the two-hundred-mile-wide band that extended a state's control over its adjacent waters, including the seabeds. The American move was particularly outrageous to treaty framers since the United States, in taking the candy and ignoring the medicine, asserted ownership over what was far and away the largest EEZ on Earth.[17]

It was a wintry day in Washington as President Reagan lifted his pen to sign the terse, 532-word proclamation that claimed resource jurisdiction over waters extending 200 nautical miles from the nation's territories and possessions. The date was March 10, 1983. The act, Presidential Proclamation No. 5030, made barely a ripple in Washington's sea of news. The President himself was vague on what it all meant. This expansion of sovereign rights, said his proclamation, would "advance the development of ocean resources and promote the protection of the marine environment." A little-noticed policy statement that accompanied the proclamation offered more detail, saying the step had been taken because recently discovered deposits of polymetallic sulfides and cobalt-rich crusts on the seafloor promised to yield minerals vital to the economy and national security.

The President's admirers were elated, calling it a master stroke that rivaled or exceeded in importance the Louisiana Purchase of 1803. After all, it cost nothing. And the new dominion was huge. The economic zone included waters off the mainland United States, Hawaii, Alaska, Puerto Rico, the Virgin Islands, and a host of little-known trust territories and possessions in the Pacific, including Samoa, Guam, Johnston Island, Wake Island, Howland Island, Baker Island, Jarvis Island, the Mariana Islands, and the Midway Islands. By the magic of the two-hundred-mile geometry and Reagan's bold assertion of seabed jurisdiction, these tiny Pacific isles added immeasurably to the size of the United States. The Hawaiian zone alone added nearly a million square miles, an area bigger than Mexico. All told, Pacific waters that lay thousands of miles from the continental mainland made up nearly half of the new economic zone. The added acreage was so vast that geographers quibbled over its exact dimensions for years. By some estimates, the new zone encompassed more than four million square miles—a tract larger than the entire U.S. land mass. By most reckonings, the United States had suddenly doubled in size.[18]

Two weeks after Reagan's move, the Minerals Management Service of the U.S. Department of the Interior announced that it would open up commercial bidding for polymetallic sulfides on the Gorda Ridge, the main seafloor spreading center within America's new economic zone. The leasing program was described as "an important element of the Administration's national strategic and critical minerals policy." And it was big, calling for the leasing of a tract of ocean bottom bigger than New York State. The targeted seabed was up to two and a half miles deep and extended in a giant swath that ran parallel to northern California and southern Oregon, its western edge abutting the EEZ limit and its eastern edge just over the horizon from the beach. In one scenario, a big seabed operation would daily bulldoze 16,500 tons of deep ore, weighing more than a big warship, and pump it to the surface. There the slurry would be treated for shipment to a coastal processing plant, which would generate nearly ten million pounds of tailings each day.

In a burst of candor, a draft environmental impact statement said the deep mining (possibly including the use of explosives) might kill sea creatures, decimate shipwrecks, hurt commercial fisheries, destroy hot-vent communities, rupture radioactive waste containers at two undersea dump sites, injure marine mammals, poison shellfish beds, mar coastal tourism, and even harm "Native Americans and unemployed persons in the coastal area," whose subsistence lifestyles centered on fishing.

Reaction to the plan was swift. Environmentalists attacked it as a catastrophe for the coastal zone and the unexplored ecosystems of the hot vents, charging that it threatened to produce a nightmare of death and pollution. State and local governments faulted the lack of consultation with them prior to its unveiling. Many academics were perplexed, saying the plan was premature at best. Most notably, industry, supposedly a main beneficiary of the plan, reacted to it with studied indifference despite early enthusiasm.

The apathy was based on a number of factors. Prices in many metals markets were dropping, making exotic sources less competitive with traditional ones on land. Virtually no exploration of the Gorda had been done, making any move to lease or mine it wildly speculative. And finally, industry had lost hundreds of millions of dollars in the first boom and was wary of repeating its past mistakes.

Nearly a year after the Reagan Proclamation, in February 1984, the Governor of Oregon and the Secretary of the Interior announced that the leasing plan was on indefinite hold. In its place, the Federal government proposed to pick up the pace of exploration on Gorda. The aim was to resolve some of the uncertainties about what the volcanic ridge harbored by way of mineral wealth and to gauge the likely environmental impacts of deep mining.[19]

The hunts were initially a bust. An *Alvin* dive of July 1984 came up empty-handed—no vents or sulfide deposits were discovered. It was a sobering moment, given all the earlier rhetoric. Things went no better in 1985. A ship towing particle and heat sensors found signs of hot vents, but two later cruises failed to spot any rocky chimneys on the seafloor. Finally, in 1986, the Navy's submersible *Sea Cliff* discovered an area of chimneys and mounds and polymetallic sulfides more than a mile beneath the waves. Analysis showed the deep material to contain copper, zinc, lead, cobalt, and silver in concentrations higher than any described before in seabed deposits. Researchers were elated. From zero, their estimates of possible metal deposits zoomed upward to as high as billions of tons. The ultimate thrill came as gold was discovered just outside the EEZ on the Juan de Fuca Ridge off Oregon, and along the volcanic spreading center that runs down the middle of the North Atlantic. At twenty-three parts per million, the concentrations there were two thousand times greater than the average in the Earth's crust and many times higher than what was routinely mined on land.

Though enticing, the glitter of the volcanic ridges was ultimately judged unattractive for mining in this century. As was the case with manganese nodules, the risks were seen as outweighing the rewards. Some indication of the extent of the uncertainties was that, from the start, the technologies meant to carry out the envisioned job were largely imaginary. No drill ships or other kinds of existing gear were seen as adequate for the job, only yet-to-be-invented devices such as seabed bulldozers. Put off by a host of such unknowns, industry showed an overwhelming lack of interest in the Federal plan. So it was that the firebrands of the Reagan Administration, for all their free-market ideology, were revealed to be no less ignorant of the ways of the business world than were the Kremlin's autocrats. The denouement came just before Ronald Reagan left office. The Minerals Management Service, after nearly five years of planning and debate, quietly ended all plans for Gorda leasing. The hot vents were no longer hot politically. As suddenly as it had arisen, Washington's gold fever was over.[20]

TODAY THE TECHNOLOGICAL speedup at the end of the cold war has rekindled interest in the mining of deep minerals, albeit with less animation than in the past. Both manganese nodules and polymetallic sulfides are being investigated. The redirection of state and military assets is catalyzing progress and lowering costs, often dramatically so. And rhetorically at least, the wave of new activity is quite moderate. No claims are being made of instant wealth and riches. Indeed, the action is often quite discreet, perhaps bespeaking a new seriousness. And it is widespread. Even such developing giants as China and India rely increasingly on their navies for aid in deep-resource appraisal and development, as issues of trade and economics supplant the exigencies of the cold war.[21]

Finally, and perhaps quite significantly for the work, given the intensifying politics of global environmentalism, the explorers are putting new emphasis on understanding what deep mining might do to abyssal creatures, a delicate issue of ecology often overlooked in the early rushes.

Once on the military's front lines, the deep submarine *NR-1* now has a wider agenda that includes searching for deep riches. For eight days in 1993, geologists from the U.S. Geological Survey dove in the nuclear-powered craft to survey manganese nodules and mineral crusts on the Blake Plateau near Florida, in total traversing a distance of more than forty miles. It was a feat of horizontal travel no manned submersible could achieve. The mineral deposits were photographed with electronic video and still cameras that produced an extensive record for later analysis. Among the chance discoveries were deep caverns that loomed up out of the rocky seabed like portals to ghostly netherworlds, as well as three mysterious circular holes, each nearly a mile wide and more than three hundred feet deep. In the end, the geologists decided the odd depressions were unrelated to the scars of cosmic impacts and were probably produced when the ocean floor gave way of its own accord, like terrestrial sinkholes.[22]

At a Pacific site rich in manganese nodules, the Russian ship *Yuzhmorgeologiya,* the same one that found the *I-52,* was hired by the National Oceanic and Atmospheric Administration to investigate the impact of nodule mining on deep ecology. It was a peace dividend—former foes working together on a neglected issue of environmental stewardship—impossible to imagine during the atomic standoff. And, as usual, the *Yuzhmorgeologiya* did the work for a song. In the early 1990s the Russian ship repeatedly towed across the seabed a fifteen-foot-long American sled that mimicked the gathering of nodules. The sled cut into the muck and raised a plume of sediments up to a height of thirty or so feet before the particles slowly redistributed themselves on the seafloor. Closely monitoring this East-West effort were Germany and Japan, whose Metal Mining Agency made arrangements to borrow the sled for its own ecological studies.[23]

The ominous question behind such research is not whether seabed mining will kill sea creatures but how great the carnage will be. The typical bottom-dwelling fauna of such abyssal regions include sea cucumbers, sea urchins, sea stars, bristle worms, sea pens, sea squirts, sea lilies, acorn worms, peanut worms, lamp shells, and probably hosts of undiscovered animals. It is widely assumed that a mining machine will destroy all organisms directly in its path as its blades probe the muck to tear loose manganese nodules and as its pumps send all the material, animate and inanimate alike, clattering up to the surface. What are uncertain are the distant effects. Clouds of sediments raised by the collector would rain down outside the mining path to bury bottom-dwelling creatures, probably killing them by suffocation. Farther away, the very gentle rain of extremely fine sedimentary particles might kill not by suffocation but by covering up meager food supplies and causing starvation.

The question, which only field studies can answer reliably, is how extensive such effects would be and how fast the new layers of bottom sediment would be recolonized. One grim study by NOAA estimated that a commercial mining operation might cut a swath of destruction sixty-five feet wide. Over twenty years of mining, the total area of environmental havoc would be about fifty thousand square miles, an area bigger than the state of Ohio. The study suggested that such destruction might be lessened by avoiding rich habitats, by raking nodules with small tines rather than large blades, and by finding a way to guide sediments back into the collector track rather than letting them drift over the dark seabed.[24]

Because hot chimneys and related structures are now known to be a far richer source of minerals than are manganese nodules, the civilian scientists working in military craft after the cold war have given high priority to the study of deep volcanism. Alexander Malahoff, the NOAA scientist who helped fuel the Reagan frenzy, was an early leader of the postwar work from an academic post he had taken up at the University of Hawaii. In 1990, the former Federal geologist and *Alvin* expert joined up with the Russians and their *Mir* submersibles to probe the depths around the Loihi submarine volcano, the youngest associated with the Hawaiian chain. It is located some fifteen miles southeast of Hawaii's big island. Its summit lies more than a half mile beneath the waves, and its sides drop precipitously to the seabed nearly four miles down. During eight dives, Malahoff's team of scientists probed deeper and deeper along the volcano's flanks, finding down more than three miles a field of extinct chimneys. The sulfide structures were apparently the deepest ever discovered, suggesting a greater global prevalence of them than previously suspected.[25]

On the American side, the fleet in search of volcanic riches includes not only such robots as *Jason* and the *Advanced Tethered Vehicle,* and such piloted craft as *Sea Cliff* and *NR-1,* but *Alvin* as well, which the military has used less and less in the 1990s, at times making only a single dive per year. By contrast, the Navy during the cold war worked the little submersible for weeks and months on end, often for an appreciable part of a season. The main beneficiary of this shift to peacetime work has been the National Science Foundation, or NSF, which in concert with the nation's universities runs an aggressive program to probe the volcanic depths.[26]

The 1993 voyage to the Juan de Fuca Ridge that I went on and its series of *Alvin* dives, funded largely by NSF, typifies such investigations. The geologists on board were struggling to gain not only basic insights into the workings of the rifts and chimneys but to weigh their financial promise as well. Two economic geologists aided the appraisal, Ian Jonasson of the Geological Survey of Canada and Randy Koski of the U.S. Geological Survey. Their job, among other things, was to assess the likelihood of an undersea gold rush sometime in the future. "The work on the ridges is really in its infancy," Randy told me over dinner one night in the ship's mess, adding that to date far less than one percent of the earth's spreading centers had been studied close up. "We've looked at the convenient ones," he said. "The Indian Ocean has 'em, too. And the Antarctic. And the Arctic. There's a great deal we don't know."

As our expedition hauled up deep volcanic rocks into the light of day,

Ian from Canada could often be seen on various decks laboring away on his hands and knees, pounding the rocky samples into tiny fragments for analysis in his laboratory. Such work is more or less constant. Each season since our expedition, a flotilla of ships and submersibles, including ones owned by the American Navy, has crisscrossed the deep sea's volcanic domains to assess their mineral potential.[27]

For the United States, the wave of military spinoffs is accelerating the hunt for deep gold. But for smaller countries, the new tools and techniques are often a first introduction to the icy darkness.

Britain is a good example. Despite its pioneering of deep studies, as exemplified by the *Challenger* expedition, and despite its leadership in many modern fields of abyssal research, the nation never developed the kind of vehicles that carry humans down to the dark seabed. For mineral exploration, this deficiency made little difference when manganese nodules were the fashion and when hauling up a handful of the black pearls took nothing more complicated than a dredge towed over an abyssal plain. But that technical want began to hurt when the focus shifted to volcanic vents and chimneys, which lie in mountainous domains and by nature are less common than the nodules and far more difficult to sample and analyze. So oceanographic Britain sat out the dance in the 1980s and watched as the Americans swarmed all over the Pacific hot vents, making discoveries and announcing a new kind of mineralogical affair.[28]

No more. The technology glut after the cold war and the collapse of East-West hostilities let the British approach their former foes, the Russians, and hire out the *Mir* submersibles, diving in them to the bottom of the Atlantic to examine a volcanic mound riddled with gold. Aboard the *Keldysh* during the expedition were thirty British experts.

The target lay fifteen hundred miles southwest of the Azores and two miles down in the darkness of the sea along the Atlantic's mountainous spine. It was an eerie hulk known as TAG, named after an early survey of the region, the Trans-Atlantic Geotraverse. Its surface aswarm with millions of crabs and shrimps, TAG is the size of the Houston Astrodome, one of the largest active sulfide structures ever discovered on the seabed. Its top bears hundreds of black and white smokers and thousands of extinct chimneys, some eighty feet high. Overall, the big undersea mound puts out enough heat to electrify London or Chicago. Its blistering waters have been measured at temperatures of up to 685 degrees Fahrenheit, more than enough to melt tin or lead. Moreover, the mound's gargantuan bulk is continuing to grow as the hot waters hit the icy surrounding sea and precipitate new layers of polymetallic sulfides in the form of active chimneys.[29]

In 1994, as the British dove on TAG, the key question for a large body of international scientists was what lay in the mound's core and lower extremities—in short, what was its structure. The mound appeared to be made of millions of tons of iron, copper, and zinc sulfides, though no one knew for sure. Gold had been discovered on its surface in concentrations of twenty-three parts per million, about five times as great as the average deposit mined on land. But no one knew the extent or depth of such glitter.[30]

To assess the hidden wealth and better understand the makeup of such behemoths, TAG was to be deeply drilled, a historic first for any volcanic mound. The drill was the *JOIDES Resolution,* a giant descendant of *CUSS I,* which, at 471 feet in length, is longer than some city blocks. The roving ship is run by JOIDES, the Joint Oceanographic Institutions for Deep Earth Sampling, an international consortium of universities, sea organizations, and government agencies, including ones in the United Kingdom. Positioned miles above TAG, its derrick twenty stories tall, the *Resolution* in late 1994 was to lower a hollow, five-inch pipe to the bottom and bore deeply into the mound's hot interior, revealing its secrets. Before, during, and after the operation, an array of deep gear was to check the mound for such things as changing patterns of heat and water flow. Such circulatory changes were seen as vital to understanding its inner plumbing. In August 1994, a team of Japanese and American researchers worked more than two miles down to set up monitoring gear.[31]

The British aboard the *Keldysh* sailed to TAG a month later, interested not only in minerals but animals as well, in line with the new environmentalism. The voyage was funded not by Britain's mining interests but by its Natural Environment Research Council. "There is a possibility that this place may change very dramatically," Adam Schultz, a Cambridge University geophysicist, told reporters before leaving. Diving in the *Mir* submersibles, the British team wired the hot mound with battery-powered instruments meant to monitor changes in its temperatures, water flows, and fauna. A camera left on the seabed was to automatically take pictures every three hours for six months. "We want to map the distribution of animals," said Paul A. Tyler, a Southampton University biologist and international expert on deep creatures, who ran the setup. A few months after the drilling, the British team picked up the instruments, again demonstrating its new independence of deep action.[32]

The drilling, done between September and November 1994, found that the mound's upper reaches were rich in polymetallic sulfides. Lower down, however, the sulfides were mixed with large amounts of anhydrite, the chalky precipitate of seawater that disappears when temperatures drop. The finding suggested the mound was blistering hot much deeper than suspected. That lessened the mineralogical allure of active mounds, since, if TAG was

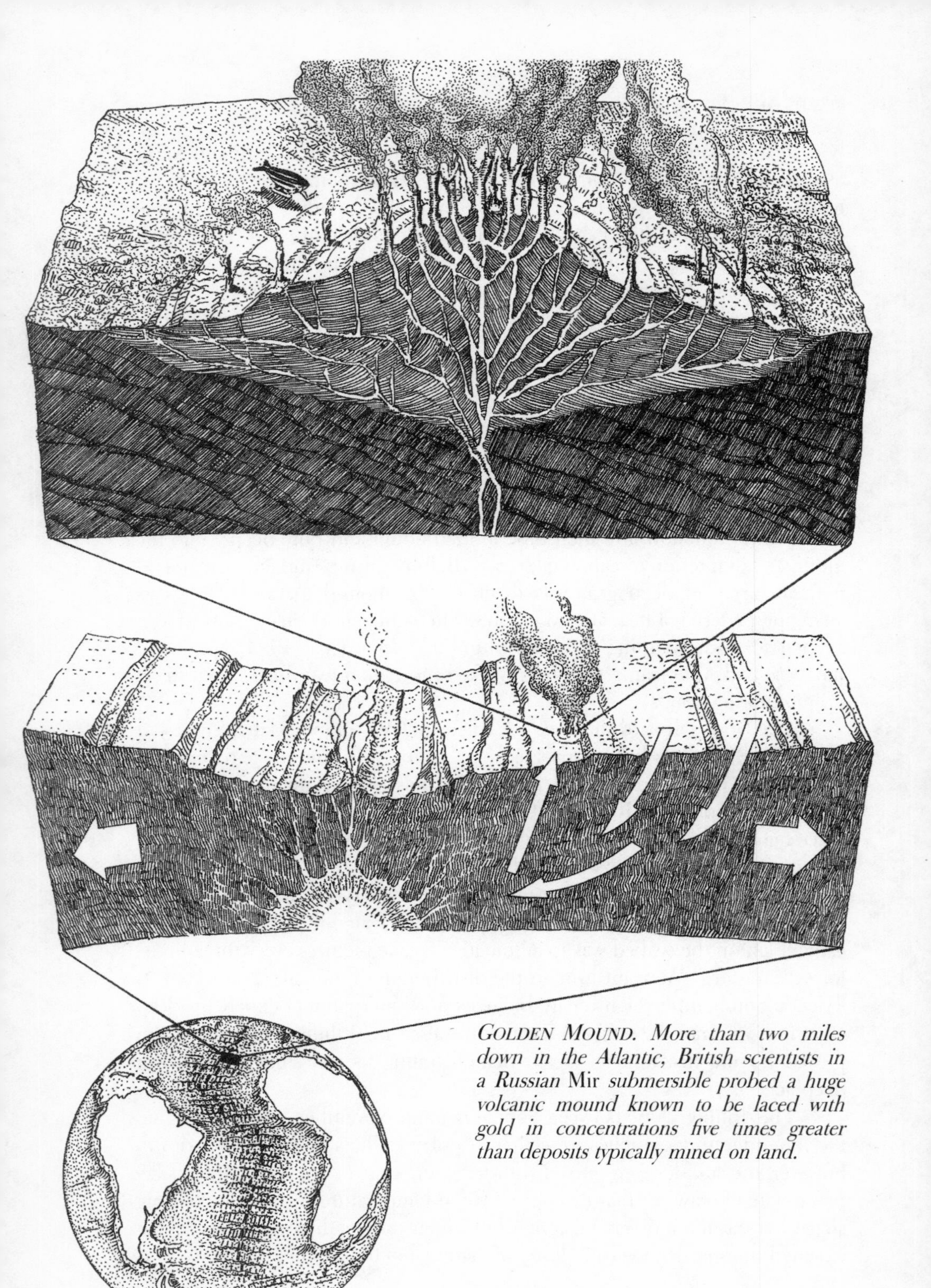

GOLDEN MOUND. More than two miles down in the Atlantic, British scientists in a Russian Mir *submersible probed a huge volcanic mound known to be laced with gold in concentrations five times greater than deposits typically mined on land.*

representative, much of their mass appeared to be tied up in uneconomic deposits. And it increased the allure of extinct formations, where anhydrite would have disappeared. Significantly, probings around TAG before and after the drilling revealed many such dead giants in the region, their tall chimneys looming cold and lifeless in the darkness.[33]

The military spinoffs, while greatly aiding small states, have produced more modest accelerations among nations that have long targeted deep gold and have worked hard to build up their infrastructure over the decades. These states include Japan and Germany, the world's leading economic powers after the United States.

Germany, despite Haber's failed hunt, is still working hard to mine the seas. Its main interests are global rather than its small North Sea and Baltic holdings, so the nation maintains a fleet of research ships that sail the world in search of deep minerals. The *Sonne,* for example, is dense with cranes, winches, robots, and underwater cameras, and is generally considered one of the world's best-equipped vessels for deep research. Germany has built an advanced system for mining and processing manganese nodules from depths of up to nearly four miles, and has tested it extensively in the Pacific fields, where it has three large claims in places where the nodules are particularly rich in nickel. Like the United States, it is also exploring what nodule mining will do to deep life. Significantly, Germany moved quickly to assess the volcanic fields when they materialized in the eighties and has remained at the forefront of such research ever since. In 1985, the *Sonne* dredged up many tons of the deep Galápagos mounds that Malahoff had hailed. And in 1994, in collaboration with Canadian geologists, it mapped a largely uncharted area of the South Pacific off the New Guinea archipelago, investigating volcanic seamounts and the submerged flanks of volcanic isles, finding deep hot springs choked with creatures and unusual concentrations of gold. One rocky sample yielded forty-three parts per million, ten times the average concentration of deposits mined on land.[34]

Around the globe, no country of late has shown more skill, persistence, and enterprise in the conduct of deep exploration and resource development than Japan, a nation of nearly seven thousand islands, where no point on land is more than seventy-five miles from the sea. Smaller than California, the nation sees the deep as a repository of biological and mineral riches and is making preparations to mine them both.[35]

It plies two large claims in the Eastern Pacific that are thick with manganese nodules, prospects around the globe for new sites, and is building the world's most advanced set of deep-sea gear, including manned submersibles for prospecting and robotic sleds for harvesting minerals.[36]

Japan learned the art of deep exploration at the feet of American masters, and Washington blessed the act. Tokyo after all was an anti-Communist bulwark in the west Pacific that Washington took many steps to strengthen. When Tokyo grew interested in the deep, especially the perilous work of piloted exploration, Washington was happy to share its undersea expertise. The lessons began in the early 1970s, just as America worked hardest at its deep-sea undertakings. In effect, Japan became the first external beneficiary of America's deep armada, the transfer taking place nearly two decades before the cold war's end.[37]

For aid Tokyo looked especially to Woods Hole on Cape Cod, which in concert with the Navy had pioneered the field of deep-maneuvering submersibles. The pride of the institution, of course, was *Alvin,* which at the time could carry three people down to a depth of 2.5 miles. A delegation of fifteen Japanese scientists and engineers visited Woods Hole in 1973 and inspected the submersible closely, taking back blueprints and many undocumented details of its development. In a remarkable step, Emperor Hirohito, a marine biologist, in 1975 paid a visit to Woods Hole as well, discussing with scientists there the kinds of discoveries *Alvin* made at great depths.

The Japanese were good students. In 1981, drawing on their American data, as well as their own strengths as master craftsmen, they launched the *Shinkai* ("Deep Sea") *2000,* an *Alvin*-type submersible that could carry people down 2,000 meters, or 1.25 miles. Cautiously, the Japanese began to survey their watery domain. A 1986 dive off the Ryukyu Islands found a cluster of volcanic mounds up to thirty feet wide and fifteen feet high, the tops and crests covered with bright orange mineral deposits. Another *Shinkai 2000* expedition some seventy miles off Okinawa dove to a depth of nearly a mile and found a jungle of dead and active chimneys, fairly hot ones, whose temperatures ranged up to 320 degrees Fahrenheit. Analysis showed the ore to contain zinc, copper, lead, silver, and gold.

The Japanese fascination with American technology and deep exploration only grew as experience whetted their appetite. When *Alvin* and its support ship visited Tokyo in 1987, fresh from making headlines by investigating the sunken hulk of the *Titanic,* the American submersible was toured inside and out by Crown Prince Akihito, later Japan's Emperor. A marine biologist like his father, he talked with Woods Hole scientists about *Alvin*'s dives to Pacific hot vents and the odd creatures the submersible found there. Eager to press ahead, the Japanese in 1990 completed the *Shinkai 6500,* which cost about $60 million to build and can carry three people down 6,500 meters, or more than four miles. That makes it the world's deepest-diving manned vehicle, surpassing the American *Sea Cliff,* the French *Nautile,* and the Russian *Mir 1* and *Mir 2.* Its proficiency gives Japan an edge in deep exploration. In 1994, as Woods Hole scientists involved in a joint program looked on somewhat sheepishly, *Shinkai 6500* set an Atlantic depth record, diving down nearly four miles to hunt for clues as to how the deep seabed had formed.[38]

Soon after that, *Shinkai 6500* joined the international assault by ship, submersible, and robot on TAG. It ferried down gear for both Woods Hole and Japanese scientists, seeking to uncover some of the hot mound's secrets. Tokyo was no longer a pupil but a peer, and in some ways a teacher.[39]

THE EXPLORATORY STIR engendered by the end of the cold war, as well as the continuing efforts of Japan and other advanced states and international consortia, appear insufficient to drive the global work of deep mineralogy from the stage of cautious inquiry to active mining, at least for now. More than a quarter century after its birth, the whole enterprise is still aimed at exploration and gaining or maintaining position for the time in the future when seabed minerals will be routinely gathered as an adjunct to land supplies. Just when that might occur is hard to predict. The politics and economics of the issue are so colossal, the financial investments so great, and the projected times so immense, with plans often drawn up on scales not only of years and decades but centuries, that the issue will undoubtedly be decided by many factors in addition to those of technological advance and redirection. Even so, such forces can influence one another in unexpected ways and perhaps hasten the era of deep mining. Indeed, it is possible that the technological stir has already served as something of a political catalyst.[40]

DECADES OF CONFLICT over the fate of more than half the Earth, its last great wilderness, came to an anticlimactic end on November 16, 1994, as the United Nations Convention on the Law of the Sea went into effect. Its activation was triggered a year earlier when Guyana, a small coastal state atop South America, became the sixtieth nation to ratify the treaty. To industrialized states, the defects of the treaty were lessened somewhat as the Clinton Administration launched a last-minute diplomatic drive to dilute the socialistic clauses on seabed mining. At an international ceremony celebrating the treaty's activation in Kingston, Jamaica, there was much backslapping and mutual congratulations on a job well done. UN Secretary General Boutros Boutros-Ghali hailed the treaty as one of the century's greatest achievements in the field of international diplomacy. "The convention is a repudiation of the pursuit of progress through a competition for spoils," he told hundreds of delegates. "It is a strong disavowal of the tactics of gun-boat development. This is a historic day."[41]

In fact, as is often the case with well-intentioned acts, the long process of negotiation produced not only legal confusion but a jungle of unforeseen troubles. In a prominent one, coastal states around the world were arming themselves to the teeth to defend their EEZs. In early 1995, Indonesia, fresh from purchasing thirty-nine old East German warships, declared that it needed hundreds more to guard the country's eastern waters. After all, its new maritime claims overlapped those of neighboring Australia, whose own size more than doubled when, in concert with the sea-law activation of November 1994, it too declared an EEZ. For states seeking to bolster their sea powers, the *International Defense Review* in 1995 ran an article highlighting the latest in warships, aircraft, helicopters, radars, radios, machine guns, and advanced arms for antiship and antisubmarine warfare. Of course, declaration of an EEZ gave a nation serious duties to police its new waters against smuggling, terrorism, piracy, and other incivilities, and gave it an incentive to maintain the area's health for fishing and other economic activities. But it also abetted the unsavory aspects of sovereignty, opening a Pandora's box of potential bullying, border disputes, and open warfare. Some effects were more subtle. Oceanographers, once free to roam the seas, found themselves shut out of many EEZs. "We are gradually being excluded from areas where oceanographic research is extraordinarily important," said C. Barry Raleigh, director of Columbia University's Lamont-Doherty Earth Observatory. "We simply cannot work" in what remains of the high seas and hope to "understand something as complex and interrelated as the oceans."[42]

In its wisdom, the United Nations gave up potential control over 42 percent of the world's oceans, encouraging a process by which this area was turned into national annexes. This is a fair bit of planetary real estate, roughly equal to the world's total land mass. The UN fostered this transformation in hopes of gaining support for the sea covenant from coastal states (the majority of the Earth's nations), which might otherwise have lost their offshore areas to a global commons. The tactic worked. But what the UN and its global constituency got in return is murky, and may remain so for decades.

Based in Jamaica, the UN International Seabed Authority, and its mining arm, the Enterprise, are ostensibly the means by which poor countries get a piece of the seabed action. But this legal and bureaucratic labyrinth could just as well keep private investors from ever taking the plunge, despite the last-minute softening of some provisions. Wesley Scholz, a State Department official who led the United States delegation to the first meeting of the Authority, warned delegates of the ambiguity and dangers inherent in their charter. Broad support was not guaranteed, he said, but would depend on whether the Authority adhered to free-market principles, gave mining interests a voice in its deliberations, and tailored its rules and regulations in response to commercial interests in deep mining.[43]

In some analyses, the heavy hand of the United Nations might simply make seabed exploitation too burdensome to bear economically and encourage the pursuit of alternatives, including recycling, conservation, metals substitution, and more terrestrial mining. The history of cobalt pricing shows how markets can be quite flexible if the need arises. In 1978, when rebels invaded Zaire's mineral-rich Shaba province, whose mines produce up to 60 percent of the world's cobalt, prices for the rare metal shot up to nearly seven times their previous levels. That boosted cobalt production in neighboring Zambia. It also encouraged global conservation and substitution. Nickel can often replace cobalt, especially if it suddenly costs less. Pratt & Whitney did exactly that kind of substitution to make some of its jet engines during the cobalt price rise. Around the world, cobalt consumption dropped sharply, suggesting considerable elasticity in demand. In a similar way, global markets might ultimately judge the costs of seabed mining too high, prompting more reliance on marginal land sources and clever ways to get around the need for rare metals altogether.[44]

On the other hand, perhaps the ostensible jurisdiction of the United Nations will be ignored, as the Reagan Administration sought to do by negotiating its own set of bilateral accords with developed nations interested in mining the sea. Or perhaps the UN's approach will be so enlightened that would-be miners will obey its rules while working in deep international waters. Even if socialistic and onerous in application, the rules may be followed by nations that are motivated by forces other than the most fundamental ones of the market. Industrial states with few mineral resources may want to develop their own supplies from the seabed as a hedge against market shortages or as leverage in bargaining with terrestrial suppliers. At a minimum, physically small, high-technology states with few terrestrial mineral resources, such as Germany and Japan, and perhaps Israel as well, are likely candidates for such activity.

Some of these political dynamics were hinted at when the 1994 activation of the sea treaty was ignored by such mineral-rich states as Canada, which refused to ratify the document, and was carefully attended to by Japan, which sang its praises. "The agreement provides an economically sound and viable framework which will improve the climate for investing in deep seabed mining," Hisashi Owada, Japan's permanent representative to the United Nations, told the Kingston delegates. Going beyond what the occasion called for by way of diplomacy, Owada added that his nation relied heavily on imports of manganese, cobalt, and nickel and was eager to diversify its supplies. "The development of deep seabed mineral resources," Owada stressed, "is of considerable importance to Japan."[45]

Then, too, the United Nations push could have the unintended consequence of strengthening the hand of coastal states exercising control of the deep seabed outside the UN's ostensible jurisdiction. Miners might see these waters as oases free of legal ambiguity and international strife, their rulers perhaps possessing a pro-business attitude.

The Cook Islands, the tiny paradise with an EEZ the size of Argentina, in late 1995 went on an advertising binge to communicate just that image. Located on the isle of Rarotonga, the national government of Sir Geoffrey Henry, prime minister of the Cook Islands, as well as his cabinet, were promoted in magazine articles as cheery alternatives to the United Nations' dim bureaucracy, eager to share their bounty of manganese nodules with the world. The feat was advertised as increasingly doable because of advances in technology and cold-war spinoffs. The leaders of the Cook Islands, a government official wrote in conclusion, "are inviting international consortia to stop their interminable planning and start harvesting." In seeking to woo investors, the government estimated that the treasure in its deep waters had a market value of at least $1,135 billion—a fair sum in paradise, or anywhere else for that matter.[46]

IT IS NO SMALL irony that the greatest excitement to date in undersea mining centers not on deep minerals—the gold of the United Nations, the munitions of the Reagan warriors, the resource that as of 1994 is ostensibly regulated on the high seas under international law—but on something beyond these beguilements and strictures. It is the mining of life, and life on a lilliputian scale at that.

Quietly, relentlessly, the explorers of the watery unknown are gathering up unusual microbes in the volcanic deeps and in so doing are speeding the genetic revolution and making some industrialists rich. By weight, these single-cell organisms are worth far more than gold. The mining of deep life was never anticipated in all the international hubbub over the divvying up of the sea's mineral wealth. Even scientists were slow to catch on. Yet the biological harvest is now happening at an increasingly rapid pace around the globe, aided by a number of the cold-war spinoffs. Overall, the action has implications not only for economic growth but, perhaps most importantly, for humanity's emerging role as an architect of life, a job that will surely test the depth of human wisdom.[47]

My eyes—and hands, as it turned out—were opened to this new industry by our expedition to the Juan de Fuca Ridge in general and by my own *Alvin* dive in particular. Early in the expedition I found myself drafted to aid the miners of life by reconditioning the titanium bottles that during explorations of the hot vents were used to gather up waters teeming with microbes. The quart bottles worked like syringes. In the deep, lashed to *Alvin*'s front end, activated when the time was right, they drew in water from the long sampling arm. The force that pushed their pistons back came from thick coiled springs. Each night after the day's dive, the bottles had to be taken apart, washed, dried, and carefully greased—otherwise they had a tendency to freeze up in the deep because of the enormous pressures. That was my job. Held together by some two dozen small screws, banded in places with gaskets to ease motion, they were quite intricate and, for better or worse, I often cleaned them in a daze late at night, only half awake in the ship's cluttered central lab. Happily, all the bottles that I reconditioned worked just fine. (My supervisor, Eric Olson of the University of Washington, jokingly offered me a job as a technician.) I was especially pleased during my own *Alvin* dive when the four bottles we carried down to the depth of a mile and a half succeeded in collecting the hottest waters discovered on our expedition.

Samples of all the hot fluids we gathered found their way into the shipboard laboratory of Jim Holden, the graduate student who worked tirelessly to isolate heat-loving microbes. Rare bugs, he told me later, were discovered in not only the 543-degree waters of Church but the 65-degree waters of the Floc site as well, probably coming there from hotter areas deep beneath the ocean floor more favorable to their growth. In all, he said, more than a dozen heat-loving microbes were uncovered.[48]

Microbes able to withstand very high temperatures, to flourish at them, to break all the rules of life, had turned out to be a treasure trove for genetic engineering. The main allure was their enzymes, which, like the organisms themselves, work at temperatures that are extraordinarily high. Enzymes are biological catalysts. They are the brokers in thousands of chemical reactions within the cells of all living things. Without them, there would be no chemistry of life. Over the years scientists had slowly transformed enzymes from natural wonders into important tools for rearranging nature, doing so increasingly with the aid of deep microbes. Enzymes from these creatures easily survive the high heats of biochemistry, unlike many conventional ones that break down relatively quickly and have to be replaced. That saves money and speeds up processes. Also, they allow some reactions to run hotter and faster, saving time and advancing new kinds of biochemistry. Most important, they allow biochemical reactions to be conducted at very high temperatures, a step that automatically kills off most terrestrial bacteria and immobilizes their constituent parts, helping ensure that genetic concoctions are pure. For these and other reasons, companies around the world are racing to isolate, clone, and sell the extremely heat-stable enzymes of deep microbes, giving biotechnology and other industries a major lift.

As I investigated the field after my voyage, I found that Jim and his adviser at the University of Washington, John A. Baross, an author of the vent-creation theory, were important figures in such work. Indeed, the microbes we collected from the Pacific seafloor might one day yield important tools by which scientists would vie to supersede the artistry of nature. As Baross told me:

> We're like kids in a candy shop. With biotechnology, we're just scratching the surface. The food and pharmaceutical industries are also starting to get into it, particularly with enzymes that modify sugars. These organisms have the potential to do lots of remarkable things, such as degrading toxic wastes. All sorts of breakthroughs are possible.[49]

The path to this biological gold rush was anything but straight. As is often the case in science, the journey was full of serendipitous twists and turns that occurred over the course of decades.

A first tentative step was taken by Thomas D. Brock, a microbiologist at Indiana University, who in 1966 paid a visit to Yellowstone National Park, home of the celebrated geyser known as Old Faithful. He was looking for uncommon microbes. While sampling hot waters in a woodsy, out-of-the-way area, alert for grizzly bears, Brock was surprised to find bacteria thriving at temperatures that were thought to be anathema to life. He named one of them *Thermus aquaticus* (Taq for short), which flourished at 70 degrees Celsius or 158 degrees Fahrenheit—at the time, an unheard-of temperature for living things. Though noteworthy, the find was a curiosity of no foreseeable practical value. Brock sent a microbial sample to the American Type Culture Collection in Maryland, a bank for such things, so that other researchers might have a chance to investigate the microbe.[50]

Many years later the curio was seized upon by the Cetus Corporation and a scientist there by the name of Kary B. Mullis, who was searching for high-temperature enzymes. In 1983, while driving along a moonlit road in the mountains of northern California, Mullis thought up a clever way to artificially duplicate DNA, or deoxyribonucleic acid, the delicate strands of genetic material that reside in the cells of most living things and control heredity as the blueprint of life. The duplication method centered on DNA polymerase. In nature, this enzyme takes a single strand of DNA and doubles it into the normal twin helix by attaching the right chemical substances in the right order, much like building a zipper by using one row of teeth as the template for the other. The enzyme is fundamental to cellular duplication and reproduction in general.

That moonlit night while driving through redwood country, Mullis hit upon a simple way to greatly amplify this process, allowing tiny smidgens of DNA to be multiplied a billionfold. The method worked by boiling DNA in a test tube to unglue its double strands, then cooling the mixture and adding

DNA polymerase so that each strand would be copied. Each hot-cold cycle doubled the amount of genetic material. Thus, the multiplication of strands had the potential to be exponential—two, four, eight, sixteen, thirty-two, sixty-four, and so on up to millions and billions of copies.

The power of the process was its repetitive nature. But therein lay the rub as well. Testing in the laboratory showed that heat destroyed the DNA polymerase (which came from *Escherichia coli,* the common bacteria of the human intestinal tract), so more of the enzyme had to be added to a test tube after each duplication cycle, slowing things down and driving up the cost. It was in this context that Cetus and Mullis seized upon *Thermus aquaticus,* taking it apart for its DNA polymerase. Unlike other enzymes, Taq polymerase retained its catalytic power through many hot-cold cycles. Suddenly, what once took days or weeks could be done in an afternoon, and cheaply. The Yellowstone microbe allowed a single strand of DNA to be copied a hundred billion times in a few hours.

It is hard to overstate the importance of this discovery. DNA is microscopic and notoriously hard to manipulate. The Mullis process, known as the Polymerase Chain Reaction, or PCR, changed all that. For the first time, scientists needed only a tiny scrap of DNA to get sufficient material to begin analyses and manipulations. The wisp of genetic material could come from a human hair, a tissue specimen, an Egyptian mummy, or a woolly mammoth frozen in a glacier. In the popular novel and film *Jurassic Park,* researchers used PCR to re-create dinosaurs. In real life, the method is playing an increasingly prominent role in murder trials, allowing prosecutors to build cases around the DNA found in blood at crime scenes. It is also setting free the unjustly accused and convicted.

More broadly, PCR is turning dusty museums into hotbeds of inquiry as ancient DNA is analyzed for the first time, illuminating the dim evolutionary past. Among its diagnostic feats, PCR paved the way for the prenatal diagnosis of sickle-cell anemia. Perhaps most important, it made possible the titanic labors of the Human Genome Project, a Federal undertaking that promises to map all human genes and to help cure many ills. In short, by multiplying tiny dabs of DNA into amounts large enough for manipulation, by revealing molecular chemistry that would otherwise be invisible, PCR changed the face of biology and medicine. Its importance was acknowledged in 1993 when Mullis won a Nobel Prize for the insight that materialized during that moonlit drive.[51]

The hunt for enzymes superior to Taq polymerase got underway even before Mullis announced PCR to the world in 1985. After all, it too broke down during the hot-cold cycles, though far more slowly than normal polymerases. *Thermus aquaticus* flourished at 158 degrees Fahrenheit. It stood to reason that microbes that withstood even higher temperatures would yield more robust enzymes, and nowhere were such organisms being discovered more abundantly than in the deep sea.

The reason had to do with pressure. At sea level, water never gets very hot (212 degrees Fahrenheit is about the maximum) because the phenomenon of boiling automatically keeps things cool. The phase transition known as boiling occurs whenever a substance turns from a liquid into a gas. Added heat goes into inciting more molecules to break free of electrostatic bonds rather than raising the liquid's temperature. But things are different in the volcanic domains of the deep sea, where gargantuan pressures allow no boiling. Water gets superheated. Moreover, the great heats have none of the violence associated with boiling, which kills many microorganisms. The vents are hot and relatively peaceful and, as it turned out, aswarm with all kinds of exotic microbes.

From a hot spring more than a mile deep in the Gulf of California, amid dense thickets of tube worms thriving in otherworldly darkness, Holger W. Jannasch, a microbiologist at Woods Hole, in 1988 isolated an Archaeal hyperthermophile of the *Pyrococcus* genus. The microbe was obviously special. It grew at temperatures of up to 104 degrees Celsius and could withstand much higher heats for short periods of time. New England Biolabs, Inc., of Beverly, Massachusetts, took the microbe, isolated its DNA polymerase, cloned it, and then sold the enzyme, beginning in December 1991. It was the first time a deep-sea microbe had been brought to market. Appropriately enough, the trade name of the DNA polymerase was Deep Vent. "Thermostability, Fidelity & Versatility from the Ocean Depths," read one of the company's ads. Deep Vent moved the field forward in important ways. At a temperature of ninety-five degrees Celsius the half-life of its catalytic powers was fourteen times as great as that of Taq polymerase, and at one hundred degrees Celsius its half-life was a startling eighty times as great. And though working longer and harder, allowing more automation of PCR amplifications and exotic kinds of process, it made fewer copying errors. Perhaps most important, its ability to work at high temperatures helped biochemists attain new levels of purity as a range of terrestrial contaminants got knocked out of action. Last but not least, Deep Vent and its kind helped break an industry monopoly and bring down prices, aiding the spread of the amplification technique.

PCR and Taq polymerase are covered by patents won by Cetus and sold as part of a $300 million deal in 1991 to Hoffmann–La Roche, Inc., the Swiss pharmaceutical giant. In theory, all scientists must buy their PCR enzymes, at rather high prices, from Roche's licensees. But in practice, dozens of companies around the world sell Taq polymerase and other enzymes for the amplification job. Conspicuously absent from their ads and literature, however, is any mention of PCR. The use is implied. The rationale for such sharp business practices, usually voiced in private, is that a PCR patent should never have been issued. It was like trying to license the microbes that ferment alcoholic beverages or leaven dough, the reasoning goes. The comeback is that it took ingenuity to dream up PCR, and that a tenet of modern scientific life is that such creativity should be rewarded financially.[52]

In exploring the field, I found that other companies were racing to extend the work of New England Biolabs, capturing and dissecting deep hyperthermophiles so their unique enzymes and other components could be brought to market. Stratagene, a major biotechnology company based in La Jolla, California, as of 1993 had already cloned bacteria from the Juan de Fuca Ridge and was searching for others, partly in collaboration with the University of Washington. "It's a growing area with lots of potential," Eric J. Mathur, director of the company's high-temperature laboratory, told me. The United States Biochemical Corporation, in Cleveland, was working with the University of Maryland's Center of Marine Biotechnology, in Baltimore, to gather abyssal microbes from such places as the Sea of Japan and deep waters off Iceland.

"The field is untapped," said Vincent Kazmer, a senior vice president at United States Biochemical. "The annual enzyme market might be $600 million. Just think about replacing chemical catalysts with these kinds of enzymes. There you start talking billions." Jannasch of Woods Hole, who helped found the field of deep-enzyme retrieval, said the raw potential was great since probably only a tiny fraction of the existing hyperthermophiles had so far been found. "Every time we go to sea we isolate new ones," he told me. "We get surprises all the time. The biology and biophysics of these organisms is completely different, which is very exciting." Commercially, Jannasch added, the field is "already a big deal." He said the utilization of hyperthermophiles was becoming a major source of income for biotechnology companies, estimating that more than a dozen firms around the world were scrutinizing or exploiting microbes from the deep, and that many others planned to do so.[53]

Today the mining of microbes is done routinely around the globe as part of most investigations of the deep volcanic fields, no matter what the sea or explorer. For instance, the British during their 1994 *Keldysh* expedition to the Mid-Atlantic Ridge gathered up microbes during *Mir* dives to two different sets of hot fields, TAG and Broken Spur, an unfamiliar site that lies about 150 miles northward. A main goal of the government's Natural Environment Research Council, which funded the dives, is to aid the development of Britain's biotechnology industry.[54]

So, too, the Japanese gave high priority to capturing hot fluids during their 1994 exploration of the TAG mound. Diving repeatedly through more than two miles of seawater in *Shinkai 6500,* Japanese scientists sampled the mound's microbe-filled fluids with a variety of instruments and syringes, capturing water as hot as 317 degrees Celsius, or 603 degrees Fahrenheit. During the series of fifteen dives, the Japanese also sampled bottom sediments, which can also harbor unknown organisms.[55]

Soon after, the unprecedented step of drilling the mound was also accompanied by a search for rare microbes. In late 1995, as the *Resolution* dug its drill bit deep into the heart of TAG's blistering-hot mass, an American biologist was aboard the ship, ready to tease hyperthermophilic life from the rocky samples hauled up from the deep. "There's a lot of companies interested" in obtaining the rare organisms, Anna-Louise Reysenbach, a microbiologist at Indiana University in Indianapolis, told me before she left on the expedition. "We could find anything. It's very cutting edge."[56]

In this business, the United States has a unique advantage by virtue of the Navy's global network of undersea microphones, a cold-war spinoff now being used to track the low-frequency vibrations made by erupting volcanic rifts and deep volcanoes. It is a stethoscope listening to the planetary heartbeat, in the process revealing where new kinds of hot microbes are erupting out of the icy seabed. By definition, sudden flows of lava and hot water from the planet's interior are seen as harboring a greater diversity of microbial life than the steady and well-sampled flows of such established sites as TAG. So Federal scientists are targeting them, seeing them as incubators for new kinds of life as well as biotechnology innovation. "These episodic events are a way into the diversity," Steve Hammond, who heads the NOAA program that monitors the volcanic ridges, told me. "They're windows into the biosphere."[57]

Early in 1996, the Navy's deep microphones revealed hundreds of seaquakes shaking the seabed ninety-five miles off the coast of Oregon. The upheaval zone, about two miles deep, was along the Gorda Ridge, the same region that the Reagan Administration had targeted for deep mining. Battling foul weather, the *McArthur,* a 175-foot NOAA research ship, raced to the scene and lowered sensitive detectors into the icy depths. On March 10, scientists discovered a huge plume of warm water that wafted over the volcanic rent like a smoky cloud six miles wide. Lowering bottles on long lines, the experts scooped up water and tiny, heat-loving microbes from the plume and took them back to the University of Washington in Seattle, where Baross eagerly studied the exotic organisms. "This is the soonest we've been able to get to an eruption site to do microbiology," he told me. During the summer of 1996, a small fleet of vessels explored the eruption zone on the Gorda Ridge, looking, among other things, for new types of biological novelty.[58]

With few exceptions, the microbial gold rush in the United States is driven by the entrepreneurial fire of such firms as Stratagene, New England Biolabs, and United States Biochemical, with interest rising fast among such giants as Pfizer, Monsanto, Eli Lilly, and Du Pont. The Federal establishment in Washington funds expeditions and makes available such gear as the Navy's undersea microphones. The surge of cold-war gear is generally speeding the harvesting work. But, beyond that, the government's microbial role has been largely symbolic, providing moral support rather than dollars. No new Federal labs or facilities have been built. It was considered big news in 1993 when United States Biochemical won a $1.5 million grant from the Commerce Department to speed its high-temperature enzyme work.[59]

By contrast, Tokyo has put substantial money into such microbial work, anticipating that it might aid a twenty-first-century boom in making new drugs and new tools for genetic engineering. Perhaps this strategy will prove wrong, as did the early push to mine deep metals. Japan is far from immune to bureaucratic error. But American scientists look upon the Japanese work with admiration and some envy, having failed to persuade their own government to make similar investments.

The centerpiece of the Japanese effort is a $43 million project known as Deepstar, an advanced plant that can mimic on land the extreme pressures and temperatures found in the sea's depths. The aim is to have an incubator for breeding unique microbes brought back to land by such Japanese craft as *Shinkai 2000* and *Shinkai 6500.* The project is something of a gamble. No one is sure what percentage of the microbes gathered in the deep at pressures hundreds of times greater than those at the surface die during the trip up to the usual terrestrial environment of one atmosphere pressure. Scientists only know about the microbes they successfully culture in their laboratories, such as the hyperthermophile of the *Pyrococcus* genus, whose cloned enzymes are sold as Deep Vent. The guess is that only a small percent of the deep microbes survive the changes in light, temperature, and pressure. Scientists say high compressions may counteract some effects of heat, stabilizing enzymes and increasing their activities by promoting processes that reduce volumes. In short, the field is full of conjecture and short on facts.[60]

Rising six stories above the industrial landscape at the port city of Yokosuka, at the southern terminus of Tokyo Bay, is a new research laboratory where Deepstar is forging ahead. With the usual Japanese penchant for big plans, it is a fifteen-year project. At the nearby dock, ships of the Japanese deep-research fleet unload pressure vessels containing samples of deep water and sediment. Then, in the lab's automated incubators, the microbes are cultured and isolated at pressures from one atmosphere to hundreds of atmospheres, at temperatures from below freezing to above boiling, and at chemical conditions from acid to alkaline. Laser beams fired into the incubators light up the microbes to reveal levels of growth. The welter of vessels can, at a minimum, create pressures of 650 atmospheres, mimicking depths of 6,500 meters or 6.5 kilometers (pressure goes up by one atmosphere for every ten meters of water), the maximum depth of *Shinkai 6500.* And Japanese scientists have hinted that ultimately it will produce much higher pressures, such as those found in the deepest ocean trenches, seven miles down.[61]

"People said I was crazy when I asked for one thousand atmospheres," said Koki Horikoshi, director of the Deepstar project. "The engineers said they had never done anything like it before."

But microbiologists see nothing crazy about such a goal. They understand. The infinitesimal bits of protoplasm that dwell in the oceanic darkness at heats and pressures that are hard to comprehend, undisturbed for eons, perhaps direct descendants of the planet's first life, are intrinsically interesting and may prove to have commercial and medical uses that no one can now foresee, much as the curious microbe that Brock discovered living in a Yellowstone hot spring helped found a new industry.

For one, Horikoshi is ready to explore this deep world and make a few waves, if he can. "Our research institute," said the director with unadulterated pride, "will be the best."[62]

THE BIOLOGICAL MINING of the deep sea involves more than microscopic dabs of life. To date, the biggest gold rush of all involves fish and crustaceans and other creatures that live in icy darkness but in growing numbers are finding their way to the table. Commercial fishermen are discovering, often to their surprise, that they are able to catch edible fare from the depths by the ton, prompting the founding of whole new industries and the hauling in of millions of dollars in profits, if not billions by this point. No global agency is keeping records.

True, the creatures of the deep are frequently unattractive by human standards and thus, when sold in stores and restaurants, are often carefully prepared and marketed to conceal their origins and unappealing looks. No slimehead is ever served to the public under that name. And many creatures of the deep are made less scary by the removal of their heads. Going further, some merchants cut away nearly all the body parts so the public sees only meaty fillets, not what engendered them.

In spite of the marketing hurdles, the industry of deep fishing is showing unmistakable signs of growth. Unlike the microbial hunt, it seeks out creatures that are widely distributed throughout the deep and relatively easy to catch. Moreover, the technology of retrieval can be quite simple, usually involving nothing more complicated than a baglike net or trawl for scooping up deep fauna, though cold-war spinoffs are making the job easier to pursue over extensive areas. In general, however, the enterprise requires no special skills other than the ones fishermen have practiced for centuries.[63]

Seafoods traditionally have come from sunlit zones. Working relatively close to shore, fishermen at first focused on harvesting the shallow seas and continental shelves that teem with life. As ships of the nineteenth and twentieth centuries improved, seamen broadened their hunt to the less-productive waters of the open ocean. But whether near or far from shore, fishermen typically pulled their nets through the lighted realms near the sea's surface. That was the mother lode. That was the place that sheltered the sea's living wealth, a bounty until recently thought of as inexhaustible.

Today, the search for deep fare is driven in part by the collapse of such shallow fisheries as the Grand Banks off Newfoundland, where fishing wars and the push for short-term profit have crippled productivity and driven such popular species as cod and haddock to the verge of commercial extinction. Worldwide, after centuries of steady growth, the total catch of wild fish peaked in 1989 and has subsequently declined. Foraging deeper is sometimes a survival strategy for fishermen forced out of traditional grounds by government regulators and pressed by creditors to make payments on costly boats.[64]

Around the globe, deep creatures now exploited or targeted as seafood include hoki, ling, skate, slimehead, sablefish, red crab, oreo dory, blue whiting, black scabbard, spiny dogfish, and rattail fish. The field is so unscrutinized and so uncoordinated that no reliable figures exist on the overall size of the global catch from the deep, though anecdotal evidence suggests it is growing rapidly. In the future, some government studies have suggested, aggressive moves to deep fishing could double the size of the global take, so the world's current annual catch of roughly one hundred million metric tons might rise to two hundred million metric tons. Whatever the actual numbers turn out to be, the issue, as David Packard suggested, seems destined to grow in importance in the decades ahead.[65]

The gathering of deep seafood, its advocates say, can be done with ecological care, and the results can be quite tasty. Many chefs, fishermen, boat makers, development experts, and pro-business government agencies back the practice. But ecologists worry that the rush could upset the rhythms of the sunless regions, which are poorly understood, threatening to tip a biological balance. Little is known about deep creatures, they emphasize, and what is known often bodes ill. For instance, all fish are cold-blooded. Their metabolisms by nature are more readily influenced by their environment than is the case with mammals and their warm bodies. Thus, fish that inhabit the icy depths are often studies in slowness, growing and reproducing at a rate that is often remarkably unhurried, a fact that makes their populations particularly vulnerable to disruption by man.

"Deep fisheries can provide a pulse of good fishing but they're often not sustainable," Jack Sobel, a scientist at the Center for Marine Conservation, a private group in Washington, told me. Bruce Morehead, an official of the National Marine Fisheries Service, said his agency was aware of the risks and was working closely with commercial fishermen to strike a balance between exploitation and conservation. "If you don't accurately assess the stocks," he conceded, "you can accidentally kill them off." And some scientists would argue that accurate assessment is presently difficult if not impossible, given the difficulties of monitoring the deep.[66]

Peter J. Auster, science director of the National Undersea Research Center at the University of Connecticut in Avery Point, told me that moderation in the new field is essential to avoiding trouble. "Some of these fisheries might be sustainable with a five-boat fleet but would be rapidly depleted if fifteen or twenty boats decided to go after them," he said. The appeal of deep fishing, Auster added, was quite understandable. "For the individual fisherman trying to pay off the boat or put his kid through school, and facing draconian rules to reduce pressure on overexploited stocks, this is a wide-open area. It's a place to go and use your wits and knowledge to try to make a living, which is what fishermen have been doing for centuries."[67]

Around the globe, deep fishing has been practiced on a limited basis for decades. But new sciences and technologies are rapidly making it more practical and efficient, even as its attraction grows because of the worldwide depletion of shallow stocks. Deep hauls that once took days or hours can now be done in minutes. From the beginning, the main tool has been a long steel line hauling a stout net or trawl, which can be deployed to cut through dark midwaters or across the bottom, its way over difficult terrain eased by rollers and wheels. Today the art has advanced widely. Ships are bigger and faster. Lines are longer, thinner, and stronger. Winches have undergone radical improvement. Their winding drums are wider, their engines more powerful, their controls often computerized, significantly easing the process of manipulating long lines and making it much safer. Most important for the dimensions of the catch, nets and trawls have grown so greatly in size that today their maws are often hundreds of feet wide, allowing them to scoop up many thousands of animals at a single pass.

Aiding the hunt in a more general way are former military technologies that have been developed in the past fifty years, including radars that let boats navigate in dense fogs, sonars that locate deep schools with great precision, and navigation satellites that pinpoint geographic coordinates so vessels can return to rich sites. The upshot of all these factors is that the bottoms of many continental shelves and shallow seas have been scraped repeatedly in the quest for deep seafood. Increasingly, the action is moving off the relatively shallow continental shelves into the inky depths.

Cold-war spinoffs are accelerating this trend. Fishermen recently began pinpointing zones of deep richness by studying seafloor maps based on formerly secret military data, which can reveal secluded habitats. For instance, the one made public in late 1995 by the National Oceanic and Atmospheric Administration instantly doubled the number of publicly known seamounts in the global deep, which are associated with upwellings of nutrient-rich water and whose flanks often harbor swarms of sea life that are remarkably dense. In a snap, the global count went from roughly six thousand to twelve thousand seamounts, giving deep fishermen vast new territories to explore. This windfall will probably take the industry decades to exploit, perhaps giving scientists time to assess the ecologic implications of the opening.[68]

Surprisingly, few people suspected the existence of the deep hordes. Decades of scientific dredging had suggested that the icy seabed was home mainly to small creatures that, though legion, were relatively feeble. Not so. Between 1968 and 1975, scientists at Scripps repeatedly lowered onto the deep seabed lures baited with dead fish and automatically photographed them every few minutes for up to two days. The fuss usually began a few minutes after touchdown as shrimps, brittle stars, and small fish tore at the bait. It picked up to include swarms of hagfish and other large animals fighting furiously for the prize. And it ended abruptly when some big creature, usually a shark, frightened off the competition and wolfed down whatever was left. The sharks were too big to be seen in their entirety. But the scientists, judging from photographic clues, estimated their length at more than twenty feet, or longer than some trucks.[69]

The commercial harvesting of deep creatures on a regular basis got underway in the late 1960s, just as science was starting to comprehend the density. In the North Atlantic, the Russians pioneered the catch of rattails, which are apparently ubiquitous in the sunless depths of the sea. Like some kind of mythological beast made of dissimilar parts, the rattail has a bulbous head and huge eyes, while the rear part of its body narrows down to become curiously thin. There is no hint of the usual caudal fin found on most fishes. The coloration of the rattail is drab, often brown or ashy gray. But the fish makes up for its dull cast by undulating back and forth in a serpentine motion that, once seen, is never forgotten. Of all the curiosities hauled up from the depths by the *Galathea* expedition, it was the rattail that drew the most attention in ports, with newsmen repeatedly photographing the specimens, fascinated by their appearance. The rattail grows more than three feet long. Its size and prevalence make it a tempting commercial target, as does its vitamin-rich liver and flavor, if reports are to be believed. The rattail is often called a grenadier in an effort to better its image.

The Russians appear to have readily welcomed the fish, problems and all. Operating mainly in the Northwest Atlantic off the Newfoundland coast, the Russian fleet began hauling up rattails on a commercial scale in 1968. The main targets were *Coryphaenoides rupestris,* also known as the rock or roundnose grenadier, which has a protruding, fleshy snout, and *Macrourus berglax,* or the roughhead grenadier, which has a prominent ridge that runs across its face from the tip of its nose to below its gills. The catch quickly peaked in 1971 with an extraordinary haul of 83,000 tons. Thereafter, the annual take steadily declined, shrinking to a paltry 4,000 tons by 1982. It is unclear whether this drop was the result of natural fluctuations or unbridled exploitation, although the latter explanation seems likely. The issue is difficult to address because scientists know so little about the rattail's growth, reproduction, movements, and lifespan, despite the heavy take by commercial fishermen over the decades. Like many creatures of the deep, the fish is assumed to grow slowly, mature late, and live an unusually long life.[70]

The practice spread in the 1970s to such species as deep-sea crabs, which grow up to a foot or more in length and look much like their shallow-water cousins. The crustacean was taken off the northeastern United States and southwest Africa. In the Pacific, the sablefish, or blackcod, was pursued at depths of up to a mile. Rather belatedly, scientists realized that it could live up to the age of seventy years, making its populations vulnerable to disruption. "The stock could become quickly depleted" in the absence of careful regulation, warned John D. Gage and Paul A. Tyler, authors of *Deep-Sea Biology,* a college text.[71]

Far and away the best seller from the depths has been orange roughy, a pug-nosed, big-mouthed predator that apparently feeds near the sea's bottom, perhaps up to a mile or so deep. Little is known of its habits. But its large eyes are apparently good at tracking luminescent krill and small squids. Its stomach often holds small fish as well, such as lantern fish, known for rows of glowing dots along their sides. This predator of the deep won its popular name from the color of its bright, reddish-orange body, which is quite meaty. Mature fish grow up to nearly two feet in length. Before roughy became a popular seafood, it was usually known by its unappealing family name, the slimehead, after the tendency of family members to bear a number of mucus-secreting cavities on their heads, especially around the eyes.

The rise of orange roughy to culinary fame began in a curiously roundabout way that illustrates the power of marine geopolitics. In 1978, New

Zealand aligned itself with United Nations strategies on the Law of the Sea by declaring an Exclusive Economic Zone. This remote nation of the South Pacific is made up of two large islands that are set close together and many smaller isles that are widely scattered across the sea. Each and every island of the archipelago, in standard EEZ fashion, exerts a territorial reach from shore that extends for a distance of two hundred nautical miles, spreading outward like an egg in a frying pan. So it was that the EEZ move instantly won New Zealand a marine area of about two million square miles—nearly ten times the size of France.

Eventually, as other nations made marine claims, New Zealand's EEZ emerged as the world's fourth largest. This watery expanse lay mostly beyond the continental shelf of the two main islands and, because of the state's outlying specks of land, extended quite a distance to the south and east. As it turned out, this area was ideal for deep fishing. The prize, lying due east, was Chatham Rise, a rocky plateau the size of Texas, which lies hidden up to a mile or so beneath the waves and drops off steeply on its sides. The rise and its flanks were found to swarm with deep sealife, including *Hoplostethus atlanticus,* a type of slimehead that until then, if known in New Zealand at all, was viewed mostly in natural-history museums and seen as a curiosity, dried and wrinkled and scary with its big mouth turned down in a permanent frown and its jaws boasting rows of tiny, sharp teeth.

Orange roughy was an immediate hit. The discovery of hordes of the animals in waters declared to be the exclusive domain of New Zealand threw the nation into a fit of development, with companies quickly building fleets for its catching and parallel operations for its global export and marketing. The culinary merit of roughy was mainly its mild flesh, which could easily be turned into thick, white, boneless fillets and substituted for nearly any other whitefish. The economic attraction was its vast numbers, which promised low prices and high profits. The only drawback for fishermen traveling out of New Zealand's easterly ports was the additional time and fuel it took to get to distant grounds and the extra cost of deep gear and boats big enough to weather the storms of the open sea.

Starting in 1978, concurrent with the EEZ declaration, trawlers from New Zealand fished for roughy on the flanks of Chatham Rise, raking the depths. Single ships sometimes caught a phenomenal fifty tons an hour. By trial and error, the fishermen found that the populations of roughy were densest and shallowest—coming as close to the surface as about half a mile, or three-quarters of a kilometer—during the summer months, as huge aggregations of the fish materialized along the edges of the deep plateau for the annual rite of spawning. Dozens of companies were set up to tap the reproductive frenzy. Factory ships were equipped with on-board freezers, pro-

ORANGE ROUGHY. Dense schools of orange roughy were fished with abandon off New Zealand until scientists discovered that the deep fishes could live more than one hundred years, their slow growth and reproduction raising the risk of commercial extinction.

cessing plants were set up on shore, and frozen fillets were exported in vast numbers, particularly to Japan, Canada, and the United States and their food-service industries.

The New Zealanders took in hundreds of millions of dollars in revenue. In America during the 1980s, roughy became a favorite as prices of cod and other traditional table fish began to soar. At $3.99 a pound (as low as $2.99 in some places), it was a cheap if bland alternative that was easy to dress up with a good sauce. And the health conscious liked its low fat content. It seems likely that few consumers, given the general disconnectedness of modern life and the smoothness of mass marketing, realized that their inexpensive dinners were the final act of a drama that began thousands of miles away in the sunless depths.[72]

At first, the gold rush around New Zealand was a free-for-all that was unregulated and largely undocumented. Then, beginning in the season of 1981 and 1982, government ministries in Wellington began a slow process of trying to regulate the size of the catch. The move was driven by worries that the boom might abruptly turn to bust and the pain of economic dislocation. Even so, the curbs were steadily loosened during the eighties as scientists observed no ill effects. Annual catches on Chatham Rise rose until they averaged 35,000 tons, and many of the annual landings were much larger, approaching the mass of the *Titanic,* which weighed 46,000 tons and was longer than a city block. In short, the haul was huge. The rule was saturation fishing, in which an area was repeatedly swept with deep trawls until few or no fishes were caught. Critics, including ecologists worried about the disruption of oceanic food chains, derided the practice as strip mining. But businessmen in many parts of the world admired the New Zealand success and sought to emulate it. In the North Atlantic, the French, Danes, and Icelanders began trawling the depths for roughy, struggling to keep the fertile zones secret from one another. The deep fishing, usually hundreds of miles offshore and outside any EEZ, was seen as an ideal way to escape tough governmental restrictions on shallow catches.[73]

Then came a surprise. It turned out that roughy, if it escaped the deep nets, could live to a remarkably old age. Hints of the animal's antiquity were gathered in the late eighties and early nineties by scientists in Wellington at the Ministry of Agriculture and Fisheries, who eventually found that the fish could live more than one hundred years, maturing sexually at around thirty years of age. Eventually, one Australian study put the age of roughy at up to 170 years. Whatever the precise figure, the fish was clearly a planetary elder. This news changed everything. Suddenly, the booming fishery around New Zealand looked quite fragile. The danger was that roughy, in growing and reproducing so slowly compared to traditional species, was at risk of commercial extinction because its stocks would keep dwindling rather than undergoing steady replenishment. By comparison, a fish like cod matures at three years of age and lives for about fifteen years. It quickly multiplies. In Wellington, ministry scientists began to draw up disaster scenarios. By their estimate, based on the sampling of deep stocks over several years, the virgin mass of the roughy fishery was about 400,000 tons. By the early nineties, however, the intense harvesting had whittled this down to about 80,000 tons, a fifth of the original size. In short, the energetic fishery was within a few years of collapse.

Agitation began for strict quotas. A sustainable level, ministry scientists estimated, would be about 7,500 tons a year. That would give a sufficient number of immature fish time to grow and reproduce, thereby ensuring the fishery's survival. But the commercial industry, having made large investments and taken large profits, fought the restrictions hard. So the politics of compromise prevailed. For Chatham Rise, the Total Allowable Commercial Catch, known as the TACC, ramped down slowly. The peak quota occurred during the 1988–1989 season, when the TACC was 38,300 tons. Thereafter, it ratcheted down during the early nineties until by 1994 it was 14,000 tons, nearly twice the recommended level.

Unhappy with the pace of reductions, the environmental group Greenpeace sued for regulatory speedup, arguing that the ministry was ignoring the advice of its own scientists. In November 1995, a high New Zealand court ruled that the fishery by law had to be managed at a sustainable level but declined to tamper with fishing quotas. No matter. The public agitation and negative publicity, combined with the continuing pressure to stave off collapse, worked to aid the ascendancy of wisdom. So it was that the TACC during the 1994 and 1995 season was lowered to 8,000 tons, just above the recommended level. As this tightening took place over half a decade, the iron laws of economics caused the price of roughy in America to soar, going from a low of $2.99 a pound up to $8.99 a pound. The cheap alternative became what for many consumers was an unaffordable luxury.[74]

"They never should have been exploited at all," Mike Hagler, a fisheries expert in Auckland, New Zealand, for Greenpeace International, told me. "People wouldn't eat rhinoceros or any other land creature that they knew was threatened by extinction. But they're eating fish like orange roughy without a clue to what's happening." The economic engine behind the fishery had no driver, he added, diminishing the chances that it could steer a reasonable course. "In deep water you get high costs, high technology, high finance, and consequently, the pressure is on to heavily exploit the stocks," Hagler said. "You're up against the wall, which pushes fisheries beyond the limits of economic sustainability."[75]

As the roughy fishery contracted to a tiny fraction of its former size on Chatham Rise, the industry responded by spreading out to search for new regions of the deep to trawl for roughy as well as for alternative deep species, such as ling and hoki. Both these fish are tapered and eel-like and vaguely resemble rattails. Retail chains found that hoki was ideal for fish and chips and other types of fried seafood, often substituting it for cod, while the Japanese turned hoki into surimi. Deep fishermen of the South Pacific also targeted the oreo dory, a creature that lives up to a kilometer or more down and previously had been thought of as a garbage fish and often discarded. It comes in four marketable varieties—black, smooth, spiky, and warty. These fishes, while sometimes sold as frozen fillets, also found their way into the more byzantine operations of the fast-food business. And on occasion they were fraudulently substituted for roughy and seized in the United States by law-enforcement authorities. As dory catches grew in size, the issue of age again arose, as with roughy. Scientists discovered that some oreo dories appeared to live more than a century, raising questions of how long the fishery would last.[76]

A worry of ecologists in all this is that deep fishermen have no idea how they are altering the complex food chains of the sunless regions—of who eats whom in a feast of life that is all but invisible. Such chains are fairly easy to understand on dry land, where the progressions are often short and readily observable. It is easy to see how a leaf is eaten by a caterpillar, which in turn is eaten by a bird, which in turn is eaten by a cat. But the ecology of the deep is mostly a riddle. No one understands the habits of the orange roughy, much less the nuances of how it fits into the larger ecological picture. By some estimates, the fish may be an important part of a food chain that ends with the giant squid, one of the deep's great wonders. If so, the decline of the roughy may threaten the giant squid as well.[77]

The rush for orange roughy was born of opportunity. That is less the case in other parts of the world, where the hunt for deep seafood is often a direct response to the closing of shallow fishing grounds.

In the United States, the investigation of deep fisheries is actively encouraged by the National Marine Fisheries Service, which aids such exploratory work with millions of dollars in grants. Off both coasts in deep water, off Alaska, and off such Pacific areas as Guam and the Marshall Islands, the fisheries service is helping industry hunt for shrimp, rattails, chimeras, sea cucumbers, orange roughy, smoothheads, slatjaw eels, blue hake, skates, and dogfish. In the case of the dogfish, the National Fisheries Institute, an industry group, has renamed it cape shark in an effort to improve its marketability. Federal officials see the push into deep waters as a potential boost for a failing industry.[78]

"We're excited because these may open up an avenue to try something new and relieve the pressure on cod, haddock, some of the flounders, and other species that are depleted at this point," Kenneth L. Beal, a manager in Gloucester, Massachusetts, with the northeast regional office of the National Marine Fisheries Service, told me.[79]

A pioneer in all this is Captain Bill Bomster, who with his three sons operates the one-hundred-foot *Patty Jo* out of Stonington, Connecticut, and regularly sells his catch of deep shrimp to the Daniel Packer Inn in nearby Mystic, where the "Stonington Reds" and "Royal Scarlets" are a hit. Some are the size of lobsters. Captain Bomster has plied the edge of the continental shelf from Block Island to Virginia, sweeping across the bottom with a trawl whose mouth is 3 feet high and 150 feet wide. The work is done at depths of up to a half mile.

"What's nice is that the water is very cold and exceptionally clean," he told me. "That makes the catch far superior to gulf or farm-raised or any other shrimp. Some of the damn things are ten inches long."[80]

• •

THE DEEP is undoubtedly cleaner than the shallows near land, where pollution from farms and cities and industries is often a problem and a major contaminant of seafoods. Even so, humans are having an increasing impact on the dark waters and the countless hordes that inhabit them as all kinds of toxins and contaminants reach into the deep, invisible and extremely hard for scientists to track and investigate. The cold-war spinoffs are easing the monitoring job somewhat. But it is still daunting. In all, mankind has managed to inject into the global environment at least seventy thousand synthetic chemicals, often with great ignorance of where they go and what they do. The issue of deep pollution is important not only because of its effects on the development of novel cuisines and the feeding of growing multitudes. In the end, the repercussions could bear on the health of the planet and all its inhabitants.

SEVEN

Tides

AS FATE WOULD HAVE IT, the first dumping of radioactive waste into the global sea began not in a barrens or a backwater but in what is indisputably a paradise. Often shrouded in fog, the Farallones Islands lie some thirty miles west of San Francisco amid one of the most productive fisheries on the West Coast. The area is not unlike Monterey Bay in the sheer extravagance of its wildlife. The rocky isles (the meaning of *Farallones* in Spanish) are home to thousands of seals and sea lions and birds, including cormorants and auklets. Save for Alaska, the jagged rocks support the nation's largest population of breeding seabirds. The surrounding waters teem with whales and porpoises and such fish as snapper, sole, salmon, striped bass, halibut, rockfish, sablefish, and herring, as well as such bottom dwellers as sea urchins, octopi, and abalone. Down below, just beyond the islands, the continental slope quickly falls away to depths of a mile or more. There fishes tolerant of cold, dark, and pressure snap up the leftovers falling from the feast above, competing with worms, crabs, brittle stars, and sea cucumbers. In all, more than one hundred different kinds of large invertebrates are known to inhabit the icy depths, an abundance in keeping with the region's general fertility.[1]

Added to this celebration of life, starting in 1946, were many thousands of barrels of radioactive waste, one of man's most deadly artifacts. The wastes were mainly leftovers from the unlocking of the atom during the Second World War and the making of the world's first atom bombs. The underlying rationale for ocean disposal was cost. No digging had to be done. No ditches, entombment, or fences were needed, as opposed to the storage of radioactive waste on land. The barrels simply and effortlessly sank into the sea, lost to the depths. And deep abandonment was judged to be safe. Isolated in deep water, far from humans, the wastes would slowly undergo radioactive decay and eventually be rendered harmless. Any leaks, it was reasoned, would be thinned to near nothingness by the global sea's seemingly limitless powers of dilution.

Why the Farallones? Out of all the conceivable oceanic dump sites on the nation's periphery, of all the wastelands, the deep waters west of these isles were chosen apparently out of expediency. They were close to naval facilities in San Francisco and they were close to the University of California at Berkeley, which played a leading role in the Manhattan Project and the running of the Los Alamos Laboratory in the mountains of New Mexico, the birthplace of the bomb. Eventually, the area also became a dumping ground for a Federal weapons center set up just outside San Francisco, which also designed nuclear arms, the Lawrence Livermore National Laboratory. The closeness of the dump bore directly on the issue of cost. It was more expensive and dangerous to ship the radioactive barrels long distances. Pitching them in the backyard took less effort and money.

Most of the waste destined for the Farallones was put in fifty-five-gallon oil drums, which were easy to obtain. Some were lined with concrete. Other drums had no embellishments at all, just their steel walls. The wastes included thorium, uranium, plutonium, cesium, strontium, and all manner of articles contaminated with radioactivity. Mostly it was low-level waste, though some was apparently considered high-level. No matter what the concentration, the material was noxious. Plutonium-239, a main fuel of nuclear weapons, has a half-life of 23,400 years and is so deadly that tiny specks invisible to the eye can cause cancer if they become lodged in soft tissues. Cesium-137 has a half-life of thirty years and strontium-90 has a half-life of twenty-eight years and a bad reputation because it binds readily with human bones. A half-life is the time it takes half of a radioactive substance to decay into atoms that are less complex and often less harmful to humans. In the Farallones, some of the wastes would be dangerous for thousands of years.[2]

After being dumped at sea, the drums sometimes refused to sink and instead would bob in the waves, threatening to drift away. So the Navy took action that was quick and practical, in line with the ethos of the day. It shot them full of holes. Water poured in and made the drums heavy, taking them down to the bottom, leaky but gone at last.

In all, more than 47,000 barrels of waste were sunk in the Farallones between 1946 and 1970. This quarter century of dumping is the longest period in which the United States continuously used a single ocean site for radioactive disposal. Experts knew that the barrels in the Farallones littered the continental slope from depths of about a half mile to a mile, spread out over hundreds of square miles. But the exact locations were a mystery, as were the conditions of the drums. They were basically out of sight, out of mind. In all, the drums at the time of their abandonment are believed to have held about fifteen thousand curies of radioactivity. A curie is the amount of radiation given off by one gram of radium and, in any nuclear material, is equal to the disintegration of 37 billion atoms per second. The nuclear accident at the Three Mile Island power plant in 1979 released about 50 curies. And the accident at the Chernobyl nuclear power plant in 1986 emitted about 50 million curies. That seems like a lot, and the Chernobyl accident took an enormous toll in terms of cancers and birth defects. But the truth is that most of the radiation was short-lived isotopes that decayed into less dangerous forms in a few months, not plutonium, with its slow decays that go on for millennia.[3]

The Farallones were the first but not the last of the nation's undersea dumps. All told, the United States disposed of radioactive wastes at twenty-nine major sites in the Atlantic and Pacific and lots of minor ones, for a total of about fifty areas. In step with growing worries over possible health effects, the dump sites over the years tended to be situated in waters that were increasingly deep. The hottest spot of all was located about 150 miles east of Delaware Bay in Atlantic depths of up to two miles. Dumped there were 14,300 barrels of radioactive waste as well as the reactor shell from the *Seawolf* nuclear submarine, which was commissioned in 1957. The *Seawolf,* the world's second nuclear sub, had an innovative reactor that reached high temperatures. But operational tests revealed safety problems and that style was quickly abandoned. Altogether in its dumping operations, the United States put into the sea about 100,000 curies of radiation, enough, if contact was direct, to maim or kill untold masses of people.[4]

America initiated the atomic age and a chain reaction of global emulation. So, too, it became a role model for disposal. The United Kingdom began dumping atomic wastes at sea in 1949; New Zealand in 1954; Japan in 1955; Belgium in 1960; France, Germany, and the Netherlands in 1967; South Korea in 1968; Italy, Sweden, and Switzerland in 1969. It was a major global trend, with no sign of where it would end.[5]

Efforts to understand the effects of the radioactivity on sea life were very slow to get started. In the Farallones, it was more than a decade after the start of dumping before Federal researchers did their first underwater tests of the site, sampling and analyzing the wildlife. Surveys in 1957 and 1960 found no obvious signs of radiation among animals. But a series of different tests undertaken in 1960 suggested for the first time the potential for trouble. In the tests, oil drums of the type that had been used to deposit wastes were slowly lowered to great depths and photographed every twelve seconds, revealing the crush of increasing pressure. Of ninety-nine steel drums examined, 35 percent showed deformations ranging from relatively minor dents to major implosions. Even intact ones, when carefully dissected, were often found to be moist or filled with water, suggesting that the radioactive contents of real waste drums would easily come in contact with the sea. Today it is unclear whether the problems were anticipated. But, once documented, they worked like acid to slowly eat away at aloof disinterestedness.[6]

The U.S. Environmental Protection Agency in the mid-1970s began to scrutinize the Farallones depths, its scientists, among other things, diving in *Alvin* to examine drums and sea life. It was an eye-opener. Robert S. Dyer in 1974 found fractures in about 25 percent of the barrels he was able to examine. Some of them were covered with sessile animals that relished hard surfaces in a wilderness of ooze. Dyer photographed a giant sponge of unidentified genus growing on a deteriorating drum, as well as crushed barrels surrounded by crabs, anemones, and deep fish swimming and rooting around in nearby sediments. EPA scientists concluded that bottom life at such dump sites was up to ten times denser than normal because the canisters made a kind of artificial habitat, just as a shipwreck in shallow waters can act as an artificial reef for swarms of fish.

The grim chain of logic was soon extended as EPA researchers found that some of the animals were radioactive. A biological survey in 1977 discovered the greatest dangers appeared to be associated with sessile filter feeders that resided permanently at the dump site owing to their immobile life style. "The sponge specimens are remarkable," scientists wrote, in that the creatures contained "readily measurable" amounts of plutonium-239 and plutonium-240, an isotope with a half-life of 6,600 years. Radiation levels were generally higher in sessile species than in fish, which presumably came and went from the dump area. Sponges had levels of plutonium-239 and plutonium-240 that exceeded the background by up to 80 times. A sea cucumber had levels of the same isotopes that exceeded the background by up to 227 times.

Overall, the analytic work was fairly crude, involving the sampling of only a few animals and the complete absence of control samples taken outside the dump site. Such controls are a standard part of scientific procedure and help rule out the embarrassment of false positives. Even so, the evidence supported the impression that the crushed, rusting barrels had delivered to the seabed and sea life a noxious legacy. Unaddressed was the question of whether the poisons could be transmitted up the food chain to man.[7]

Even before the emergence of such warning signs, the United States around 1970 ended its dumping of radioactive wastes, aware of potential health risks and growing public disenchantment with the practice. Here, too, the United States was a global leader, using its influence to engineer a policy reversal. The first steps were taken by the London Convention, a world body that regulates dumping at sea and has seventy member nations. In late 1972, it outlawed the disposal of high-level radioactive wastes, such as spent fuels from nuclear reactors laden with cesium and other deadly isotopes. This step was a formality since no nation had ever admitted to ditching such things at sea. More practically, the group also required that nations wishing to dump low-level wastes had to do so with sturdy containers at special sites where depths were greater than four kilometers, or two and a half miles. Such depths were seen as sufficient to provide a safety barrier for contamination to man. A decade later, in 1983, the group voted a voluntary moratorium on all radioactive dumping. With this step, all varieties of radioactive waste were judged as

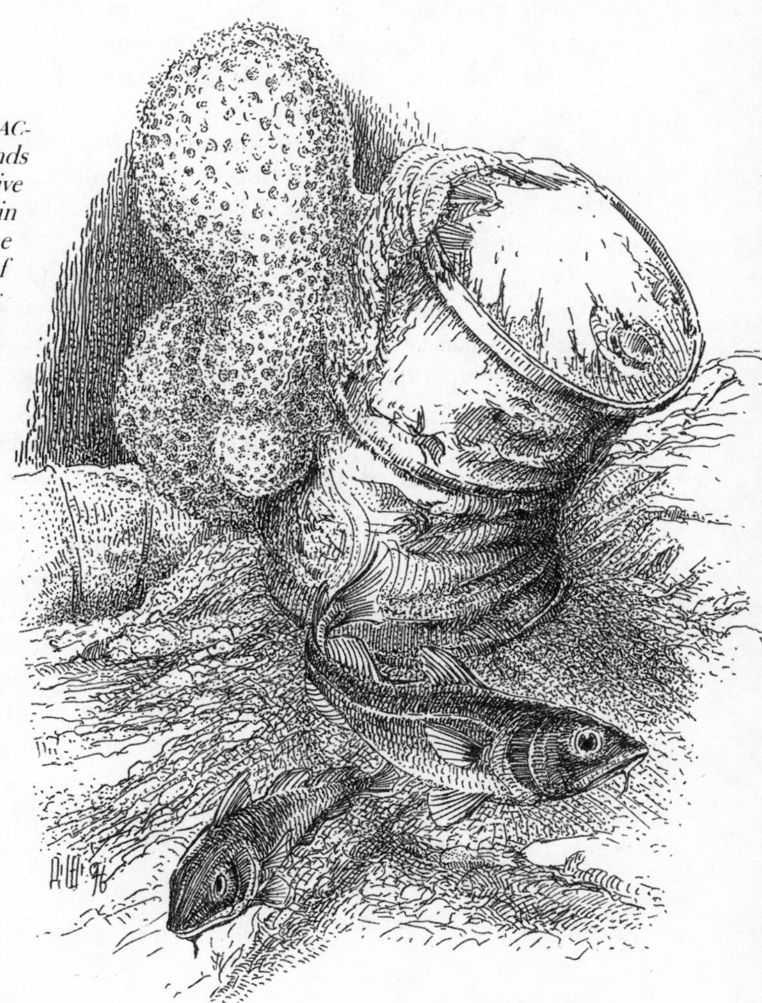

SPONGE ON RADIOACTIVE DRUM. Thousands of barrels of radioactive waste were dumped in the deep sea off the Farallones Islands of California, becoming home to deep fish and sponges and a tangle of questions about the biological effects of nuclear toxins.

too dangerous to oceanic life (and ultimately to man) to be dumped, though some member states questioned whether low-level wastes really posed a threat. But in light of the unknowns and uncertainties, the global consensus was to rule the sea off limits to mankind's most pernicious toxins, to err on the side of caution.[8]

As the moratorium withstood various political attacks over the years, a proposal arose, and was strongly backed by such environmental groups as Greenpeace, for a global accounting of what had been dumped to date. Such an inventory was eventually undertaken by the International Atomic Energy Agency, an arm of the United Nations that monitors nuclear work. Its report, published in March 1991, was based on detailed questionnaires sent to all known states that had dumped radioactive wastes into the sea. The report said that over a period of forty-five years a dozen nations dumped radioactive wastes at seventy-three different sites around the globe and put into the sea a total of 1.24 million curies of radiation. Other total estimates of the dumping were slightly higher. By any reckoning, no matter what the specifics, the disclosures marked a watershed. For the first time, the light of public accountability had illuminated one of the darkest legacies of the nuclear era. Or so it seemed.[9]

The Soviet Union in all this appeared to be a pillar of nuclear responsibility. It was a signatory to the London Convention and maintained that it had never dumped any radioactive wastes at sea and had no plans to do so in the future. Its reputation was unsullied.

The façade cracked as the union disintegrated. In late 1991, Andrei Zolotkov, a member of the old Soviet Parliament, who was an engineer with a nuclear icebreaker fleet stationed at the northern port of Murmansk, began to charge that thousands of tons of nuclear waste had been dumped clandestinely in the northern seas. In October 1992, a Greenpeace ship tried to confirm the claims by sailing to the arctic sites but was fired upon by the Russian Navy and seized. On October 24, three days after the ship's release, Russian President Boris N. Yeltsin ordered his top ecological aide, Aleksei V. Yablokov, to investigate the dumping claims. A team of forty-six experts swept through secret archives and military bases, all of which at that time were still reeling with the pains associated with the rebirth of the Russian state. Finally, in March 1993, after his presentation to the President, Yablokov publicly released a thick report, a riveting indictment of how the Soviet Union broke international laws, as well as its own internal regulations, "consciously and frequently," as Yablokov put it. Comrade Zolotkov, it turned out, knew only the tip of the iceberg.[10]

The dumping had begun in 1959. In the north it spread across the White, Barents, and Kara seas, and in the Far East across the Sea of Japan, the Sea of Okhotsk, and the Pacific Ocean. One type of radioactivity dumped into the sea was liquid, often composed of fluids for cooling nuclear reactors that had become contaminated over time. The volume of such discharges was more than eighty million gallons, enough to fill a lake. The discharges were not only historical, the report said, but ongoing. The Russian Navy was still dumping liquids because it lacked processing plants for them on land. This mingling of radioactive fluids with the sea was totally at odds with Western practice, which emphasized putting wastes in some kind of container or package.

Another part of the Yablokov report focused on solid waste. Some of this was packed in drums, seventeen thousand of them, while larger pieces were sunk individually or enclosed in special ships consigned to the deep. Echoing the Western practice, drums of radioactive waste were occasionally shot full of holes to make them sink, the report said. Most remarkable of all, the Yablokov report detailed how derelict Soviet submarines and a crippled icebreaker had bequeathed the sea a total of eighteen nuclear reactors—something apparently never done or contemplated in the West, the *Seawolf* episode having involved only a reactor shell. Sixteen of these power plants were cast into the waters of the Kara Sea, an arctic region not far from major northern fisheries, from puffins and polar bears. Seven of them were heavy with spent radioactive fuel, the deadliest of radioactive wastes. (Atoms split in nuclear fission produce dozens of isotopes, some radioactive for a million years or more.) The report said no spent fuel had been removed from these reactors because of "the damaged condition of their cores," a vague reference to meltdowns and other serious accidents.

The dumpings of hot reactors took place between 1965 and 1981. In a nod to safety, all were packed in cement or other kinds of protection before abandonment. Even so, none were disposed of in the kind of deep water recommended by the London Convention. Remarkably, they were ditched at depths of feet and meters, not miles and kilometers. Near the large arctic isle of Novaya Zemlya, the reactors (and sometimes whole submarines) were dumped in waters ranging in depth from twenty meters to three hundred meters, or sixty-five to about one thousand feet. Land was quite close by. One photograph in the report showed a damaged submarine powered by two nuclear reactors being scuttled in September 1981 off Novaya Zemlya, its bow in the air, the rugged shoreline looming close behind. Half a world away in the Sea of Japan, things were only slightly less appalling. The two reactors dumped there were unfueled and were sunk in waters up to three kilometers deep, still a violation of international norms, but less gruesome than the Kara episodes.

All told, Moscow's decades of secret dumping injected into the sea some

2.5 million curies of radiation—nearly twice what was previously thought to have been dumped at sea during the whole of the nuclear era by a dozen nations, including the United States. It was a time bomb ticking away on the seabed, mostly in the Kara. This forsaken gear, these decaying cores of nuclear submarines, were located in a shallow burial ground swept by fierce storms and ice packs for much of the year. To keep people from stumbling upon hazards that washed up on shore, the authorities each year conducted a visual inspection of Novaya Zemlya's eastern coast for suspicious objects. In 1984, they found an unidentified metal item emitting high levels of radiation—over one hundred roentgens an hour, many times the acceptable dose for nuclear workers. The very hot object was said to be part of a reactor fuel rod washed to the surface from below. It was disposed of very carefully.

At first, Moscow kept a relatively close watch on its atomic netherworlds at dozens of ocean sites around the nation's periphery. But after 1967, no monitoring ship got closer than fifty or so miles away, keeping back as dumping continued and radiation levels soared. No direct monitoring, the Yablokov report said, had been done for twenty-five years. It noted that metal containers fail in seawater after about ten years, and concrete ones after about thirty years. Distant monitoring over the decades had so far shown no dangerous rise in radiation levels. But the Yablokov report said monitoring had been "unsatisfactory" and noted that a 1992 expedition to the Kara Sea found that concentrations of cesium-137 unambiguously rose as the sampling got deeper, data it called "alarming." Overall, it said, reliable monitoring was needed urgently. If the ecological dangers were eventually found to be high, the report said, then the only reliable solution would be to raise large pieces of the radioactive waste for reburial on land.[11]

"Nobody knows what the situation is now," Yablokov told a Moscow audience in April 1993, saying the Kara was particularly suspect. "Nobody has studied the concrete place where sixteen reactors have been sunk. It's ridiculous, but there has been no study."[12]

Global reaction to the report verged on panic in some places. Many Western experts, especially those involved in nuclear studies, sought to clarify the danger. "Any potential problem would be a local one and would pose no threat on a global scale," a large team of international scientists meeting at Woods Hole said in a written statement that was widely publicized. But Greenpeace, at a congressional hearing in September 1993, charged that team had been unscientific and misleading. "Pertinent data is sparse to nonexistent," Clifton Curtis, oceans adviser to Greenpeace International, told the congressmen. "Monitoring surveys or risk assessments have not been conducted."[13]

Other experts at the hearing noted that Moscow's waste was steadily

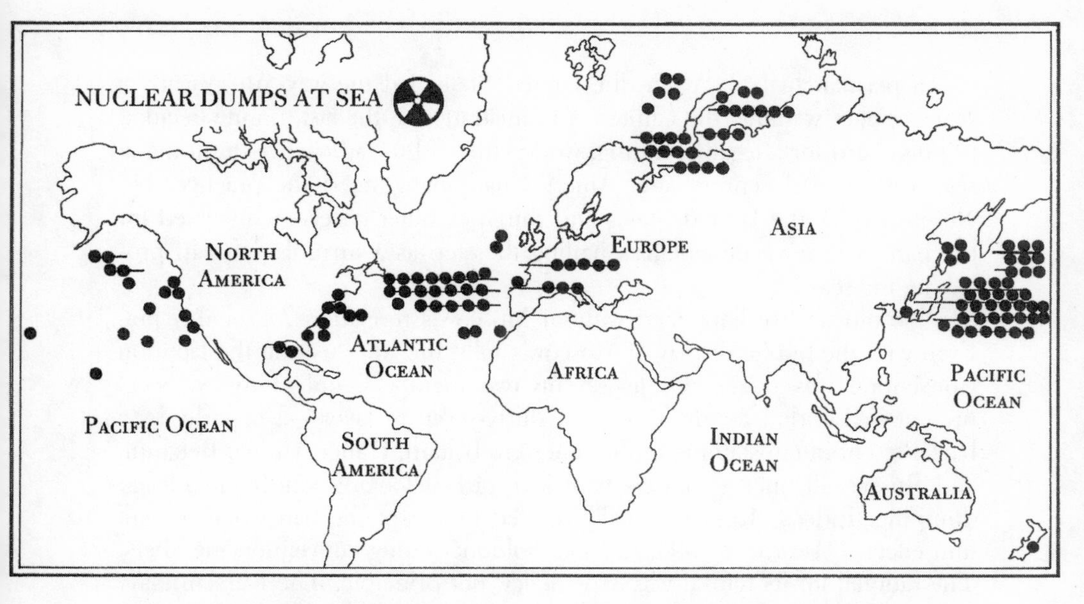

NUCLEAR DUMPS AT SEA. Following the example of the United States, a dozen countries put their radioactive wastes into the sea over the decades. The Soviet Union did so clandestinely in amounts that dwarfed the nuclear refuse of all other nations.

weakening because of radioactive decay, just as the architects of sea dumping had originally envisioned. A Federal team led by Mark E. Mount, a physicist at the Livermore National Lab, the bomb center near San Francisco, estimated that the 2.5 million curies of radiation had already dropped in potency to 0.5 million curies, and after two more decades of decay would further decline to about 0.3 million curies. So the vast majority of the danger had already vanished. Still, a half million curies was a lot of radioactivity. It was nearly half of what a dozen other nations had acknowledged dumping into the sea over the decades. But it was within the realm of experience—a monster, perhaps, but one that was slowly diminishing in size.[14]

Japan was generally of the Greenpeace school, reacting with alarm to the Yablokov report. News of the sunken reactors in the depths off its coasts startled Tokyo and prompted it to petition Moscow for details. Apprehension soared as a Russian naval ship in October 1993, a half year after the report's release, dumped hundreds of tons of low-level nuclear waste into the Sea of Japan, touching off a diplomatic row between Tokyo and Moscow just after they had declared a new era of cooperation.[15]

The final act of the drama came in November 1993 as the members of the London Convention, shaken by the Yablokov revelations and the spectacle of continued Russian dumping, worked to turn the voluntary moratorium into a permanent ban. Again, the United States led the way. After nearly a year of policy waffling, the Clinton Administration at the last minute decided to press hard for a legally binding, worldwide end to radioactive dumping at sea nearly a half century after America had inaugurated the practice. On November 12, the United States and thirty-six other governments voted for the ban, with environmentalists hailing the step as a turning point in protecting the seas.

Would it have happened without Moscow's revelations? Probably not. Even with the public display of Moscow's folly, the consensus at the London Convention was fragile. Of its seventy-two members, only forty-two were present in London for the vote, with thirty-seven in favor—a majority by a hair. Present but abstaining in the vote were Britain, France, China, Belgium, and Russia, all nuclear powers with a record of looking kindly on nuclear dumping. Indeed, Russia formally refused to accept the ban when it went into effect in February 1994, the sole holdout among convention members. The rational for its refusal was expediency, not principle. The Russian Navy was running out of storage space on land for radioactive wastes and Russian officials apparently wanted to use the crisis, and the threat of renewed dumping at sea, as a way of pressuring the West for financial aid. It was diplomatic blackmail, pure and simple.[16]

Save Russia itself, Norway and its rich fisheries lay nearest to the most dangerous of Moscow's dumping sites. So this northernmost nation of Scandinavia was eager to know exactly what had transpired in the past and what threats were likely to arise in the future from the radioactive hulks. In 1993, Norwegian researchers were allowed to visit some relatively minor sites close to the fjords of Novaya Zemlya. The breakthrough came in 1994, when the Russian military dropped its objections and opened up the hottest of the nuclear dumps to foreign experts for the first time. Sailing to Abrosimov Bay, where most of the radioactive wastes lay, scientists of Norway's environment ministry lowered a remote-controlled robot on a tether to film the wilderness of sunken reactors and radioactive waste tanks. The debris was strewn across the bottom in the gloomy half-light like a forgotten world, a lost enterprise of metallic shapes and looming shadows, as a British Broadcasting Corporation documentary later revealed. Sometimes when the robot drew close, its lights and cameras showed the bright orange of rust and the jagged edges of gaping holes. Corroded items that the robot touched often dissolved in a cloud of debris.

"The containers are in bad shape," said Lars Foyn, a scientist at the Norwegian Institute of Marine Research. "I doubt that it's possible to take them up without destroying them." Strapped to the top of one sunken barge were two big cylinders, each about ten meters around and two meters wide.

The metallic top of one flaked apart at the robot's slightest touch. What the big cylinders held was a mystery, although the Norwegians speculated that the contents might be reactor parts. Eager to document any leakage, the Norwegians had the robot carefully and repeatedly scoop up bottom ooze for radioactive analysis. It showed conspicuous leakage close to the waste sites but, to the surprise of Norwegian scientists, the radioactive contamination was apparently staying put, at least for the moment.[17]

The United States, too, struggled to investigate the dumping's aftermath, if only because of the threat to Alaskan waters. Each year America harvests nearly one billion pounds of cod, pollock, herring, salmon, and crab from the Bering Sea, and lesser amounts from the Chukchi Sea, with the arctic fisheries together constituting a multibillion-dollar industry. These fisheries are far from the Russian dumps—thousands of miles, not hundreds or dozens—but the poisoning of so important a resource, however unlikely, had to be guarded against. So Washington made a major effort to monitor for arctic contamination. Unlike the Norwegians, the Americans were barred from hot spots by the Russian military (a planned trip in 1993 was abruptly canceled), but on other joint cruises with Russians scientists were able to sample around the zone's perimeter. The Food and Drug Administration in 1994 began to monitor seafood originating in arctic waters for radiation, including strontium-90. All such monitoring efforts reportedly uncovered only very low levels of contamination, which were attributed to radioactive fallout from the atmospheric testing of nuclear weapons in the early days of the atomic age rather than from the dumping of Moscow's nuclear wastes at sea. But doubts lingered, given the sketchiness of the sampling and the potential size of the problem.[18]

No freeing up of military gear and personnel at the end of the cold war can reverse a half century of atomic dumping. But it is increasing the scrutiny of Moscow's dump sites as well as many others scattered around the globe, and is beginning to aid cleanups as well. In one case, American Navy attack submarines manned with civilian researchers were put onto the job of sampling a wide range of arctic waters for radioactive contamination. Such sampling by submarine was done in 1993 by the *Pargo* and in 1995 by the *Cavalla.* The Russians, meanwhile, had a team of a half-dozen top scientists develop a new type of detector meant to better track undersea radiation. Its unveiling took place in 1996, with the Russians saying it would yield "significant progress" in the art of atomic surveillance.[19]

As a test of rehabilitation, Moscow lavished attention on the *Komsomolets,* a Russian submarine that sank in 1989 in North Atlantic waters a mile deep. Its nuclear torpedoes began to corrode and threatened to leak plutonium and other radioactive debris, to the peril of nearby fisheries. Between

1989 and 1995, the *Mir* submersibles on forty-nine occasions dove down through dark waters to investigate the deteriorating wreck, monitoring its radioactivity and sealing up a number of holes in its hull in an attempt to stem leakage.[20]

The forerunner of all atomic dumps at sea, the Farallones, is also getting new scrutiny. Investigations of the paradise were spotty during the 1970s and '80s, with the gathered data inconsequential compared to the region under surveillance. That was true even after the Federal government in 1981 recognized the rich diversity of the Farallones ecosystem by proclaiming the area a national marine sanctuary, despite the hidden drums of radioactive waste. Beginning around 1990, however, the Navy began working closely with civilian scientists to study the atomic legacy. In 1991, 1993, 1994, and 1996 the Navy's *Sea Cliff* submersible dove with Federal experts and scientists to investigate the deteriorating drums as well as sea life. "The biology down there is poorly known," Ed Ueber, the sanctuary's manager, told me after assessing his protectorate in a dive. "One bottom sample had forty unidentified species, including worms and a new kind of clam. There's a lot down there we don't know about."[21]

In terms of radioactivity in the Farallones, as with so much of the global dumping, the decades of investigations and dives are inconclusive. The repercussions for marine life and humans are still largely a mystery. The only solution, according to Federal officials, is closer scrutiny of the drums and their surroundings, animate and otherwise.

"We haven't detected high levels of radiation," Ueber said. "We haven't found it in the fish. We haven't found it in the sediment. And we haven't found it in other places." However, he added, "it's just kind of like painting a room. We've barely opened the paint can."[22]

CARSON CANYON is a yawning fissure in the North Atlantic seabed just off the Grand Banks of Newfoundland, not too far from where the *Titanic* lies. It drops away precipitously from the continental shelf and plunges to a depth of about two miles. Lying some two thousand miles northeast of New York City, it is far removed from any major port or urban center, a deep wilderness if ever there was one. Its recesses hide such inhabitants as *Coryphaenoides armatus,* untold thousands of them. These fishes are among the most ubiquitous of the hundreds of known species of rattails and are close cousins of the roundnose grenadier, which the Russians in the 1970s hunted so intently as seafood. In retrospect, that might have been unwise.

While probing the depths of Carson Canyon in the 1980s, American scientists caught several *Coryphaenoides armatus* and subjected them to careful analysis. The ominous results hint at the chemical pall that is quietly descending on the deep and illustrate that not all recent advances of abyssal inquiry are the result of military spinoffs. The analytical strides took place during the Reagan military buildup at the height of the cold war, and had little to do with naval gear or personnel.[23]

John J. Stegeman is a senior biologist at Woods Hole who is interested in understanding the effects of pollutants on deep life and more generally in maintaining the Earth's habitability. His logic is straightforward: large volumes of sewage and oil and other contaminants are pouring into coastal waters around the globe, and inevitably some of that miasma must make its way into the deep. The implications for shallow waters are already clear. Increasingly in urban harbors, scientists find that bottom-dwelling fish are rife with cancer and worry about the risks to humans from eating contaminated seafood. It is Stegeman's goal to see whether the planet's tides and currents are quietly spreading such dangers into the deep. He is particularly interested in a ubiquitous family of industrial chemicals known as polynuclear aromatic hydrocarbons, which cause cancer and genetic mutations. Other targets of his are polychlorinated hydrocarbons, which promote tumors and interfere with reproduction. Such compounds often abound in the coastal waters of many nations and readily taint sea life because they are soluble in fat, often concentrating in liver and flesh.[24]

Stegeman found a clever way to hunt for clues of an animal's biochemical reaction to such poisons. His analyses were deemed better than chemical ones because they showed biological effects rather than simply the presence of contaminants, which can be ambiguous. When under chemical attack, Stegeman found, fish produce an enzyme he named cytochrome P450E, which helps wage biochemical war on the intruder. He then developed a sensitive way to measure the enzyme. As specimens of *Coryphaenoides armatus* from Carson Canyon were examined, signs of the enzyme emerged in the form of dark bands on light-colored strips of analytical paper. The deep fish were clearly under siege. These results were then compared to ones from specimens of *Coryphaenoides armatus* that had been taken closer to urban centers, specifically, a little more than one hundred miles out to sea from New York City, in the depths of the Hudson Canyon, nearly two miles down. For these fish, the bands on the paper strips were much darker, indicating a greater reaction to contaminants. Such warning signs, Stegeman wrote, provide "the first direct evidence that some chemicals may already be causing biological change in the deep ocean." His argument grew stronger as he later found similar signs of distress in fish from coastal regions of North America and Europe, as well as in whales. The rivers of pollution that run into the

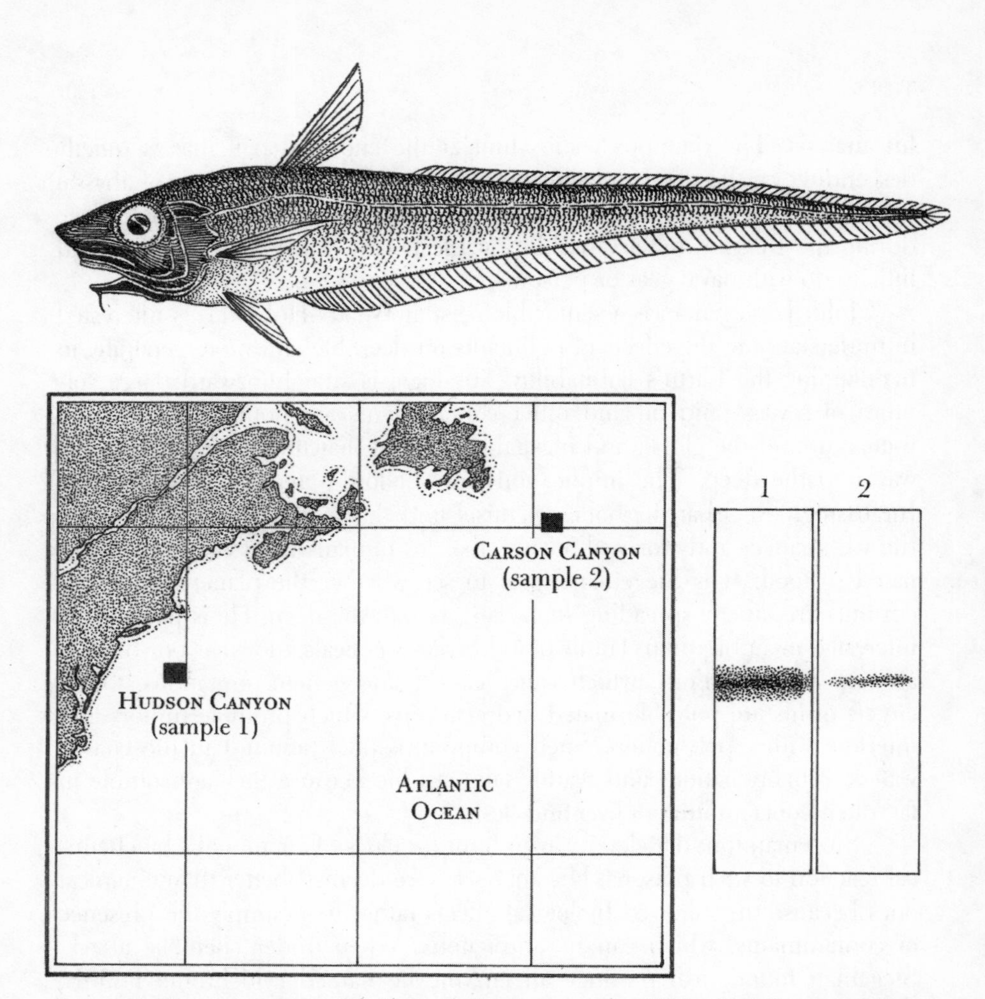

Tainted Rattails. Rattail fish caught in deep waters off Newfoundland and New York City were found to harbor signs of contamination by polynuclear aromatic hydrocarbons, a ubiquitous family of industrial chemicals that cause cancer and genetic mutations.

sea, he concluded, hurt not only residents of the sunlit zone but are beginning to haunt the inhabitants of the darkness as well.[25]

A brain coral the size of a small whale sits at the bottom of Exuma Bay in the Bahamas. Once it was a festival of life. Now it is bleached and barren. Over the years, a thermometer at its base has revealed a trend to rising temperatures, as have other thermometers monitored by the Caribbean Marine Research Center. The situation is the same for coral reefs from Belize to Australia, many of which are showing signs of disease and decline linked at least in part to temperature rises.[26]

Monterey Bay in the Pacific has no coral reefs, but its coastal zone is nonetheless undergoing a major shift of temperature and fauna. Between 1931 and 1994, southern types of snails, limpets, anemones, and rock barnacles grew in abundance while northern types of invertebrates went into decline, according to a study by MBARI and Stanford scientists. During the six-decade period, the mean shoreline temperature rose 0.8 degrees Celsius, and peak temperatures in the summer rose a remarkable 2.2 degrees. In Fahrenheit, that is 1.4 and 4.0 degrees, respectively, a degree of warming that in theory could be noticed by swimmers and beachcombers.[27]

Further south off the coast of California, changes in sea life have been even more dramatic as average temperatures in the region rose up to 1.6 degrees Celsius, or 2.9 degrees Fahrenheit, in the period between 1951 and 1993. The study area extended out three hundred miles from Point Conception in the north to San Diego in the south, some fifty thousand square miles of sea that constitute the southern third of California's coastal waters. Examining data from 222 cruises, scientists from Scripps found that populations of zooplankton dropped an astonishing 80 percent during this time of rising temperatures. Zooplankton are tiny animals such as copepods that drift in currents and are a founding link of the oceanic food chain, feeding sardines, anchovies, hake, rockfish, mackerel, and, indirectly, all kinds of large animals that feed on smaller ones.

Coincident with the plankton drop, landings of commercial fish in California fell by more than 30 percent. For seabirds it was worse. Populations of the sooty shearwater, a zooplankton feeder, once the area's most abundant seabird, fell by a stunning 90 percent. "We found that a small temperature change resulted in a very large biological change," said John McGowan, one of the Scripps scientists. If the sea off California continues to warm, he added, "we can anticipate a collapse of fisheries."[28]

What all this portends for the deep is unclear. Scientists assume that major drops in surface productivity will hurt abyssal residents who live on leftovers falling from above. But no one has studied that prospect. Relatively little is known about the creatures of the darkness, much less how food reductions could affect their health and behavior. Moreover, it is intrinsically hard to measure temperature changes in the deep, in contrast to the relative ease of doing so at the surface. Despite the difficulty, scientists have found hints of worrisome trends, ones that have implications beyond the well-being of deep fauna. The abyss is so enormous and so influential in the regulation of the global heat balance that even minor changes in temperature have the potential to unsettle all the planet's habitats and inhabitants.

The foundation for the hints was laid in 1957 during the studies of the International Geophysical Year, a global assembly of scientists that among other things tried to take the temperature of the deep Atlantic. A second set of observations was made in 1981. Comparison raised the possibility of oceanic warming. In 1992, as part of the quincentennial celebration of the new world's discovery, an oceanographic ship manned by Spanish, British, and American scientists sought to elucidate the trend. Following the same route that Columbus took, the Spanish naval vessel *Hesperides,* named after the Pleiades, a cluster of bright stars, departed from the Canary Islands in July 1992 and stopped every thirty-eight miles, making a total of 101 stops before reaching the Bahamas in August. At each stop, scientists dropped a weighted probe into the deep that recorded temperatures continuously from the surface down to the seabed.

The cruise was costly and complicated. But it shed light. Overall, a warming of the abyss over a period of thirty-five years—between the voyages of 1957 and 1992—was found in waters ranging up to 2.5 kilometers deep, or about 1.5 miles. The effect was greatest at a depth of 1.1 kilometers, where temperatures had risen since the fifties by as much as 0.32 degrees Celsius, for an extrapolated overall rate of one degree per century. (On the Fahrenheit scale, the equivalents are, respectively, 0.6 degrees and 1.8 degrees.) The measured temperature rise in the deep Atlantic was roughly in line with predictions of global warming, according to the research team. "That's a huge signature and certainly warrants further monitoring," said one team member, Robert Millard of Woods Hole.[29]

While the warming of everything from Pacific bays to Atlantic deeps is intriguing—and perhaps alarming to some people—it should also be clear on reflection that it is ambiguous, at least in terms of human culpability. The problem is that no one knows whether the temperature shifts and life upheavals are natural or unnatural. The warming could mean that rising levels of carbon dioxide in the atmosphere are heating the planet by means of the greenhouse effect, trapping sunlight that otherwise would be reflected back into space. Such heating is predicted to damage continental breadbaskets with droughts, coastal regions with flooding, and the whole planet with intensifying storms. Or it could mean we are simply experiencing the same old climate swings that over the ages have pummeled the Earth with periodic cycles of fire and ice, producing wide ecological shifts and species extinctions. Or it could be some combination of the two factors, intermingled and difficult to isolate. No one knows.[30]

So ambiguous is the current situation that some experts dispute whether global heating is occurring to any significant degree at all. The whole Earth, based mainly on land measurements, is said to have warmed a total of 0.54 degrees Celsius, or 0.97 degrees Fahrenheit, between 1881 and 1993. But critics dismiss such data as meaningless. The increasing heat of cities is known to magnify the readings taken near urban areas, they say, which climate scientists concede is true but try to take into account. Critics further argue that major changes in instrumentation have undercut the accuracy of the historical record. Despite such problems and skepticism, experts advising the world's governments on climatic issues in 1995 concluded for the first time that the warming of the global atmosphere was very real and that human activity is probably one of the culprits. As evidence, the panel cited improved computer models of the atmosphere that are better able to extract trends from noisy data. Even so, the finding was promptly denounced by reputable experts as erroneous, which is no great surprise. In science, as in other disciplines, the intensity of debate is often inversely proportional to the quality of the evidence.[31]

In the warming dispute, the sea and its temperature readings are uniquely authoritative, at least in theory. The reason is that the ocean is a sponge not only for carbon dioxide, as MBARI and others groups are studying. It also soaks up heat. All water does so with little effort compared to most other liquids and solids. It absorbs heat with a minimum rise in temperature, and holds on to it tenaciously. This fact is driven home every day as different parts of the Earth pass from day to night. When darkness falls, the sea's surface cools down a bit, a degree Celsius or so. In contrast, the interiors of continents undergo a drop of up to thirty degrees Celsius, or more than fifty degrees Fahrenheit. Islands and coastal areas have milder climates and temperature swings because of the moderating effect of the nearby sea. The sea's heat capacity is so great that small changes in its usual temperature patterns can throw the planet's whole atmosphere into turmoil, as happens periodically when an El Niño change of ocean temperatures in the South Pacific triggers floods in California and droughts in Africa.

The reverse never happens. The sea is imperturbable over the course of months and years, no matter how great the temperature changes above. Beyond a depth of about two hundred meters, it experiences virtually no seasonal changes at all, even in places where the summers are very hot and the winters very cold. The sea is a colossus unto itself, absorbing and storing huge amounts of solar energy and over time averaging out the violent swings of atmospheric weather.[32]

For all these reasons, the sea is considered one of the best places to look for clues of a global temperature rise over the course of years and decades, especially its dark regions, down low, far from the vicissitudes of the surface, despite the difficulty of taking measurements there. To be really authoritative, the readings would have to be more accurate and reliable than those made by the *Hesperides* and other ships that have tried to measure the temperatures below. But no reasonable scientist could dismiss the steady warming of the deep as mere static in the Earth's climatic din. The trend would be real.

Such was the backdrop to a $35 million test in the mid-1990s of a radically new way to take the temperature of the sea, one made possible by the military spinoffs of the cold war. The allure was so great that seven nations signed up to participate, including Russia. The experiment was run by Scripps and led by Walter Munk, an icon of oceanography who had helped pioneer studies as diverse as tide mechanics and plate tectonics. It envisioned sending bursts of sound from underwater speakers across the Pacific at least once a day, every day, for at least a decade. Sound travels faster in warm water than in cold. And it can go quite far, as the Navy discovered during the cold war. In theory, the test would reveal temperature shifts as subtle as a few thousandths of a degree Fahrenheit. The experiment was designed to distinguish between natural variability and greenhouse warming. Over a decade, the warming produced by people was expected to cause the infinitesimal speedup of sound waves so that their overall travel time across the Pacific would be cut by 1.5 seconds. Thus, the experiment could check the computerized forecasts of global peril with unprecedented rigor.

The monitoring effort was known as Acoustic Thermometry of Ocean Climate, or ATOC (pronounced A-tock). Its measurements would be unique by virtue of their geographical reach across thousands of miles of the world's largest ocean and by virtue of their sheer quantity. ATOC promised not only great reliability compared to terrestrial readings but a virtual deluge of reliable data compared to the trickle that *Hesperides* achieved during its monthlong voyage or, for that matter, any other effort at deep temperature taking. The experiment would probe the average temperature of an immense body of water, rather than a finite number of readings over many miles of ocean by surface ships or buoys, readings that might easily fail to accurately represent the complexities of the whole. For all these reasons, it was applauded by many scientists as ideal for investigating the warming of the Earth, probably the best experimental setup possible for addressing one of the knottiest environmental problems of the age.

If a preliminary yearlong run was successful, the definitive answer on warming would take longer to register, perhaps a decade or two or more. But to many scientists, the prospect of a long run was reasonable given the magnitude of the issues and the elegance of the setup. And though seemingly costly, the test in reality was economical, since it relied in large part on the $15 billion system of underwater microphones that the Navy had used to spy on enemy ships and submarines.

The Navy undersea microphones that were to record the temperature signals were located up and down the West Coast, off Alaska, around Hawaii, and in Guam. In addition, civilian microphones were to be scattered throughout the Pacific basin and operated by Australia, Canada, France, Russia, New Zealand, Germany, and South Africa. Loudspeakers, each three feet wide and five feet tall, were to be moored off California and Hawaii at a depth of a little less than one kilometer. They would rumble with low-frequency noises for twenty minutes, repeating that signal up to six times a day. In theory, the monitoring of the undersea rumble would open the concluding chapter of the warming debate, settling it once and for all.[33]

That was the plan. But ATOC and its scientists ran into the whale lobby, a political force to reckon with in the late twentieth century, especially in nature-loving California. The debate became extraordinarily bitter, at times hysterical. Rock stars wailed. The Santa Monica City Council condemned the experiment. Environmental groups—including the Humane Society, the Sierra Club, Save the Whales, Save Our Shores, the Great Whales Foundation, the Enemies of Pollution, the Environmental Defense Fund, the American Oceans Campaign, the Earth Island Institute, and the Natural Resources Defense Council—fought ATOC hard. A wave of angry letters crashed down on attentive congressmen and government bureaucrats. The charge was that ATOC threatened to harm whales and other marine mammals, to render them deaf in a world where hearing is synonymous with survival, with finding food and avoiding becoming someone else's meal. "A deaf whale is a dead whale," Lindy Weilgart, a scientist who led the protests, said with characteristic bite. "It's not worth the risk."[34]

In truth, how mammals use sound in the world below is poorly understood, much less how artificial noises might hurt the process. The hearing ranges and diving depths of whales and other marine mammals at best are known only in outline, and at worst are mysteries. Logic alone suggests that hearing must be vital in undersea gloom, where eyesight is virtually useless, as is the case with bats flying through pitch darkness. But it clearly works, and has done so for ages, amid a natural din of seaquakes, thunder, breaking waves, cracking ice, and erupting undersea volcanoes, as well as a symphony of mammalian grunts, warbles, and squeals, some of which can be quite piercing. However, marine mammals now face a rising clamor. Growing numbers of sonars, ship engines, ocean drills, oil rigs, icebreakers, supersonic jets, industrial underwater blasts, and countless other sources of human noise are creating a growing wave of undersea sound pollution. Preliminary research suggests that the cacophony, if nearby, can temporarily or permanently injure the hearing of some marine mammals, damaging their ability to feed, communicate, and navigate.

It was against this backdrop that ATOC was born. As is often the case in emotional conflicts, the ambiguity and uncertainty worked only to harden the battle lines.

The trouble began in February 1994 just as ATOC scientists were getting their final permits and preparing their gear for the start of the sound experiment. Weilgart and Hal Whitehead of Dalhousie University in Halifax, Nova Scotia, both specialists in sperm whales, which are toothed and dive deep in search of prey, posted messages on the Internet claiming that ATOC's transmissions could deafen the big mammals. With lightning speed, the rumor mill of electronic mail on the global computer network amplified and distorted the charges until the ATOC researchers were portrayed as inhuman monsters ready to risk deafening more than a half million whales, dolphins, and endangered creatures, crippling and killing them. The electronic outcry caught the attention of reporters, whose stories lent credence to the fears. One story reported that the sound would be like an explosion ten million times louder than ones known to damage human and mammalian ears.[35]

Munk and the ATOC architects fought a rear-guard action, saying the fears were unfounded, that numbers and words had been misconstrued and taken out of context, that no harm would be done—to no avail. In April 1994, the project was put on indefinite hold, pending reviews by a gauntlet of governmental agencies, including the National Marine Fisheries Service, an arm of NOAA that enforces the Marine Mammal Protection Act.

As things cooled down a bit, the voices of ATOC scientists and their allies began to be heard publicly, though often just barely. No, the low-frequency hum was highly unlikely to damage marine mammals. The frequency of the sound, sixty to ninety cycles per second, was chosen because it seemed to be practically inaudible to seals, sea lions, and toothed whales (the kind that swim deep). No, the sound did not erupt as an explosion but began softly and grew louder, giving mammals time to swim away if they found it unsettling. No, the low rumbles were not louder than large ships and other common sources of ocean noise but, with the speakers moored three thousand feet down, would be far away. No, most ocean animals do not appear to dive that deep. Yes, the ones that do and are sensitive to the sound would have to come up very close to the loudspeaker for a long time to damage their hearing. No, human divers could never hear the sound unless they swam right above the loudspeaker, and probably not even then. Sure, the ATOC scientists said in conclusion, what they proposed to do was somewhat uncertain. That is the nature of experimentation. But it was a question of balance, of very small risks and very big rewards. The experiment might discomfort a few whales but produce a wealth of insights about the fate of the planet, which not

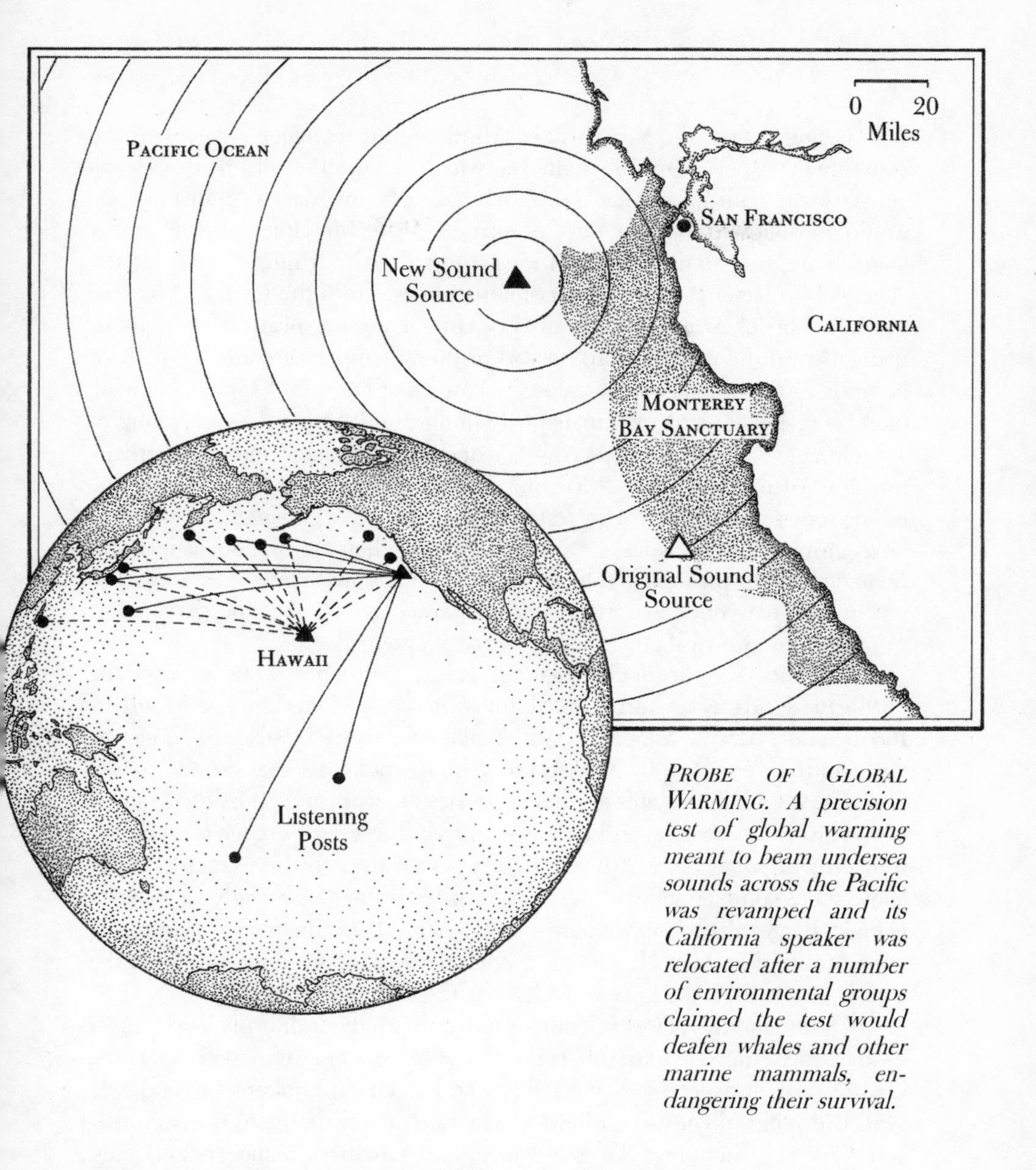

PROBE OF GLOBAL WARMING. A precision test of global warming meant to beam undersea sounds across the Pacific was revamped and its California speaker was relocated after a number of environmental groups claimed the test would deafen whales and other marine mammals, endangering their survival.

incidentally included all the sea's whales. The opportunity was too good to pass up.[36]

The plea fell on deaf ears. No whales should suffer, the activists shot back. No migrations should be impeded. No hint of harm should be tolerated. It was a question of integrity and fairness, the activists said. Whales had been around a lot longer than people, more than sixty million years longer. They had rights and deserved respect rather than being treated like road kill on the highway of progress.

Facing defeat, the Vienna-born Munk agreed to major changes in the experiment. A new plan was drafted in which the number of broadcasts was cut back dramatically. In lieu of daily broadcasts, two days of transmissions would be followed by four days of silence. More fundamentally, the main focus of the test would shift from measuring global warming to gauging the effects of underwater noise on marine mammals. From the start, ATOC had a sizable part of its program devoted to animal studies, nearly $3 million of them. But in the revised plan, instead of doing temperature and whale tests in tandem, the biologists won exclusive control of the sound source, allowing direct investigations of mammalian discomfort and injury. The gathering of any climate data would be secondary and opportunistic and still perhaps revealing even though the setup compared to its predecessor was weak. The biologists wanted the four-day resting period between transmission cycles so areas around the loudspeakers would return to their normal acoustical and biological state, presumably allowing any animals that were scared away to return and new ones to arrive. And all parties agreed that the experiment would be shut down if any evidence of harm emerged.[37]

The most significant change of all, at least symbolically, was to move the California loudspeaker further offshore so it was no longer in the Monterey Bay National Marine Sanctuary, a holy place for many West Coast ecologists and nature lovers. Run by NOAA, the marine sanctuaries that dot the coastal waters of the United States and its territories are as near as possible to idylls of nature. No oil or gas can be tapped, no sand or gravel dredged, no wastes dumped, no areas damaged, no shipwrecks torn apart. The sanctuaries do allow some commercial activities, like kelp harvesting and nearly all types of fishing. Bigger than Connecticut, the Monterey Bay sanctuary is one of the most beautiful of them all, including such parts of the California coast as Big Sur. It is where MBARI does much of its research.

Originally, the ATOC scientists wanted to put the California loudspeaker in an area of the sanctuary off Point Sur. The reasons were good acoustics, logistic ease, and nearby Navy facilities and electrical hookups for undersea gear. But politically it was a blunder. The sanctuary is home to many endangered species, including the gray whale, and sanctuary managers and their allies are dead serious about marine conservation. Critics of ATOC saw the sound experiment as a stark violation of that ethic. So in the new plan, the loudspeaker was moved some fifty-five miles offshore and into seas that were rougher. The distance and waves forced the cancellation of animal studies from small boats. And setting up the experiment so far from land cost more. But the acoustics were good and the logistics tolerable.[38]

So, after a year and a half of fights and delays, and after the whale lobby gave its reluctant blessing, the switch was thrown and the apparatus in December 1995 began to broadcast its deep rumblings.

No spasms. No seizures. No rushings away. No signs of any discomfort or change whatsoever—at least as far as could be determined by methods that, while advanced, gave only the most cursory overview of the potential drama in the wilderness below. From boats and planes and satellites, biologists from the University of California at Santa Cruz closely watched the whales and other marine mammals of the region, hundreds of them. Despite great efforts, the experts could find no evidence that the broadcasts were having any biological or behavioral effects at all. If that apparent harmlessness held up over a year or so of transmissions, the plan was for the ATOC scientists to seek to operate the speakers according to their own schedule, understanding, of course, that the work could proceed only with the permission of the whale lobby. With that go-ahead in hand, however, the scientists would embark upon the most sensitive test yet devised to measure the dimensions of global warming, though nearly three years after they intended to do so, a period during which public debate over the warming topic became increasingly confused and contentious.[39]

WE LEFT BOSTON early one morning in October 1995, just as the city was waking up, ready to address a riddle of undersea ecology. The fog was thick. The NOAA ship *Ferrel,* a Louisiana mud boat adapted for research, repeatedly sounded its low horn as we made the twenty-five-mile run from Boston Harbor to Stellwagen Bank, an area between Cape Ann and Cape Cod that is famous for its whales. The sandy bank lying invisible beneath the waves is some twenty miles long. It and its rocky flanks abound in plankton and fish and the cetaceans that feed on them, including porpoises, dolphins, pilot whales, minke whales, fin whales, right whales, and humpback whales. Indeed, the bank is a mainstay of the Massachusetts economy—not for whaling, of course, which is mostly banned after hunters pushed many species to the verge of extinction, but for tourism.

The whale-watching industry attracts more than a million visitors to the bank each year. Or it did until the numbers of whales went down. Fleets of tour ships still sail out from Newburyport, Gloucester, Boston, Plymouth, and Provincetown. At the peak during the late 1980s, forty-two companies sailed regularly. Now, late in 1995, there were thirty-five companies and questions about how many would be around next season. The big tourist attractions were humpbacks, acrobats of the sea famous for hurling themselves out of the water and slapping the surface with flippers and tails. Of the scores of different kinds of whales that ply the seas, Melville called humpbacks "the most

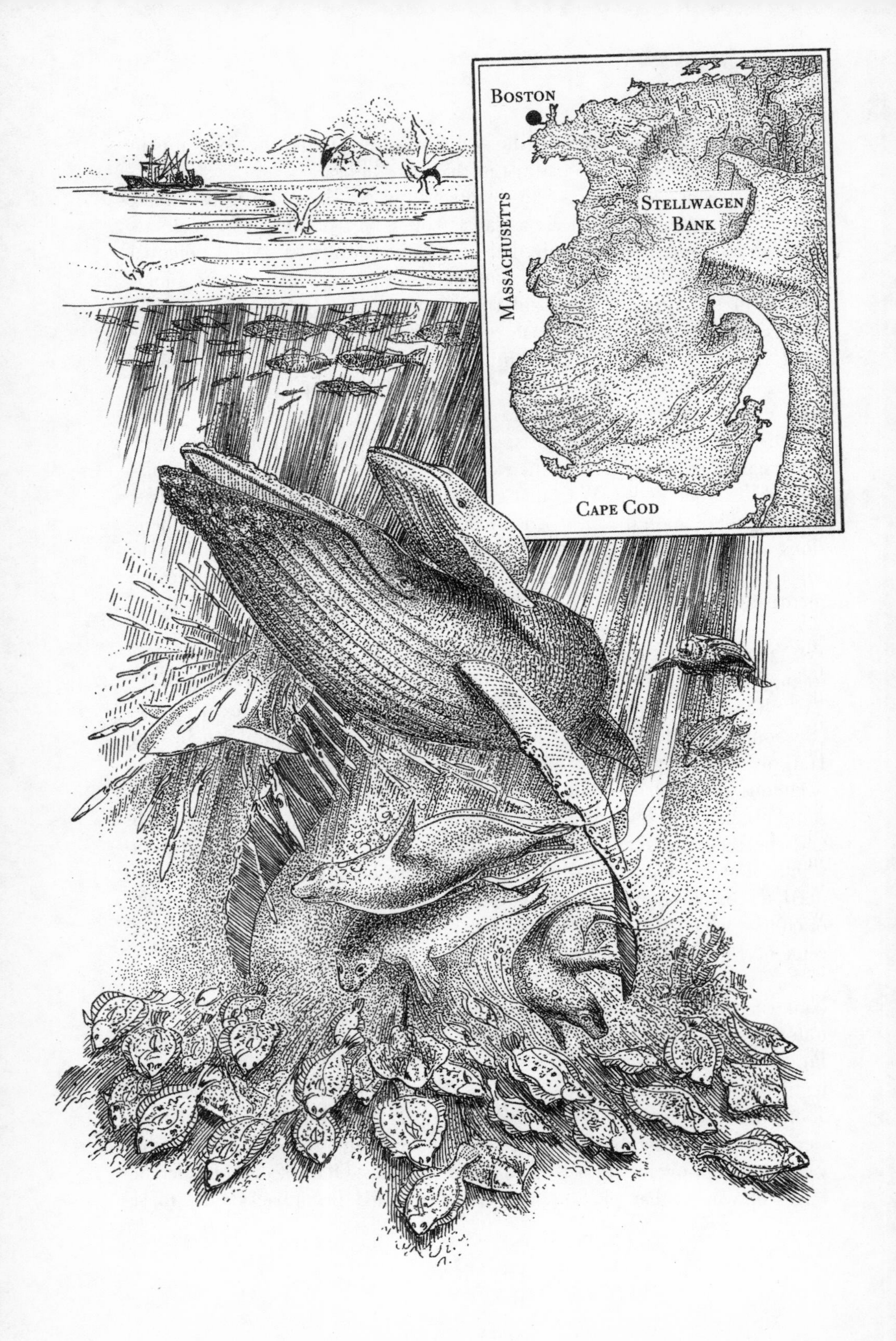
Boston
Massachusetts
Stellwagen
Bank
Cape Cod

gamesome and lighthearted." The Center for Coastal Studies in Provincetown published a poster featuring the rich tapestry of Stellwagen's bottom fauna and just above, dominating the picture, a mother humpback caressing her calf. In fact, the numbers of humpbacks at Stellwagen had fallen precipitously, down from hundreds to dozens or fewer. What caused their disappearance was a mystery. It was also something of an embarrassment, since Stellwagen had been declared a National Marine Sanctuary largely on the basis of its whales, mainly humpbacks. The issue was seen as possibly offering warning signs about the ecological health of the whole region. In the murkiness of the undersea world, whales are often considered sensitive barometers of environmental change.[40]

The scientific riddle was complicated by the tangle of possible factors at work. Red tides of toxic microorganisms, fed by city sewage and agricultural runoff, had periodically befouled Cape Cod Bay. But no bodies of humpbacks were now washing ashore as they had in 1987. Another factor was the general dirtiness of the whole Massachusetts Bay region, Boston Harbor having achieved the grim distinction of possessing the nation's highest coastal concentrations of polynuclear aromatic hydrocarbons, the industrial chemicals that cause cancer and genetic mutations. But cleanup efforts were underway. Other suspicions centered on a dump site for radioactive waste just outside the harbor, where state and federal contractors in the 1950s had taken more than four thousand containers packed with atomic refuse and ditched them in waters just three hundred feet deep. The radioactive site was the nation's hottest after the *Seawolf* area and the Farallones, yet nuclear decay was slowly making it and its cousins less deadly each year. The most plausible culprits were deemed to include changes in water quality and declines in the fishes that the humpbacks feed on, including herring and sand lance. Raked by nets and trawls, Stellwagen is part of the New England fishery and, like the whole business, is beset by disruptions. Catches were down and the industry was gearing up for a battle.[41]

As we plowed through the fog, Bradley Barr, Stellwagen's bearded manager, told me of a proposal to close down a large part of the bank to commercial fishing, an experiment to see if that would help restore the ecological balance. The industry, he said, was fighting the plan. It wanted to keep taking whatever fish it could, even if that hurt things in the long run.

A subtler test than the fishery's closure was already in progress. On the *Ferrel*'s rear deck was a sleek, dolphin-shaped robot some eight feet long

STELLWAGEN FAUNA. Humpback whales, the acrobats of the sea, once frolicked on Stellwagen Bank. But their numbers have dwindled, prompting scientists to search for ecological changes that may be responsible for the exodus.

known as *Odyssey.* All week long, the yellow device had probed Stellwagen bank and its flanks, recording the chemistry of its waters and the ecological makeup of its seabed. Unlike most robots, it had no tether. Battery powered and packed with computer chips, *Odyssey* was smart enough to guide itself on its own missions and journeys.

To my surprise, I discovered that its overseer was none other than Bob Grieve, the *Alvin* pilot who had taken me down to Pacific depths and displayed such good humor, his feet warmed by animal-claw slippers. After careers in the Navy and Woods Hole, and after a total of 245 deep dives, Bob, in his early thirties, had quit the world of piloted exploration for that done by machines. He told me that he had surveyed the industry's trends and decided to cast his fate with *Odyssey.*

"It's humble now but this is the wave of the future," he said as he worked on the small craft, screwdriver in hand.

Unlike Bob, the undersea robot was no spinoff of the nation's naval establishment. Its military genesis was more direct than that, as I learned from its creator, James G. Bellingham, a lanky physicist from the Massachusetts Institute of Technology in nearby Cambridge. Jim and most of the dozen or so people who work for him in the Underwater Vehicle Laboratory that he runs at MIT were in their twenties and thirties, and at that young age were already grizzled veterans of the cold war's unexpected end. Jim himself had worked on secret projects for the Navy and had received his doctorate from MIT in 1988, eager to join the war against Moscow. MIT was famous for its intellectual aid to that contest. For his part, Jim dreamed of becoming a Navy astronaut. But the sudden end of the war threw his future into doubt. Bearing a new Ph.D. in solid-state physics but seeing a sudden drop in military prestige and research monies, he searched for a new career, which he decided lay in the creation of small, smart, low-cost robots to probe the deep, mainly for the tracking of environmental change.

"We're all refugees," Jim said of himself and his team as he leaned on the ship's rail. "The cold war ended and it was clear what direction things were going—that national security was not going to be solved by military force, but by education and economic competitiveness and understanding the environment. I had a real tough time at first looking for jobs."

He succeeded not only in finding a job but in launching an innovative line of robots. They were tiny compared to Navy behemoths such as *AUSS.* In fact, the Navy at first showed no interest in the devices, considering them mere toys. But Jim had recently won a five-year contract with the Office of Naval Research. It had seen the light, as it had decades earlier with the lumbering bathyscaphs. Like many vehicles, *Odyssey* was meant to dive six kilometers down, or nearly four miles (the same depth specified decades ago by the Navy in Craven's twenty thousand feet). But it did so for a pittance, its parts costing not tens of millions of dollars but $50,000. One low-cost secret was its casing. Relatively flimsy, made of polyethylene, it actually flooded with seawater as the robot sped downward from the surface. Inside the case, key parts and components were kept dry inside a set of either two or three glass spheres—the same kind that housed the electronic pingers that we deployed on the Juan de Fuca Ridge. The number of seventeen-inch spheres was adjusted to suit the particular requirements of the dive. The robot was pioneering yet used many parts that were off the shelf rather than custom made, which kept costs low.[42]

I inspected the device, which was split open for servicing. Two glass spheres in the bottom half of its casing looked like peas in a pod. The spheres were full of bright wires and circuit boards, and one held the chassis of a Sony Handycam to make video recordings.

"This is the brains, this is the brawn," said Bob, pointing out the front sphere, which held the computer, and the rear one, which held batteries and other devices.

Odyssey's brainy independence made it a true robot. Most of the other unpiloted machines that prowled the deep were remotely operated vehicles directly controlled by people, usually over the long tethers. Even *AUSS,* the Navy's torpedolike *Advanced Unmanned Search System,* was no true robot, since its deep searches could be guided by people over an acoustical link to the surface. But *Odyssey* made its own way. The advantages were several. A ship dropped the robot off and picked it up hours, days, or even months later. Human overseers did what they wanted in the interim rather than having to constantly provide guidance. And deep, wide searches were easier to conduct since there were no long cables to drag around. The problem was reliability. Robots were unable to solve problems and outwit crises. Yet *Odyssey,* despite problems inherent in any experimental vehicle, had so far lived up to its name, always managing to find its way home.[43]

Two years hence, the robot was to take on an ecological job that would tax its brains far more than any probing of Stellwagen, which was part of a program of preparatory runs. At the bottom of the Labrador Sea, between Newfoundland and Greenland, the robot would sit for six months at a depth of 3.3 kilometers, or 2.1 miles, activating itself periodically to fly around and sample the deep waters in wintertime. The Labrador Sea had some of the roughest winters of the North Atlantic, a sizable deterrent to researchers working in ships. Yet the area is vital to ecological studies because it is one of the "lungs" where the ocean breathes oxygen into the global deeps and thus energizes all kinds of life and chemical reactions in the frigid darkness. The breathing occurs when surface waters in high latitudes near the Earth's poles become rich in oxygen but also become very cold and dense and sink to the bottom in plumes and gentle currents that eventually wander all over the seafloor. The Labrador Sea is thought to play an important part in this oxygenation. A grim, long-term uncertainty is whether global warming could conceivably slow or stop this descent of seawater that is icy cold and highly oxygenated, killing deep life. It is a question that requires better knowledge of the status quo. *Odyssey* would shed light on the hidden process, recording in detail the descent of the cold plumes.[44]

I wandered into the ship to learn something of what the robot was currently doing and was shown one of the videos it had taken in the preceding days. Flying about two meters off the seabed at a speed of two knots, *Odyssey* and its camera had revealed a blur of changing bottom environments and occasionally a cloud of silt kicked up by a rising flounder. From this perspective, Stellwagen Bank had no whales but instead consisted of shells, sands, gravels, and the dark burrows of fish and crustaceans, a patchwork of patterns that were constantly changing.

Peter J. Auster, a fisheries ecologist from the University of Connecticut, who was the mission's scientific director, said that such information was important for understanding how biological productivity of the bank differed from area to area. Over time, it would also reveal how fishing nets and dredges were changing things, usually for the worse. "If we harvested deer the way we harvest fish, dragging nets, eventually we'd knock down most of the woods," he remarked.

Today's run, he said, aimed at probing the chemical and temperature characteristics of the thermocline—an area some thirty meters beneath the ocean's surface at Stellwagen, where the temperature dropped suddenly, marking a dividing line between upper and lower waters. The thermocline was important to the bank, he said, because it was an area of high biological productivity, rife with phytoplankton and the animals that feed on the tiny plants. The robot was programmed to dive down to a depth of fifty meters, then back up to five, and to keep cycling back and forth between those depths, taking measurements as it repeatedly crossed the thermocline. In all, *Odyssey* was to go about four kilometers, traveling from relatively deep water to shallow areas on the side of the bank.

Back outside, on the rear deck, I found Bob and another team member hunched over *Odyssey*, working to link its upper and lower halves with dozens of screws. The robot's upper casing bore a large serious-looking circular seal, OFFICE OF NAVAL RESEARCH. On a lighter note, the handiwork of Bob's label maker was evident all over the robot. NO STEP, read a small label on a fin not much bigger than a shoe. READ INSTRUCTIONS FIRST. THIS IS NOT A TOY, read another. BEWARE OF BLAST, read a label with a tiny arrow that pointed toward the robot's propeller. In a final touch, Bob had used some kind of tape to apply a pair of eyes to the robot's front, making the sleek machine look like a shark.

Its halves joined, the robot was lifted delicately by a crane and placed in the water, which was nearly smooth. *Odyssey* looked right at home, its eyes just above the waterline. Three team members in an inflatable boat then towed the robot out a hundred meters or so over the water. One of them, Brad Moran, carried a laptop computer to communicate with the robot over a cable up to the moment of its release and a walkie-talkie to transmit progress reports back to the ship.

The fog was beginning to lift. Hints of sun were overhead. No whales were visible. None had been sighted during our five hours at sea.

Inside the ship, the electronics lab was crowded with people and computers and a large navigation monitor. The screen was lit with a green icon for the ship's position and a yellow one for *Odyssey.* Jim Bellingham and other members of his team eyed the monitor, fidgeting.

Jim picked up a walkie-talkie.

"We're tracking," he radioed to Brad in the inflatable boat.

No response.

Finally, the walkie-talkie came to life.

"Five, four, three, two, one," Brad's voice crackled above the static.

That was it, a team member said. No other act marked the start of the mission. The robot was on its own.

Bob, fresh from the launching, ran into the lab. He locked his eyes on the monitor. The icon of the robot was pulling away from the ship, signaling that *Odyssey* was diving beneath the waves.

"She's off and running," he observed, his face tense.

Epilogue

THE AGE of global exploration that began a half millennium ago was driven by more than curiosity and restlessness. Religious zeal motivated part of the enterprise. Nerve and muscle had their place. But even the thirst for economic gain, while important, was obviously no means of establishing remote empires and acquiring great fortunes. Columbus, da Gama, Magellan, and all the rest were backed by increasingly rich cultures and patrons and, as important, by a wave of increasingly subtle tools and sciences that let the explorers probe the unknown and possess some hope of a safe return. Artisans improved the magnetic compass. Mercator drew his ingenious maps. Astronomers and mathematicians, finding new order in the heavens, came up with all kinds of navigational aids. Kepler's star tables were a bible. The astrolabe and quadrant, octant and sextant, all gave increasingly precise celestial readings and all aided in the calculation of latitude, which eventually allowed mariners to find their north-south position to within a league or so. The early challenges were great enough to attract the likes of Galileo, Newton, and Halley, all of whom sought ways of calculating the longitude at sea, the knottiest of the navigational problems. Whole institutions, including the British Royal Observatory at Greenwich and the French Academy of Sciences, were founded with the specific goal of aiding the mariner and over time produced a wholesale improvement in charts, maps, almanacs, and tables.[1]

It is no accident that the exploratory age coincided with the start of the scientific revolution. The two were entwined. It was rigorous application of scientific method that allowed naval surgeon James Lind to show that fruits and vegetables could cure and prevent scurvy, the dread disease of long voyages. It was love of experimentation that led Benjamin Franklin to discover that pointed metal rods could channel lightning away from buildings and other structures—an invention that reduced the danger of storm-induced fires at sea. It was the British Royal Society, that pioneer of scientific patronage, that funded John Harrison's work on the perfection of a precise seagoing clock, the achievement of which finally allowed for the accurate calculation at sea of longitude and the finding of east-west positions. And not just clocks advanced. Steady improvement marked all kinds of machines and devices, the most important being ships. Caravels were equipped with lateen sails that caught the wind from a variety of angles, allowing mariners to travel in almost any direction and to make progress in shifting winds. Deep keels eventually gave way to shallow-draft hulls studded with nails or covered with copper, reducing drag and discouraging worms and barnacles. Big tillers that required several men were replaced by pulley systems that allowed a single helmsman at a wheel to control the rudder.

Over the centuries, this tide of material improvement and scientific comprehension transformed man's horizons. Myths died. Peril and fear abated. Circumnavigation became routine. Europeans learned that all seas are one and that any country in the world with a coastline could be visited in time. The Americas were so accessible that the Spanish Crown, habitually overextended, regularly pledged to bankers the riches of its treasure fleets years in advance of their arrival. Arab middlemen who moved oriental goods over land were eliminated in favor of European ships heavy with silks, rugs, jewelry, steels, porcelains, and spices. Russia, eager for a share of the wealth, built the port of Saint Petersburg and amassed a large navy. Voyages to the wilds of the North Pacific became commonplace, setting off a contest for its control among London, Paris, Washington, Tokyo, and Saint Petersburg. In the annals of Western history, the age of discovery has few rivals for sheer consequence and profound change of world view. The sea had always been a barrier. Now, rather abruptly, it became a bridge. The thunder of waves no longer signified an end but a beginning.[2]

We are now in the midst of a similar kind of transformation, only this time the push is downward, not outward. Perhaps we are at a point analogous to the voyages of the eighteenth century before Cook charted New Zealand and discovered Hawaii. Perhaps we are at an earlier stage. No matter where we stand on the exploratory curve, it is clear that activity is increasing, that new advances are in the works, and that much remains to be done. All the evidence suggests that, relative to the challenge, we have barely begun to slip beneath the waves.

Consider the aims of Graham Hawkes, the inventor whose shop overlooks San Francisco Bay. In October 1996, he and his winged submersible *Deep Flight* disappeared into the gloom for which he has such fondness, inaugurating the craft and, in so doing, moving a step closer to his dream of probing the craggy recesses of the Challenger Deep, nearly seven miles down, the sea's deepest spot. No human can now reach that domain, nor the bottom of any of the Earth's great ocean trenches. In contrast, people have climbed Mount Everest in the Himalayas more than one thousand times. And rockets have cast people into space more than seven hundred times. Whether or not Hawkes achieves his goal, the mere fact of his planning the dive says much about the field's acceleration. In modern times, large sovereign states had a near monopoly on the deep, turning its frontiers into battlefields. Now the abyss is increasingly within reach of small nations, firms, and even individuals such as Hawkes. We are entering an age of civil inquiry.[3]

My own first glimpse of the deep illustrates the surprising richness of the worlds below and hints at how much remains to be done. A mile and a half down, in a darkness once thought to be devoid of life, our quick samplings at the base of the Beard chimney gathered a total of 236 animals of fourteen different species, including worms, limpets, copepods, snails, sponges, and sea spiders. And it turns out that the white bottle brushes and gooey pink colonies that we saw and retrieved were in fact unusual types of deep corals. Given this menagerie, one wonders what a more leisurely sampling of that wilderness would have revealed.[4]

Perhaps the most intriguing finding of our 1993 expedition was microbial. Beard and its cousins turned out to be aswarm with hyperthermophiles. And, surprisingly, so were some of the more tepid vents sampled miles northward along the ridge. To John Baross, the biologist at the University of Washington, the finding suggested that the vents had common roots deep within the seabed where heats were higher and better suited to hyperthermophile growth and reproduction. His view gives weight to the extraordinary idea that the planet harbors a hidden biosphere made up of microbes living primarily in hot, wet rock—a third domain of life distinct from land and sea. If such a bizarre world exists, it is one I was privileged to glimpse as clouds of microbial life swirled out of Beard chimney into the gloom.[5]

That monolith and its kin turned out to be an old vent field reinvigorated by the volcanic eruption, as we suspected. The next summer, in 1994, a return expedition led by Bob Embley of NOAA discovered even larger chimneys in the area, including a giant dubbed Mongo. Northward along the ridge, at the Floc site, Embley's team found that lush bacterial fields had been invaded by tube worms and other creatures, giving a first glimpse of how virgin deep vents are colonized. More dives in 1995 revealed that the tube worms had undergone a growth spurt so that their lengths measured up to four feet. The creatures were massed in dense thickets around warm vents, thousands of them.[6]

The repeated visits to this dark world have given us our first insights into how the volcanic fury that occurs with such regularity in the sea's depths can rapidly turn a barrens into a jungle, fertilizing the water, enriching a hidden realm. The advance was made possible by the formerly secret Navy microphones that monitored the eruption in June 1993 and by the team of universities and Federal agencies that scrambled to exploit the opportunity. As we rush to eliminate Federal budget deficits, it is worth reflecting on the scientific benefits of relatively small public investments.

We know the sea is the birthplace of all terrestrial life. But we are just learning that it is also our lifeblood, that the planet's deep volcanic heartbeat renews and circulates the sea and helps make this a living planet rather than a dead one. In a remarkable twist of geopoetry, we may ultimately learn that the Earth's heartbeat gave rise to our own.

The serious work of deep exploration is just getting underway, aided by such things as the new gravity maps that reveal so many of the seabed's hidden features. It will take decades, if not centuries, for the deep vents and kindred structures to be mapped globally and investigated up close—a labor that perhaps will reveal the true nature of the Garden of Eden and, as new kinds of hyperthermophiles are discovered, give us genetic tools to help us cure diseases and shape our own evolutionary destiny.

As I found, the civilian age is also accelerating the study of the planet's dark midwaters. The Monterey Bay Aquarium Research Institute, eager to advance its robot explorations, in 1996 won access to a Navy microphone anchored off Point Sur at a depth of more than a mile. The institute's scientists are now working to discover the identities of such riddles of the deep as the Echo, Carpenter, and Woof-Woof, named after analogous sounds on land. Perhaps one day Robey will discover that the mysterious vocalizers are some of the large animals he dislikes referring to as monsters.[7]

So, too, the hunt is heating up for the giant squid, the greatest of the sea's legendary beasts, still unobserved in its lair despite any number of human efforts. Early in 1997, a team of American experts from three institutions set out for New Zealand, whose midwaters are thought to harbor the creature. *Odyssey,* the robot that speeds independently through the deep, was to lead the probing of the icy darkness. Any sightings of the giant are expected to shed light on its behavior, habitat, and perhaps how to help save the secretive animal from environmental harm.[8]

The new age is illuminating not only the dark abyss but all the sea's provinces, deep and shallow. In murky waters off Long Island, where scuba divers are virtually blind, Federal investigators and their contractors in 1996 lowered a laser line scanner to hunt for wreckage from TWA flight 800. Flashing its beam through the gloom, the laser found much of the jumbo jet and disclosed even large pieces in rivet-by-rivet detail. The pictures in one case were so clear that investigators were able to identify first-class tags on luggage.[9]

The seabed is a vast repository of failed ambition that contains not only crash debris but millions of ships and treasures, arms and artifacts have have been lost over the ages—everything from jeweled tiaras to nuclear warheads. It is in some respects the greatest, and deadliest, of all museums. And it is opening rather suddenly as people use deep technologies to uncover a wealth of human worlds that had been presumed gone forever.

Done right, the excavation of this archive will strengthen us as a species and give us new depths of self-knowledge. Even a superficial analysis will rewrite textbooks, if only by delineating surprising trade routes and contacts between ancient peoples. If it is done wrong, we will miss a unique opportunity that is never to be repeated. This wave breaks but once. It has been approaching for thousands of years and its first agitations are upon us. If Paul Tidwell, the finder of the lost Japanese submarine *I-52,* is correct in his assessments, the exploratory wave in the next century will shake up thousands of deep wrecks. For better or worse, we are at a turning point in the history of history.

The most celebrated of the lost ships, the *Titanic,* once imperturbable at a depth of nearly two and a half miles, already lies damaged in the icy darkness, a victim not only of the *Mir* crash but of Tulloch's bid to raise the hull section. In late August 1996, after hauling the piece to the surface, he watched in disbelief as equipment failures and rough seas sent it tumbling back into the depths, lost a second time. Perhaps these are the normal setbacks of bold ventures. But the *Titanic* is such a cultural icon that the jolts should sound alarms and spur discussion.[10]

Part of the problem is that the rush into the deep has upset the uneasy balance that long existed between commercial salvors and marine archaeologists, between private and public efforts to uncover civilization's past at sea. For decades both groups dove mainly in scuba gear, their hunts limited by equipment to shallow waters. Now the power has shifted into the hands of commercial teams and treasure hunters. Only they can afford deep explorations that cost $10,000 to $35,000 a day. Only they can promise investors big paybacks, sometimes quickly. In most cases, archaeologists are left out, as the frustrated ones of Portugal discovered. Commercial teams, aware of rising public interest in sunken history as well as fear of its destruction, are starting to adopt codes of ethics and to add archaeology to their deep agendas. But self-policing remains unproven and some scholars argue that commercial ventures by nature imperil the past.[11]

In light of the stakes, and the risks, it is important to address these emerging issues in a serious way and perhaps to strengthen the public sector so it can better guide the private one. Given the squeeze on governments everywhere, we also need new kinds of philanthropy. The opening of the deep museum is a wonderful opportunity for Bill Gates and other silicon tycoons. Their love of technology makes them ideal patrons at a critical juncture, and they already have a flattering role model in David Packard and his bankrolling of the midwater explorations. Whatever comes of such deliberations, the time to act is now, not ten or twenty years from now when it is too late.

Beyond the realms of booty and archaeology lie all kinds of natural treasures now facing their own varieties of promise and peril. A rush is developing for deep fish and microbes, oil and gold, perhaps foreshadowing resource wars in the next millennium. In time, perhaps inevitably, the long decades of deep mineral inquiry will give way to extraction. Indeed, that day may arrive sooner than we think as Asian nations grow rapidly and become increasingly hungry for natural resources.[12]

A grim harbinger of ecologic harm is the coelacanth, the foremost of the living fossils, a creature dating to the age of the dinosaurs. Intense fishing appears to be pushing it to the brink of extinction.[13]

The benefits of wide exploitation of the deep have to be weighed carefully against such risks as deep pollution and species eradication, which are very difficult to anticipate given our general ignorance about the deep sea. If history is any guide, a lopsided race will be run between development and the comprehension of its repercussions.

But history cannot be the guide. The deep is the one place on the planet where consequences are inescapable. There is nowhere else to go, no other terrestrial wilderness to rush off to after the excitement of pillage and plunder is over. On the other hand, we must resist the temptation to romanticize the sea and to sanctify the idea of naturalness. Every day the hot vents fire many tons of heavy metals and noxious chemicals into the icy darkness, in some respects dwarfing the toxic wastes of civilization.

Above all, we need a balanced understanding of environmental matters that disentangles real from imaginary threats. ATOC and *Odyssey* and other gifts of technology can help us better understand the state of the sea, yet budgets for such endeavors are continuously under fire.

This new age is the best and worst of times for government and university scientists. They are often unable to fully exploit the wealth of new opportunities for lack of money, unlike growing numbers of firms and nations and entrepreneurs. In the United States during the past fifteen years, Federal funds for ocean research have dropped by almost half. In 1996, a concerted effort got underway in Washington to reverse this trend, inspired in part by the military spinoffs. A rise in funding is considered especially important for deep environmental studies, which have been shunned by the free market and are a natural for government, with its long-term interests in human welfare.[14]

Deep environmental work is vitally important yet very difficult to do, as suggested by a comparison to tropical rain forests. It is relatively easy to show that the Earth's lush jungles are under siege and that their residents are going extinct at an alarming rate. But what of the darkness below? We currently have only the sketchiest understandings of its structure and inhabitants and food chains, much less the alterations being made by deep fishing and planetary heating and the seventy thousand or so synthetic chemicals that we have managed to inject into the global environment. By any measure, our ignorance is almost as boundless as the deep.

The Russians, for all their focus on economic survival, have shown themselves to possess a degree of ecological sensitivity by virtue of repairing some of the damage to the *Komsomolets,* the leaky nuclear submarine that lies a mile down in the deep Atlantic. Theirs is a small step toward addressing the scores of nuclear relics from the cold war that lie scattered throughout the deep, including reactors and warheads. Given the growing exploratory powers of small states and private companies, and the growing possibility of atomic plunder, it seems inevitable that these highly sensitive sites will have to be discussed more openly and cleaned up if such action is warranted.

A prudent step in the direction of ecological stewardship would be the creation of deep sanctuaries, of whole regions placed off limits to deep fishing and mining and development. The international move to coastal sanctuaries is insufficient. Given how little we know of the deep repercussions of our activities, and given the overall stakes and global pressures, it would be best to keep large parts of the wilderness as intact as possible for the sake of ourselves and future generations.[15]

By informal consensus, deep investigators have already set aside a place in the darkness of the eastern Pacific where a tower of lava rises nearly three stories. Top to bottom, it is surrounded by tube worms, the creatures packed densely and arranged as if in a formal garden, their red plumes facing outward like flowers. Struck by the surreal beauty, the scientists have pledged only to look, not to touch.[16]

It is folly to assume that the ecological fate of the sea is bleak and inevitable. Our decisions, individual and collective, make a difference, and those decisions will grow more important in the next century as the human population of the planet swells, along with its army of developmental pressures. Already, two-thirds of the people in Earth live near an oceanic coastline, and that fraction is growing. In the whirlwind of progress, we will either destroy the sea through ignorance or save it through knowledge. And we will feel the outcome either way. Practically every day science finds new evidence that the health of the planet is intimately tied to the state of the ocean and that our own fate and that of all living things revolves around the habitability of this global commons.

Lastly, a word about surprises. This book has focused on a few worlds of the deep, on those that are known or starting to be explored. In all likelihood, there are many others out there waiting to be discovered. Just consider the phenomenon of life. The majority of the living things on this planet dwell in the sea. Yet we know little about their representatives in the unilluminated depths, an area that makes up the vast preponderance of the biosphere.

Our deep explorations to date have tended to focus on things that we are comfortable with, on things we can conceive, on ourselves, on our past. If this book has dwelled at some length on lost ships, it is because they are attracting a lot of attention. Indeed, the action is more extensive than it seems, since commercial secrecy in some cases has come to replace the military secrecy of the cold war.

The foibles of our anthropocentric focus are driven home by the whole history of deep exploration. At almost every juncture, our imaginations and intuitions failed us. Generations of scientists dismissed the abyss (a dismissive word in some respects) as inert and irrelevant, as geologically dead and having only a thin population of bizarre fish, mainly scavengers living off a rain of detritus from above. They were wrong, impressively so. The surprises include black pearls, giant clams, hyperthermophiles, cold seeps, black smokers, living lights, gelatinous drift nets, sound channels, volcanic rifts, tectonic chasms, gulper eels, invertebrate swarms, hot vents, tube worms, volcanic gold deposits, and cobalt crusts, to name just a few of the more celebrated finds. Weighing the evidence, and remembering the unexplored vastness of the deep, it seems fair to conclude that much more of interest lies out there in the darkness, unknown to us.

One indication of the potential richness is the continuing high rate of discovery. Whenever biologists are lucky enough to descend to unexamined parts of the seabed they tend to find literally hundreds of new species, producing droves of questions about how the animals live and fit into the immensity of the dark ecosystem.[17]

The surprise factor was driven home to Tidwell during his search of the *I-52*'s debris field, more than three miles down. He found the seabed dotted not only with bits and pieces of the old Japanese warship but with puzzling dark blobs a bit larger than coffee cups. Aboard the *Yuzhmorgeologiya,* old hands conferred among themselves and decided that the mysterious objects were fuel globules from the *I-52* that somehow had congealed over the years and been preserved by the deep's intense cold and pressure. But eventually, one blob was seen moving. Close inspection showed that it was alive. At the end of the voyage, Tidwell resolved that, during any return visits, he would bring along seasoned biologists to study the living globules and any other unusual creatures that the site might conceal.[18]

Some day we will visit the stars. Until we do, out in our own backyard is the universe below.

Chronology of Deep Exploration

1818	Sir John Ross lowers a line more than a mile into the North Atlantic and hauls up worms and a large sea star.
1843	Edward Forbes proposes that no substantial life can exist below three hundred fathoms.
1858	The first transatlantic telegraph cable comes to life, its laying preceded by deep seabed surveys.
1859	Darwin's *Origin of Species* implies that the deep is a sanctuary for living fossils.
1864	Norwegians haul up from the deep a sea lily—a living fossil previously found only in rocks 120 million years old.
1870	Jules Verne's *Twenty Thousand Leagues Under the Sea* depicts no life in the ocean's deepest regions.
1872–76	British ship *Challenger* sails the globe while lowering dredges and other gear into the deep, finding long mountain chains, puzzling nodules, and hundreds of animals previously unknown to science.
1892	Prince Albert of Monaco starts to probe the sea's dark midwaters, discovering new kinds of eels, fish, and squid.
1920	Alexander Behm sails the North Sea and bounces sound waves off its bottom, advancing a new method of depth measurement known as echo sounding.
1925	Fritz Haber launches the German *Meteor* expedition in a bid to extract gold from seawater.

1934 — William Beebe and Otis Barton descend in a tethered sphere to a depth of a half mile, where they glimpse a previously unseen world of living lights and bizarre fish.

1938 — Fishermen off South Africa pull up an ungainly five-foot fish identified as a coelacanth, a living fossil thought extinct since the days of the dinosaurs.

1948 — Auguste Piccard dives in his bathyscaph, the first untethered craft that carried people into the deep.

1950–52 — Danish ship *Galathea* lowers dredges into the sea's deepest trenches and hauls up swarms of invertebrates.

1951 — British ship *Challenger II* bounces sound off the bottom, and near Guam finds what appears to be the sea's deepest chasm, its lowest point nearly seven miles down, subsequently named the Challenger Deep.

1952 — Marie Tharp, studying echo soundings, discovers that the Mid-Atlantic Ridge conceals a long rift valley, which turns out to be part of a hidden volcanic rent that girds the global deep.

1953 — Auguste Piccard and his son Jacques enter *Trieste,* an improved bathyscaph, and dive to a depth of nearly two miles.

1958 — American Navy buys *Trieste* and begins to strengthen its steel personnel sphere.

1960 — Jacques Piccard and Don Walsh dive in *Trieste* to bottom of Challenger Deep, seven miles down.

1961 — American ship off Mexico lowers a pipe through more than two miles of water and drills into the rocky seabed, a first that advances the fields of deep geology and mining.

Robert Dietz, studying echo soundings, proposes that the seabed's mountainous rifts are invisible scars where molten rock from the Earth's interior wells up periodically and spreads laterally to form new ocean crust, a process he calls seafloor spreading.

1963 — *Thresher,* America's most advanced submarine, sinks in waters a mile and a half deep with the loss of 129 men.

Trieste finds the shattered hulk of *Thresher* on the bottom after five months of searching.

1964 — American Navy founds the Deep Submergence Systems Project to develop new gear that can better probe the deep sea's darkness.

Navy launches submersible *Alvin,* the first piloted craft able to roam the deep with relative ease.

1965 — Navy tests its first underwater robot.

Navy develops *Halibut,* a submarine that can lower miles of cables bearing lights, cameras, and other gear to spy on enemy armaments and matériel lost on the bottom of the sea.

1966 — *Alvin* and Navy robot probe the deep Mediterranean and retrieve a lost American hydrogen bomb.

Halibut spies on Soviet warheads abandoned to the deep.

1967 Geologists, after fierce debate, agree that seafloor spreading involves a dozen or so huge plates that form the Earth's crust and move slowly over time, rearranging the land.

1968 Soviet submarine sinks in the deep Pacific, littering seabed with secret code books and nuclear warheads.

In stealth, *Halibut* examines the lost Soviet sub.

American Navy sub *Scorpion* sinks in the Atlantic, killing ninety-nine men and surrendering to the depths two torpedoes tipped with nuclear arms.

1969 *Trieste II,* a new Navy bathyscaph, probes the *Scorpion*'s wreckage more than two miles down and recovers the sub's sextant.

1971 Navy launches first of two piloted craft that hitch rides atop submarines and dive deep for rescues and espionage.

1973 Navy begins to design a type of tetherless robot, eventually known as the *Advanced Unmanned Search System,* for wide hunts of gear lost at depths up to nearly four miles.

1974 Disguised as a seabed miner, American ship *Glomar Explorer* lowers a giant claw to grab a Soviet sub lost on the Pacific floor.

United Nations Law of the Sea conference proposes to tax seabed miners as a way of enriching poor nations.

French-American team dives to Mid-Atlantic Ridge and unexpectedly finds its rift valley paved with lava.

1977 American team dives in *Alvin* to a volcanic rift in the Pacific and discovers warm springs teeming with undescribed species of life, an ecosystem new to science that includes tube worms, snakelike creatures standing upright in long tubes.

1979 American team exploring Gulf of California with *Alvin* finds mineral chimneys that blow clouds of black smoke and discharge water hot enough to melt lead.

1980 Scientists propose that the seabed's hot springs are the birthplace of all life on Earth.

1981 Ronald Reagan becomes President and begins an arms buildup, including new classes of deep craft and new kinds of deep espionage.

1982 Volcanic seamounts in Pacific are found to be covered with rare metals, including cobalt.

United Nations Law of the Sea Treaty is finished and opened for ratification, saying deep minerals belong to the world's people.

1983 Reagan proclaims Exclusive Economic Zone around the United States, effectively doubling the nation's size and fueling a burst of exploration in deep waters.

1984 Robert Ballard tows tethered Navy craft *Argo* over the *Thresher,* scanning the lost sub's corroding wreckage with an array of video cameras.

1984 (cont.) American researchers diving off Florida in *Alvin* discover life swarming in cold springs, another new kind of deep ecosystem.

Mikhail Gorbachev emerges as Soviet leader, starting conciliatory East-West policy.

1985 Ballard lowers Navy craft *Argo* and discovers, more than two miles down, the *Titanic,* broken in two, many of its fixtures and artifacts scattered on the icy seabed.

Graham Hawkes's *Deep Rover* submersible reveals a riot of midwater life in the depths of Monterey Canyon, helping inspire billionaire David Packard to fund deep explorations.

1986 New Navy robot *Jason Junior* probes the interior of the *Titanic,* and in secret missions explores the twisted wreckage of two sunken American submarines, *Thresher* and *Scorpion.*

1987 American firm hires the French *Nautile* submersible to begin *Titanic*'s salvage, hauling up thousands of items, including children's marbles and a lady's wristwatch.

First East-West treaty is signed that reduces nuclear arms.

1988 Treasure hunters searching off South Carolina more than a mile down find the remains of the *Central America,* a wooden ship that sank in 1857, heavy with tons of California gold.

Ballard tows Navy craft *Argo* over Mediterranean deep and discovers a graveyard of ancient ships, including a fourth-century Roman craft.

1989 Ballard lowers *Argo* nearly three miles down in the Atlantic and finds German battleship *Bismarck,* a mass of deteriorating guns and fading swastikas.

Jason, Ballard's top robot for the Navy, debuts and recovers from the deep Mediterranean dozens of artifacts from lost Roman ships.

Berlin Wall crumbles.

1990 Navy begins giving civilian researchers wide access to *NR-1,* a deep-diving nuclear submarine with lights, windows, and wheels.

Japan finishes *Shinkai 6500,* the world's deepest-diving piloted craft.

Russians in *Mir* submersibles probe Monterey Canyon, one of the first in a wave of post-cold-war dives for foreign customers.

1991 American Navy agrees to share with civilian scientists a fleet of deep exploratory craft, including robots and submersibles.

Mir submersibles dive more than two miles down and film *Titanic* wreckage for Canadian IMAX movie.

Soviet ship *Yuzhmorgeologiya,* which once spied on the submarines of the United States Navy, is hired by American government to do studies of deep ecology.

Soviet Union ceases to exist.

1992 Scientists, after a large seabed survey, conclude that the deep may hold ten million species of life, far more than known on land.

Ballard lowers Navy submersible *Sea Cliff* and Navy robot *Scorpio* to examine fourteen ships lost during World War Two at the battle of Guadalcanal.

CIA director Robert Gates tells Russian President Boris Yeltsin that *Glomar Explorer* recovered remains of six Soviet sailors, who were subsequently buried at sea.

American Navy adopts a new strategy in which fighting forces target shallow waters and regional conflicts, reducing the need for deep expertise.

Businessmen hire an American Navy contractor to dive on *Titanic* for commercial salvage.

1993 Two American companies unveil laser cameras, formerly secret Navy tools for seeing long distances in the deep.

Federal scientists listen to Navy deep microphones and hear a deep volcanic outburst on the Pacific's Juan de Fuca Ridge, prompting a number of expeditions to study how heat on the dark seabed can beget a jungle of life.

Japan begins testing *Kaiko,* the world's deepest-diving robot.

French submersible *Nautile* dives on *Titanic* to recover artifacts.

Ballard lowers Navy *Jason* robot in Celtic Sea to probe the deteriorating remains of *Lusitania,* torpedoed by Germany in 1915.

1994 American Navy agrees to share its attack submarines with civilian scientists for arctic studies.

Navy turns over the *Advanced Unmanned Search System,* an early tetherless robot, to private industry.

Shinkai 6500 sets an Atlantic depth record for a piloted vehicle, studying deep geology.

Russians in *Mir* submersibles carry British scientists down to Mid-Atlantic Ridge to study a huge volcanic mound laced with gold.

Nautile dives on *Titanic* to recover artifacts.

United Nations Convention on the Law of the Sea goes into force.

1995 *Kaiko* dives to bottom of Challenger Deep, finding the icy darkness alive with small animals.

Paul Tidwell arms himself with naval spinoffs and finds in Atlantic waters more than three miles down the lost Japanese submarine *I-52,* which sank in 1944 heavy with tons of gold.

Ballard dives in Navy's *NR-1* to map a field of deep Mediterranean wrecks, some more than two thousand years old.

Mir submersibles film *Titanic* for a Hollywood movie.

American Navy releases seafloor gravity data, which civilian oceanographers turn into the first good public map of the global seabed.

1995 (cont.) Civilians start broadcasting deep sounds across the Pacific and listening with Navy microphones for changes in travel time, seeking to measure global warming.

1996 Federal scientists listening to Navy microphones hear fury on the Pacific's Gorda Ridge, prompting new studies of seabed volcanism.

The robot *Jason,* one of the Navy's top deep projects of the eighties, makes its debut for a Federal scientific group, its first expedition probing hot vents on the Mid-Atlantic Ridge.

Navy widens access to its deep microphones, prompting the development of private acoustic observatories meant to listen for volcanic eruptions and whale songs.

The advanced robot *Tiburon* debuts at Packard's institute, ready to explore down to a depth of four kilometers, or two and a half miles.

Nautile dives on the *Titanic* to film the shattered hulk and recover artifacts, including a large section of the liner's hull.

Deep Flight makes its debut, taking Hawkes a step closer to flying into the Challenger Deep, seven miles down.

1997 American experts use the robot *Odyssey* to search the dark waters off New Zealand for the giant squid, the greatest of the sea's legendary beasts.

Measure Equivalents

0 degrees Celsius	=	32 degrees Fahrenheit
100 degrees Celsius	=	212 degrees Fahrenheit
degrees Celsius	=	9/5 (+32) degrees Fahrenheit
0 degrees Fahrenheit	=	−17.8 degrees Celsius
98.6 degrees Fahrenheit	=	37 degrees Celsius
degrees Fahrenheit	=	5/9 (−32) degrees Celsius

centimeter	=	0.39 inch
fathom	=	6 feet, or 1.83 meters
foot	=	0.3 meter
inch	=	2.54 centimeters
kilometer	=	3,280 feet, or 0.6 mile
league	=	roughly 3 miles
meter	=	39.37 inches, or 1.1 yards
mile	=	5,280 feet, or 1.61 kilometers
mile, nautical	=	6,080 feet, or 1.85 kilometers
story	=	10 feet, or 3.05 meters
yard	=	0.91 meter

knot	=	1 nautical mile per hour

square centimeter	=	0.16 square inch
square foot	=	929 square centimeters
square inch	=	6.45 square centimeters
square kilometer	=	0.39 square mile
square meter	=	10.76 square feet

square mile	=	2.59 square kilometers
square yard	=	0.84 square meter
kilogram	=	2.2 pounds
pound	=	0.45 kilogram
ton	=	2,000 pounds, or 0.9 metric ton
ton, metric	=	2,205 pounds

Glossary

ABYSS A synonym for the DEEP SEA.

ABYSSAL PLAIN The floor of the deep sea extending outward from a continental rise or an oceanic trench.

ACORN WORM A type of marine INVERTEBRATE whose head resembles an acorn, found down to depths exceeding three kilometers, or nearly two miles. Not a true worm but a member of the PHYLUM Hemichordata, these animals burrow in the sand and mud. Lengths range from inches to over six feet. Some acorn worms are brilliant in oranges, reds, and yellows.

ACOUSTIC THERMOMETRY OF OCEAN CLIMATE A project begun by SCRIPPS in the 1990s to measure the overall temperature of the Pacific and any changes therein by broadcasting sound waves across its breadth and measuring subtle changes in their travel time over years and decades.

ADVANCED TETHERED VEHICLE A 20-foot Navy robot that can dive to depths of 6 kilometers, or 3.7 miles. It has three arms and many still and video cameras, which send images up FIBER-OPTIC cables to shipboard operators at the surface.

ADVANCED UNMANNED SEARCH SYSTEM One of the world's first complex robots able to roam the seabed without a tether, *AUSS* was begun by the American Navy in 1973 for wide-ranging hunts for lost gear down to depths of 6.1 kilometers, or 3.8 miles. Seventeen feet in length, it made 114 successful dives for the Navy before being turned over to private industry in 1994.

AKADEMIK MSTISLAV KELDYSH The 422-foot mother ship of Russia's *MIR* twin submersibles and one of the world's largest oceanographic research vessels.

ALBATROSS A large, web-footed seabird.

ALGAE An ancient group of primitive plants ranging from unicellular PHYTOPLANKTON to KELP forests.

ALVIN The world's first SUBMERSIBLE able to roam the deep seabed with relative ease, completed by the American Navy in 1964 and run by WOODS HOLE. It was originally rated for a depth of 1.8 kilometers, or 1.1 miles, but was eventually strengthened so it could descend down to 4.5 kilometers, or 2.8 miles. The 25-foot craft can carry up to three people.

AMPHIPOD A small type of marine CRUSTACEAN, swarms of which can move across the deep seabed like insect hordes, seeking to feed on carrion.

ANGLER A group of deep-sea fish in which a thin rodlike projection near the mouth acts as lure, its tip often glowing to attract prey. About 110 known species. Anglers are often small, but reports occur of specimens up to four feet in length. The fishing rod usually projects from an area of the head between the eyes and can be moved about to attract prey. The teeth of anglers are often daggerlike.

ANHYDRITE A white or grayish mineral, calcium sulfate, that makes up a large part of the structural matrix in the deep's active volcanic CHIMNEYS. The granular mineral forms only at high temperatures, and dissolves back into the sea if temperatures go down. Since it dissolves quickly, extinct chimneys and the cool surfaces of live ones usually have no anhydrite. Conversely, thin, young chimneys are made principally of the crumbly stuff.

APOLEMIA A genus of deep SIPHONOPHORE.

ARCHAEA A large branch on the tree of life for MICROBES of ancient pedigree that thrive in extreme environments, including places that are very salty and very hot. The discovery of these organisms in the 1970s prompted scientists to draw up a basic new classification of the living world, with the new group called Archaea. It is a major kingdom, alongside eubacteria (normal bacteria) and eukaryotes (which covers all higher organisms, including plants and animals and humans). The name Archaea implies an ancient origin, and many biologists believe that these organisms are the ancestors of the first forms of life to flourish on Earth.

ARCHAEAN HYPERTHERMOPHILE An archaean MICROBE that thrives in hot spots, usually springs on land or beneath the sea, often doing so at temperatures near or above the usual boiling point of water.

ARGO A 15-foot American robot built by Ballard's team at WOODS HOLE and owned by the Navy. Towed on a long line from a ship, with no maneuvering power of its own, its video cameras beaming up images over a FIBER-OPTIC line, *Argo* debuted in 1984 and is best known for its 1985 discovery of the *TITANIC* more than two miles down.

ARTHROPOD A phylum of segmented INVERTEBRATES with jointed legs, including the insects, CRUSTACEANS, and TANAIDS.

ATLANTIC OCEAN The second-largest body of water on Earth, after the Pacific. Its area is 33 million square miles, or about 17 percent of the planet's surface. Its average depth is 11,730 feet, equal to 2.2 miles, or 3.6 kilometers.

ATLANTIS II The 210-foot mother ship of *ALVIN*, retired in 1996.

ATOC See *Acoustic Thermometry of Ocean Climate.*

AUSS See *Advanced Unmanned Search System.*

AZOIC THEORY An early nineteenth-century idea of great influence holding that the DEEP SEA is lifeless or nearly so.

AZORES An archipelago in the Atlantic Ocean where the isles are 900 to 1,200 miles west of Portugal. The islands are peaks of undersea volcanoes, whose flanks drop away steeply to deep water.

BACTERIA Unicellular organisms with no distinct CELL nuclei.

BALEEN An elastic material that grows in lieu of teeth in the upper jaws of certain whales and forms a series of thin, parallel plates on each side of the palate to filter PLANKTON from the water.

BALLAST Heavy material in a ship's hold to lower the center of gravity and provide greater stability when the ship carries little or no cargo.

BARNACLE A modified CRUSTACEAN that attaches firmly to ship bottoms or rocks and gathers food from the sea with a beating, feathery arm.

BASALT Rock formed of cooled LAVA that characteristically is dark in color, fine-grained, and rich in silica, iron, and magnesium. Though little known to most people, it is the predominant rock of the global seabed and, for that matter, of the planet's surface.

BATHYMETRIC MAP A map that shows the bottom contour of the sea, ocean, or large body of water.

BATHYMETRY The measurement of the depth of the sea, ocean, or large body of water.

BATHYSCAPH The world's first SUBMERSIBLE, invented by Piccard in the 1940s and used extensively for deep-sea exploration. Unlike the BATHYSPHERE, it could move freely if clumsily in all directions. Instead of having a surface ship counteract the heaviness of its steel personnel sphere, the bathyscaph gained BUOYANCY with an enormous upper tank filled with gasoline, which is lighter than seawater.

BATHYSPHERE A metal diving sphere tethered to a ship by a long cable in which scientists were lowered in the 1930s, mainly to study midwater life.

BENTHOS Organisms that live on the seabed.

BIOLUMINESCENCE The production of light by living organisms, including species of bacteria, marine INVERTEBRATES, and fish. It results from a chemical reaction either within certain CELLS or organs or outside the cells in some type of excretion.

BIOSPHERE The part of the Earth's CRUST, waters, and atmosphere where life can subsist.

BLACK PEARLS A nickname for MANGANESE NODULES.

BLACK SMOKER A volcanic spring of the DEEP SEA that belches hot waters at temperatures of around 350 degrees Celsius, or 660 degrees Fahrenheit, the blackness of its waters produced by particles of metallic SULFIDES.

BLAKE PLATEAU An undersea shelf extending along the east coast of Florida from the Bahamas to Cape Hatteras, varying in depth from 200 meters to 1,000 meters.

BLUE WHITING A deep fish looking less monstrous than many abyssal residents, with dorsal and ventral fins and a wide caudal fin typical of surface dwellers. It is beginning to be fished globally.

BOILING POINT The temperature at which a substance changes state from a liquid to a gas at a given pressure.

BOLLARD A stout metal post on a ship, submarine, or wharf used to secure mooring lines.

BRACHIOPODS An ancient group of bivalves in which the upper and lower shells are of unequal size, with the lower being larger. Also known as lamp shells, since they supposedly look like ancient Roman oil lamps.

BRIDGE A raised structure or platform, often enclosed to form a pilothouse, that lies toward the bow of a ship. It has a clear view ahead and is the place from which a mariner steers a powered vessel.

BRIG A two-masted sailing vessel that is square-rigged on both masts.

BRISTLE WORM A relative of the common earthworm that lives in the sea and has hairs that extend from each segment. Also known as polychaete worm.

BRITTLE STAR A type of ECHINODERM with a small dish-shaped central body and five thin arms that can be quite long, making it look like a skinny kind of SEA STAR. About 2,000 known species. They are scavengers and PLANKTON feeders that live on the seabed at almost every depth.

BRYOZOAN A member of a PHYLUM that forms branching, encrusted, or gelatinous colonies of many small POLYPS, each having a circular or horseshoe-shaped ridge bearing ciliated tentacles. These filter feeders, also known as moss animals, are found down to at least 6 kilometers, or 3.7 miles.

BUOYANCY The ability or tendency to rise or float in a liquid.

CALCAREOUS Containing CALCIUM CARBONATE.

CALCIUM CARBONATE A structural compound of animals found in bones and shells and corals that is also the major component of LIMESTONE.

CAMBRIAN A geological period dating from 510 million to 570 million years ago that saw an explosion in the quantity and diversity of the earliest forms of life, which were restricted to the sea. Among the organisms that thrived were algae and such marine INVERTEBRATES as TRILOBITES.

CARAVEL A two- or three-masted sailing ship with broad beam, high rear deck, and LATEEN rig, used by the Spanish and Portuguese in the fifteenth and sixteenth centuries.

CARBOHYDRATE An ORGANIC COMPOUND consisting of CARBON, hydrogen, and oxygen, including sugars and starches.

CARBON The ELEMENT essential to all living things and present in the Earth's environment as a free element in the form of coal, graphite, and diamond. It is a unique element in that it can form an almost infinite number of chemical combinations as carbon atoms bond to one another, making possible all kinds of chemical rings and chains.

CARBON DIOXIDE A colorless, odorless gas present in the atmosphere at a concentration of less than one percent. Produced by combustion as well as by the respiratory processes of animals, consumed by plants during PHOTOSYNTHESIS. Rising atmospheric concentrations of the gas are seen as threatening to warm the Earth through the GREENHOUSE EFFECT.

CATALYST A substance that induces or accelerates a reaction while itself undergoing no changes.

CAUDAL To the rear of an animal's body.

CELL The basic structural unit of most organisms, usually microscopic in size. Within its walls are found all the chemical mechanisms of respiration and reproduction, including the heredity molecule DNA.

CEPHALOPOD A type of MOLLUSK with well-developed eyes and a ring of tentacles around the mouth. Examples include the SQUID, OCTOPUS, and NAUTILUS. The shell is absent or internal on most members.

CESIUM-137 A radioactive ISOTOPE of the ELEMENT cesium. It is a common by-product of NUCLEAR FISSION in bombs and reactors and thus is slowly accumulating as spent-fuel waste. Cesium-137 has a half-life of 30 years. In humans, it is absorbed rapidly and is distributed throughout the body, where it enters into reactions that normally involve potassium, which are fairly common. Acute exposures can result in death.

CETACEA An order of marine mammals that includes the whales, dolphins, and porpoises.

CHALLENGER A 226-foot British CORVETTE that circled the globe between 1872 and 1876 to probe the deep sea, finding 4,717 new species on land and sea and basically founding the field of deep investigation.

CHALLENGER DEEP The deepest-known spot in the global sea, situated at the bottom of the MARIANAS TRENCH east of the Marianas chain of islands in the western Pacific. Its floor is about 35,840 feet deep, which is 6.8 miles, or 10.9 kilometers. The Challenger Deep is a yawning fissure on an already deep convergent PLATE boundary where the huge Pacific slab meets the smaller Philippine plate and plunges beneath it into the Earth's hot interior.

CHEMOSYNTHESIS The formation of ORGANIC COMPOUNDS in which energy is derived from INORGANIC substances such as SULFUR and hydrogen, as opposed to PHOTOSYNTHESIS, in which the driving force is sunlight.

CHIMERA A group of bizarre deep-sea fish that look as if they were cobbled together from other animals, including birds, sharks, and RATTAILS. About 30 known species. Many have grinning mouths and long pointy noses. Most are about 3 feet in length, much of that a long whiplike tail. Chimeras may have evolved from early sharks.

CHIMNEY Nickname for tall mineralized mounds that grow atop volcanic VENTS at the bottom of the DEEP SEA, their tops and sides usually alive with hot gushing water rich in minerals. The hottest ones are known as BLACK SMOKERS.

CILIA Short hairlike structures common on lower animals. Beating in synchrony, they may be used for locomotion or to create water currents that carry food toward the animal's mouth.

CLAM A MOLLUSK that is one of the common bivalves. Clams range in size from ones that can barely be seen to giants that weigh more than 500 pounds. Large beds of clams are common along some deep-sea volcanic RIFTS, living in symbiosis with CHEMOSYNTHETIC bacteria.

COBALT A hard silver-white metallic ELEMENT used in steel alloys, especially those meant to resist high temperatures. In the deep sea it is often a constituent of MANGANESE NODULES.

COELACANTH A primitive fish that first appeared during the DEVONIAN Period, some 400 million years ago, and was thought to have gone extinct at the end of the CRETACEOUS Period, 65 million years ago, the same time the DINOSAURS died off. But in 1938 a South African fishing boat, dredging deeper than usual, pulled up a living coelacanth, astonishing the scientific world.

COELENTERATE A member of a PHYLUM (also known as Cnidaria) characterized by simple radial symmetry and a body in which a single internal cavity serves for digestion, excretion, and most other functions; includes CORALS, SEA ANEMONES, and JELLYFISH. They are some of the planet's oldest residents, having flourished since pre-CAMBRIAN times, perhaps more than a billion years ago.

COLD SEEP A cold spring on the seabed often rich in METHANE or HYDROGEN SULFIDE and harboring a diverse FAUNA similar in many respects to the hot-VENT communities.

COLD WAR The economic, military, and political struggle between the United States and the Soviet Union, as well as their allies and clients, that lasted from about 1945 to 1990. The long stalemate was characterized by the buildup of atomic arsenals and the threat of nuclear war.

COMB JELLIES Animals of the PHYLUM Ctenophora that are often spherical or teardrop-shaped and have rows of CILIA that are fused together and beat synchronously to propel the animal.

CONTINENT A mass of rock that rises above the deep-sea floor and the waves to be exposed to the atmosphere. Continents are composed primarily of granite, an igneous rock of lower density than BASALTIC ocean crust. They account for 28.8 percent of the Earth's surface. The only other dry land on the planet is composed of volcanic isles, which account for 0.4 percent of the surface.

CONTINENTAL SHELF A shallow submerged plateau that surrounds most continents, on average about 50 miles wide and having a depth of 400 feet, or 120 meters. The bottom slopes gently away from the shore until the angle of descent suddenly increases at the shelf break, plunging into the ABYSS.

CONTINENTAL SLOPE A steeply sloping surface lying seaward of the CONTINENTAL SHELF.

COPEPOD Small, SHRIMPlike member of the ZOOPLANKTON and benthic fauna, with about 9,000 known SPECIES, found in fresh and marine waters. The most numerous of the CRUSTACEANS. A square foot of muck from the depths of the central Pacific will usually swarm with copepods, among other small invertebrates.

COPPER A metallic ELEMENT important to many ENZYME reactions in living things and to the electronics industry, which uses the metal to make copper wire.

CORAL A type of COELENTERATE that secretes a hard outer CALCAREOUS skeleton and lives in a colony that helps form the framework for warm-water reefs in shallow waters. The free-swimming MEDUSA life stage of coelenterates in the case of corals has been abandoned for the bottom-dwelling POLYP stage. Best known for living in sunlit shallows, corals are nonetheless also found in icy deeps, both as solitary individuals and in mound-like reefs up to a height of 18 meters, or 60 feet.

CORVETTE A class of warship in the age of sail having a flush deck and usually one tier of cannon.

CRAB A CRUSTACEAN with a broad, flat body and five pairs of walking legs. Crabs range in diameter from less than an inch to 12 feet. In the deep sea, they are found down to depths of at least 3.5 kilometers, or 2.3 miles, and often live among the diverse fauna of the hot volcanic springs.

CRETACEOUS A geological period between 146 million and 65 million years ago, characterized by the rule of giant reptiles and DINOSAURS that died out at the period's end.

CRINOID A type of marine animal that flourished at the time of the DINOSAURS and was thought largely extinct until new types were discovered alive in the nineteenth century. Usually has a cup-shaped body to which are attached radiating arms that collect food from seawater. Includes SEA LILIES and FEATHER STARS.

CRUST The uppermost part of the Earth, which is composed largely of BASALT under the oceans and granite in the continents.

CRUSTACEAN Member of a class of primarily marine INVERTEBRATES with paired jointed appendages and a hard outer skeleton. Includes lobsters, CRABS, SHRIMPS, and COPEPODS. Found throughout the DEEP SEA to the bottoms of the deepest trenches.

CTENOPHORE A PHYLUM of gelatinous animals that are often spherical or cylindrical in shape and characterized by eight rows of ciliated combs for locomotion and often by two tentacles for catching prey. Also known as COMB JELLIES.

CURV An acronym for *Cable-controlled Underwater Recovery Vehicle,* the Navy's first undersea robot, completed in 1965 and initially used for recovering torpedoes.

CUSS-1 The world's first deepwater drill ship, its acronym formed from the first letters of the companies that owned it, Continental, Union, Shell, Superior. Strengthened and modified by the American government, the ship in 1961 lowered a long pipe through more than 2 miles of water and succeeded in drilling into the seabed, advancing the fields of deep geology and mining.

DEEP FLIGHT A one-person winged submersible designed by San Francisco inventor Graham Hawkes that is meant to fly into the sea's depths, eventually into the CHALLENGER DEEP, seven miles down.

DEEP SEA The deeper parts of the ocean, all of which lie in darkness or where sunlight penetrates too weakly to support photosynthesis. Often thought of as the depths beyond the CONTINENTAL SHELVES, the deep is usually estimated as covering about 65 percent of the Earth's surface, or nearly two-thirds of its exterior.

DEVONIAN A geological period from 409 million to 363 million years ago when fish dominated the Earth and amphibians were beginning to arise and to invade the land.

DINOSAUR Popular name for the giant lizards that lived on land and sea during the Triassic, Jurassic, and CRETACEOUS periods, which date from 245 million to 65 million years ago.

DNA An acronym for deoxyribonucleic acid, the main chemical building block of all life, found chiefly in the nuclei of CELLS and containing in its chains of different kinds of chemicals all the genetic information that is transmitted from one generation to the next. Exists as a double helix until it undergoes a copying process prior to cell division.

DNA POLYMERASE An ENZYME that binds to DNA. In nature, DNA polymerase takes a single strand of DNA and doubles it into the normal twin helix by attaching the right chemical substances in the right order, much like building a zipper by using one row of teeth as the template for the other. In the laboratory, it is the backbone of the POLYMERASE CHAIN REACTION.

DOGFISH A member of one of the most familiar groups of sharks, at maturity about three feet long.

DORSAL Relating to the back or upper surface of a fish or other animal.

DORY A deep fish caught around the world as human food, living to depths of about 400 or 500 meters. Large eyes, head, and mouth, growing in overall length to about two feet. A popular type is the oreo dory.

DRAGON FISH A group of deep fish occurring in most oceans and living down to a depth of a kilometer or more. Close cousins of the viper fish. Dozens of known species. Long and thin, black or dark brown in color, sea dragons get their name from fanglike teeth in long jaws, accented by a long barbel dangling down from the chin that at its end has a luminous bulb or filaments. The heads and bodies of dragon fish are often covered with BIOLUMINESCENT organs arranged in rows.

DREDGE A scoop or dragnet on a long line that is towed across the bottom to sample rocks, ooze, and organisms. With a heavy metallic mouth and dense bag, dredges tend to weigh more than TRAWLS, their mouths cutting deeply into bottom muck.

DSRV Initials for Deep Submergence Rescue Vehicle, a 50-foot Navy SUBMERSIBLE with robot arms that can dive as deep as 1.2 miles, or 2 kilometers, for saving lost submariners and gathering intelligence. DSRV-1 is known as *Mystic* and DSRV-2 is known as *Avalon.*

EARTHQUAKE A quick motion or trembling in the Earth that occurs as slowly accumulated crustal strain is suddenly released by volcanic activity, or by rock slabs sliding past one another.

ECHINODERM A member of the PHYLUM of marine animals that have no heads and a radial arrangement of body parts, including SEA LILIES, SEA CUCUMBERS, SEA STARS, SEA URCHINS, sand dollars, and BRITTLE STARS. From the Greek for spiny skin.

ECHO SOUNDING A way to find the depth of water beneath a ship by measuring the time it takes a sound wave to travel to the bottom and back. The speed of sound in water is roughly a mile per second (some five times faster than in air). So in 1951, when a ship first intentionally bounced a sound wave into the CHALLENGER DEEP, seven miles down, the trip to the bottom and back took about fourteen seconds.

ECOSYSTEM The organisms in a community and the system by which they interact with the nonliving environment.

EEL A usually scaleless, bony fish with an average length of about one meter, though some species grow up to 3 meters, or ten feet. About 500 known species. Common in the deep. Snipe eels are long-nosed, deep-sea dwellers found down to depths of 4 kilometers, or 2.5 miles.

EELPOUT A large family of deep fishes occurring throughout the global sea. About 150 known SPECIES. They have long bodies with small heads and eyes, some of them looking like EELS overall. Bodies are often banded or blotched. Most eelpouts appear to be bottom feeders, living on INVERTEBRATES.

EEZ See EXCLUSIVE ECONOMIC ZONE.

ELEMENT A chemical substance that cannot be decomposed into simpler parts by means of chemical reactions. Examples are hydrogen, iron, gold, and uranium. More than 90 chemical elements exist in nature, with all of them found in seawater, often in trace amounts.

ENTERPRISE The mining entity established by the United Nations LAW OF THE SEA Convention to benefit developing nations.

ENVIRONMENT All of the factors—physical, chemical, and biological—that bear on the life of an organism or community.

ENZYME A PROTEIN made in living cells that is able to produce chemical changes in organic substances by CATALYTIC action, as in digestion.

ERUPTION The process by which volcanic action ejects solid, liquid, and gaseous materials into the atmosphere or sea.

EUPHAUSIID A SHRIMPlike CRUSTACEAN, found down to at least 3.5 kilometers, or 2.2 miles. About 100 known species. Most are BIOLUMINESCENT. Euphausiids, commonly known as krill, are a main food of BALEEN whales.

EVOLUTION The process by which organisms change over time, first described by Darwin in 1859. Evolution occurs over generations as random changes in organisms help or hurt them in the life struggle. Beneficial changes that aid survival and reproduction are passed on to descendants, with the accumulation of such alterations eventually able to produce great physical changes and new species.

EXCLUSIVE ECONOMIC ZONE A coastal zone 200 nautical miles wide where a coastal nation claims jurisdiction over mineral resources, fishing, and such things as pollution control.

EXTINCTION The total disappearance of a particular species of life.

EXTINCT VOLCANO One that is not presently erupting and is not likely to do so for a long time.

FANTAIL A fan-shaped deck that overhangs the stern of a ship.

FAULT A crack or fracture in the Earth's CRUST where geological movement can cause EARTHQUAKES or let underlying melted rock rise to the surface.

FAUNA The animal life of any particular area or of any particular time.

FEATHER STAR A free-swimming CRINOID.

FIBER OPTICS Glassy conduits at the core of advanced tethers that use pulses of light to send operator commands in one direction and relay data and crisp images in the other. Fiber-optic cables were pioneered decades ago for undersea work by the American Navy and over the years have sped the development of all kinds of gear, especially tethered robots. The glassy cables can carry up to billions of bits of digital information per second, far more than copper wires or coaxial cables.

FILTER FEEDERS Types of animals that live mostly on the ocean floor or among the PLANKTON that sift food from currents.

FISH Marine or freshwater animals that have spines and breathe by means of gills.

FISH Informal name for a reconnaissance probe or scientific instrument lowered from a submarine or a surface ship into the depths.

FLATFISH Bottom-dwelling fish with both eyes on the same side of the head, such as sole, flounder, or halibut.

FLORA The plant life of any particular area or of any particular time.

FLYING BRIDGE On a ship, an open-air platform near the BRIDGE, sometimes with a duplicate set of steering controls.

FOOD CHAIN The progression of organisms, from small to large, in which each is food for the next member in the sequence.

FOSSIL Any remains, trace, or imprint of an organism that has been preserved in rocks.

GALÁPAGOS ISLANDS A group of volcanic islands about 1,000 kilometers, or 650 miles, west of Ecuador.

GALATHEA A 266-foot Danish ship that sailed the world from 1950 to 1952 to probe the sea's deepest trenches with ECHO-SOUNDERS, TRAWLS, and DREDGES.

GALLEON A large sailing ship of the fifteenth to eighteenth centuries that had three or more masts, square-rigged on the foremast and mainmast and generally LATEEN-rigged on one or two of the aftermasts. Used in both war and commerce.

GANOID Pertaining to an ancient group of fishes, now mostly extinct, that have hard, smooth scales. Includes garpikes and sturgeons.

GASTROPODA A class of MOLLUSKS, most of which have a one-piece shell and a large flattened foot. Includes snails, LIMPETS, and sea slugs.

GLOMAR EXPLORER A 618-foot spy vessel built by the CIA in the guise of a deep-mining ship for the industrialist Howard R. Hughes. In 1974, the ship conducted a secret mission to raise a sunken Soviet submarine from the Pacific depths, trying to do so by lowering a long pipe capped by a giant claw.

GORDA RIDGE A divergent PLATE boundary off Oregon but within the EEZ of the United States, whose volcanic RIFTS have attracted the interest of mineral and microbe hunters.

GRAND BANKS A very large CONTINENTAL SHELF southeast of Newfoundland that is famous for its rich fisheries and whose average depth is between 30 meters and 100 meters, or 100 feet and 330 feet.

GRAY WHALE A Pacific BALEEN whale that feeds in Arctic seas and migrates long distances to breed and calve in the warm lagoons of Baja California and in Japan.

GREENHOUSE EFFECT The theorized heating of the Earth as water vapor and CARBON DIOXIDE in the atmosphere absorb infrared radiations from the surface that otherwise would go back into space, blocking them much as glass traps heat in a greenhouse.

GRENADIER A synonym for RATTAIL, apparently one of the most common fish of the DEEP SEA.

GULPER EEL A deep fish with an immense, pelicanlike mouth and a sinuous tail. Up to more than three feet in length. Found globally down to depths of 2.8 kilometers, or 1.7 miles. Little is known of what it feeds on. Analysis of stomach contents have revealed prawns and small fish. It has been suggested that, to lure prey in undersea darkness, this very flexible fish may wave the luminous tip of its long tail in front of its enormous gaping mouth.

HABITAT The place where a plant or animal naturally lives and grows.

HAGFISH A primitive animal that has a cartilaginous skeleton and a sucker mouth with many teeth but no jaw. It lives in burrows and feeds on INVERTEBRATES, detritus, and dead or dying animals. Hagfish have the ability to coat their finds in an unappetizing slime that deters some rivals.

HAKE A fast-swimming, carnivorous deep fish that is caught commercially off California, Canada, and New Zealand. A member of the prolific cod family.

HALF-LIFE The time required for half the atoms of a radioactive substance to decay to atoms of another element.

HALIBUT A 350-foot American submarine that, beginning in 1965, was converted into a spy vessel able to lower long cables laden with lights and cameras and other gear to reconnoiter objects on the seabed miles below. After a number of top-secret feats, it became the first of a new class of vessels for undersea espionage.

HARBOR SEAL The common or spotted seal, frequently found near shores or in harbors.

HATCHET FISH A relatively small midwater carnivore living down to depths of about a kilometer, its belly dotted with light organs.

HESPERIDES A Spanish naval vessel that in 1992 sailed across the Atlantic from the Canary Islands to the Bahamas, stopping 101 times to measure the temperature of the deep in an effort to gauge the reality of global warming.

HOKI A deep fish caught for human consumption in the South Pacific, tapered and EEL-like and vaguely resembling a RATTAIL.

HUDSON CANYON A cleft extending southeast of the mouth of the Hudson River to the edge of the CONTINENTAL SHELF in the North Atlantic, in all about 55 miles long.

HUMPBACK WHALE A stubby BALEEN whale with a hump on its back and long, thin flippers, hunted almost to extinction.

HYDROCARBON A relatively simple organic compound consisting solely of hydrogen and carbon, for instance methane. Petroleum is a mix of many hydrocarbons.

HYDROGEN SULFIDE To humans, a poisonous gas that smells like rotten eggs, its molecule composed of two atoms of hydrogen and one atom of sulfur. People exposed to low levels of hydrogen sulfide experience irritation of the eyes, nose, and throat. Moderate levels can cause headache, dizziness, nausea, and vomiting, as well as coughing and breathing difficulty. Higher levels can cause shock, convulsions, coma, and death. In the deep, the compound is emitted at hot vents and cold seeps, where whole ECOSYSTEMS depend on it. The primary producers in these ecosystems are deep microbes, both free-living and SYMBIOTIC, that break down the energy-rich bonds of hydrogen sulfide to make materials in a CHEMOSYNTHETIC way of life, much as plants use the energy of sunlight in PHOTOSYNTHESIS.

HYDROPHONE An underwater microphone used to listen for sounds from ships and submarines as well as such natural noises as whale songs. Also refers to a communications system that uses underwater microphones and speakers to transmit human speech to and from the deep.

HYDROZOA A large class of COELENTERATES that includes bell-shaped MEDUSAE, hydroids, fire CORALS, and SIPHONOPHORES. About 2,700 known SPECIES. Animals can appear both as free-swimming MEDUSAE or bottom-dwelling POLYPS.

HYPERTHERMOPHILES Heat-loving MICROBES found in deep springs and VENTS that are dormant at temperatures hot enough to kill most other forms of life but wake up when temperatures approach 100 degrees Celsius, or 212 degrees Fahrenheit. Their ancestors are seen as perhaps the first forms of life on Earth.

IFREMER An acronym for Institut Français de Recherches pour l'Exploitation des Mers, the French National Institute of Oceanography, based in the ancient port of Brest on the west coast of France. IFREMER runs *NAUTILE.*

INORGANIC COMPOUNDS Ones having a composition unlike those associated with living things or HYDROCARBONS or their derivatives.

INVERTEBRATE An animal without a backbone.

ISOPOD A small type of arthropod in the class CRUSTACEA, usually up to an inch and a half long, at times with insectlike legs and often with crushing jaws. Found everywhere on the seabed, down to the deepest trenches. About 4,000 known SPECIES.

ISOTOPE One of several atomic forms of an ELEMENT that has a different number of neutrons in its nucleus and therefore a different atomic mass.

JASON A seven-foot tethered robot designed by Robert Ballard and owned by the Navy. Built as part of the Reagan Administration's military buildup, it debuted in 1989, just as the cold war ended. With a mechanical arm and many lights and cameras and maneuvering motors, communicating to the surface over a long FIBER-OPTIC tether, *Jason* was used by Ballard in April 1989 to reconnoiter an ancient Roman wreck at the bottom of the Mediterranean, from which he recovered more than fifty of the ship's artifacts, including ten amphoras, a piece of cedar deck planking, iron anchors, a grindstone, a cooking pot, and a copper coin from the reign of Constantius the second (A.D. 355 to 361).

JASON JUNIOR A three-foot prototype of the *JASON* robot that Ballard designed and the Navy owns. Basically a flying eyeball on a short tether, it debuted in 1986 when Ballard carried it aboard *ALVIN* down to the *TITANIC,* flying the tiny robot deep into the wreck's interior.

JELLYFISH A bell-shaped COELENTERATE with a soft, gelatinous structure, especially one with an umbrellalike body and long, trailing tentacles, which often bear stinging cells. True jellyfish are widely distributed in the deep down to depths of at least two kilometers, and range in size up to 8 feet wide.

JUAN DE FUCA RIDGE A divergent PLATE boundary off the coast of Washington and Oregon.

KAMCHATKA PENINSULA A long peninsula that juts out from Siberia between the Sea of Okhotsk and the North Pacific.

KARA SEA A body of water on the margin of the Arctic Ocean east of Novaya Zemlya and south of Franz Josef Land, its average depth about 400 feet and its maximum depth about 2,000 feet. It is believed to be the Earth's youngest sea. Major activities on and around the Kara are commercial fishing and hunting, which tend to vary greatly depending on the arctic weather.

KELDYSH See *Akademik Mstislav Keldysh.*

KELP Any of several large, brown ALGAE, including the largest known algae.

LABRADOR SEA A deep body of water bounded by the coast of Labrador on the west, Greenland on the east, and Baffin Island to the northwest. Most of the basin drops to depths of 3.5 kilometers, or 2.2 miles. The climate is polar—cold, dry, and windy, with the icy water at the surface often getting colder than in the depths, causing it to sink.

LAMP SHELLS See *brachiopods.*

LANTERN FISH Relatively small midwater fish living down to depths of a kilometer, usually with rows of light organs on their undersides and sometimes a pair of lights on their heads.

LATEEN SAIL A triangular sail extended by a long tapering yard, the lower end of which is brought down to the deck.

LATITUDE Location on the Earth's surface based on the degrees of angular distance north or south of the equator.

LAVA Melted rock that flows from an opening in the Earth's surface, or the same material after it solidifies.

LAW OF THE SEA Begun in Geneva in 1958 as a series of United Nations conferences to standardize national claims to territorial waters, it went into force in 1994 with a set of rules meant to reconcile the conflicting maritime interests of many countries and to govern virtually all aspects of humanity's relationship with the sea. Among other things, it covers pollution control, seabed mining, dispute arbitration, the right of free passage on the high seas, and the EXCLUSIVE ECONOMIC ZONES that stretch for 200 nautical miles beyond the shores of coastal states.

LEOPARD SHARK A small shark with patterned gray-and-black skin that grows up to four feet long, and whose habitat is the shallow waters of the Eastern Pacific. Considered harmless to humans, it feeds on fish and invertebrates.

LIMESTONE A type of sedimentary rock consisting mainly of CALCIUM CARBONATE, often produced over ages of geologic time as the shells or skeletons of marine creatures accumulate on the seabed after the organisms die.

LIMPET A small type of marine GASTROPOD with a low conical shell that is open on the underside, making the creature look like a flattened snail. The animal is found adhering to rocks or deep creatures such as TUBE WORMS.

LING A deep fish with a long body and broad DORSAL and VENTRAL fins, up to four feet in length or longer. A member of the cod family and apparently global in distribution.

LONGITUDE Location on the Earth's surface based on the degrees of angular distance east or west of the Greenwich Meridian, or Prime Meridian, which is located outside London at the Greenwich observatory.

MANGANESE A hard, brittle, grayish-white metallic ELEMENT used chiefly as an alloying agent in steel to give it toughness.

MANGANESE NODULES Blackish rocks that occur over much of the global seabed and sometimes occur so abundantly they look like cobblestones. They have a high content of

MANGANESE and nickel and other metals and minerals and are viewed as a likely target for deep mining.

MANTLE The bulk of the Earth, a mass of hot plastic rock lying between the cool CRUST and the fiery core.

MARIANAS TRENCH A deep-ocean fissure in the western Pacific lying east of the Marianas chain of islands, harboring what is believed to be the deepest spot in the global sea, the CHALLENGER DEEP.

MBARI See *Monterey Bay Aquarium Research Institute.*

MEDUSA A free-swimming, bell-shaped COELENTERATE with a mouth at the end of a central projection and tentacles around the periphery, as in the JELLYFISH. The MEDUSA is one of two forms of all adult coelenterates, the other being the SESSILE POLYP.

METHANE A colorless, odorless gas that is obtained commercially from natural gas and is a product of ORGANIC decay, being a main constituent of swamp gas. The MOLECULE contains a single CARBON atom bonded to four hydrogen atoms. In the deep sea, methane is metabolized as a source of energy by some types of CHEMOSYNTHETIC bacteria.

MICROBE A small, usually microscopic form of life; the term is used mostly in reference to bacteria.

MID-ATLANTIC RIDGE A slow-spreading divergent PLATE boundary running north-south to bisect the Atlantic Ocean.

MINERAL A naturally occurring INORGANIC material having distinct physical properties and a composition that can be expressed by a chemical formula. Minerals are the basic components of rock.

MIRS A pair of 26-foot submersibles run by Shirshov Institute of Oceanology in Moscow, which debuted in 1988 and can carry three people down to a depth of 6.1 kilometers, or 3.8 miles. *Mir* means *peace* in Russian.

MIZAR A 256-foot cargo ship that the Navy in the 1960s converted into a unique vessel for probing the depths. Through its well-like central opening, crew members could lower miles of cables, whose ends were heavy with lights and various types of cameras.

MOLECULE The smallest particle of an ELEMENT or compound that, in the free state, retains the characteristics of the substance.

MOLLUSK Animals belonging to a very large PHYLUM of mostly marine INVERTEBRATES that are characterized by soft, unsegmented bodies and mantles that often secrete a CALCAREOUS shell of one, two, or more parts. Representatives include mussels, oysters, chitons, snails, CLAMS, SQUIDS, NAUTILI, and OCTOPI.

MONTEREY BAY AQUARIUM RESEARCH INSTITUTE A private scientific center in coastal California that does oceanographic research, particularly in the DEEP SEA. Founded in 1987 by Silicon Valley pioneer David Packard, a former deputy secretary of the U.S. Department of Defense, MBARI under Packard's guidance became a user of military spinoffs from the end of the cold war, including personnel, ship designs, and networks of undersea microphones.

MUSSEL A common bivalve MOLLUSK.

NADIR The 183-foot mother ship of *NAUTILE.*

NATIONAL OCEANIC AND ATMOSPHERIC ADMINISTRATION An arm of the Commerce Department of the United States that, among other things, runs the National Weather Service and conducts civilian deep research. After the cold war it became a main beneficiary of the American Navy's deep gear, methods, personnel, and data.

NATIONAL SCIENCE FOUNDATION A Federal agency of the United States that is the nation's major source of funding for basic scientific research, the monies being distributed mainly to universities. NSF is the main sponsor of *ALVIN*'s scientific expeditions.

NAUTILE A 26-foot French SUBMERSIBLE that debuted in 1985 and can dive with three people to a depth of 6 kilometers, or nearly 4 miles.

NAUTILUS A CEPHALOPOD MOLLUSK with upward of 90 tentacles, which is found most frequently in the South Pacific and Indian Ocean, the last remnant of a race that once ruled the seas. Apparently lives to depths of 400 meters, or 1,300 feet.

NOAA See *National Oceanic and Atmospheric Administration.*

NR-1 A 136-foot SUBMARINE with viewports and wheels that the Navy completed in 1969 as part of its abyssal buildup. Powered by a nuclear reactor, *Nuclear Research-1* can dive unusually deep for a submarine and can stay down for weeks or months on the seabed, unlike SUBMERSIBLES, which usually go down for hours.

NSF See *National Science Foundation.*

NUCLEAR FISSION The power behind atom bombs and nuclear reactors in which heavy atoms such as URANIUM and PLUTONIUM are split in two in chain reactions, releasing bursts of energy as well as creating such RADIOACTIVE WASTES as CESIUM and STRONTIUM.

OCEANIC CRUST A mass of rock with BASALTIC composition that is about three miles thick and underlies the ocean basin. It is created at the divergent PLATE boundaries.

OCTOPUS An eight-armed CEPHALOPOD MOLLUSK.

OKHOTSK, SEA OF An arm of the North Pacific whose average depth is about 775 meters, or 2,350 feet, and whose maximum depth is more than 3,370 meters, or 10,000 feet. Ice covers the sea for about half the year.

OPHIOLITE A rocky complex rich in BASALT and POLYMETALLIC SULFIDES thought to have formed at divergent PLATE boundaries in the sea and later to have been pushed up onto dry land.

ORANGE ROUGHY A pug-nosed, big-mouthed type of SLIMEHEAD that is caught for human consumption, with the fishery especially large around New Zealand and Australia.

ORGANIC COMPOUNDS Ones existing in or derived from plants and animals as well as many others that include the element CARBON.

ORTHOCONE An extinct ancestor of the NAUTILUS that arose in the late CAMBRIAN Period and whose giant shell reached lengths of 30 feet, with the animal's long tentacles extending farther. At the time, they were probably the largest animals in existence.

OXYGEN MINIMUM A zone beneath the sea where respiration, decay, and obscure factors have reduced dissolved oxygen to a minimum, usually at a depth between 800 and 1,000 meters. In Monterey Bay, the minimum typically occurs between 600 and 800 meters.

PACIFIC OCEAN The largest and deepest body of water on Earth. Its area is 64 million square miles, or about a third of the planet's surface. Its average depth is 12,925 feet, equal to 2.4 miles or 3.9 kilometers.

PCR See *polymerase chain reaction.*

PECTORAL Referring to a pair of fins usually situated behind the head, one on each side, and corresponding to the forelimbs of higher vertebrates.

PHOTIC ZONE The layer in seawater that receives enough sunlight to support PHOTOSYNTHESIS in plants and PHYTOPLANKTON, its lower boundary usually considered to be a depth of about 100 meters. Deeper areas can sustain amounts of photosynthesis that are relatively minor. Since the global sea has an average depth of 3.8 kilometers, that means the photic zone is a very thin layer of the sea's surface, accounting for less than 3 percent of the sea's average depth. The rest is darkness that becomes increasingly black with progressive depth.

PHOTOSYNTHESIS The process by which plants use the energy of sunlight and the green pigment chlorophyll to convert water, CARBON DIOXIDE, and INORGANIC substances into ORGANIC ones, such as sugars. Oxygen is released as a by-product of the photosynthetic reaction.

PHYLUM A broad category for the grouping together of varying individuals of the living world, consisting of one or more closely related classes, which in turn are composed of orders, families, genera, and SPECIES. The lower down on the taxonomic ladder, the greater the degree of likeness. Above the phylum level is the kingdom, a broad category that divides living things into such groups as bacteria and higher organisms.

PHYTOPLANKTON Plant forms of PLANKTON, mostly microscopic.

PILLOW LAVA Interconnected lobes of LAVA formed under water that are shaped like pillows and sacks.

PLANKTON The floaters and drifters of the sea, including both plants and animals. The definition is usually stretched to include organisms that are weak swimmers.

PLATE In terms of PLATE TECTONICS, a rigid segment of the Earth's CRUST that moves about in relation to the dozen or so other plates.

PLATE TECTONICS The theory that says the Earth's CRUST is broken into a dozen or so cool PLATES that float on the planet's hot interior and move in relation to one another, rearranging the CONTINENTS over the ages and provoking EARTHQUAKES and volcanic eruptions around the globe. Colliding plates build mountains on land and dig deep trenches in the sea, while diverging plates allow hot melted rock to ooze up from the Earth's deep interior, forming new ocean crust over the eons.

PLUTONIUM A radioactive ELEMENT usually made in nuclear reactors by a kind of modern alchemy involving natural uranium. Its ISOTOPE plutonium-239 is a main fuel of atomic and hydrogen bombs and some types of reactors. It has a half-life of 23,400 years and is so deadly that tiny specks invisible to the eye can cause cancer if they become lodged in soft tissues.

POINT LOBOS A 110-foot research ship belonging to the Monterey Bay Aquarium Research Institute and the mother ship of the robot *Ventana.*

POLYMERASE CHAIN REACTION A laboratory method known as PCR that is used to multiply tiny bits of DNA a billionfold or more, allowing enough of the genetic material to be accumulated for subtle analyses and manipulations.

POLYMETALLIC SULFIDES A mix of differing SULFIDES, such as pyrite (iron sulfide), chalcopyrite (copper sulfide), and sphalerite (zinc sulfide). On land, mixtures of sulfide ores, including ones bearing gold, have been mined for millennia. In the 1980s, scientists stumbled on their birthplace at the bottom of the sea along the volcanic RIFTS, setting off a feverish round of planning for deep mining. To date, no one has attempted such activity, although the sites are explored globally to assess their economic promise.

POLYNUCLEAR AROMATIC HYDROCARBONS A family of ORGANIC COMPOUNDS that are derivatives of anthracene, which consists of three benzene rings in a row. PAHs are found in coal, tar, and petroleum, and are emitted by some types of combustion. A wide contaminant of coastal waters, this family of chemicals also appears responsible in part for the cancer-causing properties of cigarette smoke.

POLYP A COELENTERATE attached to the bottom either as a solitary individual or as part of a colony. One of the two forms taken by all adult coelenterates, the other being the MEDUSA.

PRAYA A genus of deep SIPHONOPHORE that grows lengths of up to 130 feet, longer than the blue whale, which is often considered the Earth's largest animal.

PRECAMBRIAN All geologic time from the beginning of Earth history to 570 million years ago, ending just before the CAMBRIAN explosion in the quantity and complexity of life. Single-celled organisms dominated its early stages, while multicellular ones arose at its end.

PRIVATEER A privately owned warship under license to a government.

PROTEINS The complex ORGANIC COMPOUNDS that make up a large percentage of the dry weight of all living organisms and are the building blocks for such things as ENZYMES and cellular tissues. They stand in contrast to fats and CARBOHYDRATES.

RADIOACTIVE WASTE Radioactive materials no longer useful for their intended purpose.

RADIOACTIVITY The spontaneous breakdown of the nucleus of an atom resulting in the emission of radiant energy in the form of particles or waves or both.

RATTAILS A large family of ubiquitous deep fish that are related to the cods. About 260 known species. Also known as GRENADIERS. They can grow up to two or three feet in length and seem to be almost everywhere on the seafloor, their heads huge and ugly by human standards, their bodies tapering to long tails that are remarkably sinuous.

RED CRAB One of the largest of the deep CRABS, some nearly two feet across, their claws often quite large.

RESOLUTION A 471-foot roving drill ship run by JOIDES, the Joint Oceanographic Institutions for Deep Earth Sampling, an international consortium of universities, sea organizations, and government agencies.

RIFT The trough or valley that cuts lengthwise through an oceanic ridge. It is the place where tectonic PLATES diverge and molten rock wells up to form new crust.

RIGHT WHALE A slow, portly whale that was the favorite target of early whalers and eventually became the great whale most in danger of extinction, which remains the case today.

ROYAL SOCIETY Founded in 1660, it is the oldest scientific group in Great Britain and one of the oldest in the world. Members included Isaac Newton and Edmund Halley. Among other endeavors, the society sponsored the 1768 voyage of James Cook, his first to the Pacific, during which he charted New Zealand, Australia, and the Indian Ocean.

RUSTICLES Fragile reddish-brown stalactites of rust, hanging down as much as several feet on the *Titanic.* Produced by iron-eating bacteria.

RV Initials for *reentry vehicle,* the cone-shaped part of a missile or space vehicle meant to reenter the earth's atmosphere during the fiery end of its flight and survive intact until impacting on land or sea. Military ones often contain one or more nuclear warheads or, during test flights, mock warheads.

SABLEFISH Also known as blackcod, a deep fish that is found in the Pacific down to depths of 1.5 kilometers, or nearly a mile.

SCABBARD A deep fish of the Atlantic, with sharp teeth, prominent eyes, long, thin body, and petite caudal fins. Often known as black scabbard. Up to three or four feet in length.

SCHOONER A tall ship with two or more masts, in which the lower parts of the masts are rigged with sails fore and aft.

SCRIPPS Shorthand for the Scripps Institution of Oceanography, based in La Jolla, California, an arm of the University of California at San Diego. Founded in 1903 by the Scripps newspaper dynasty, the institution grew into one of the world's preeminent centers for oceanic study, including its depths.

SCUBA An acronym for Self-Contained Underwater Breathing Apparatus, a device built around an air tank, valve, and hose that allows a diver equipped with a mouthpiece to breathe normally under water. It can be used only at relatively shallow depths, normally no more than 100 feet or so.

SEA The body of saltwater that covers 71 percent of the Earth's surface, its average depth 12,465 feet, which is equal to 2.36 miles or 3.80 kilometers.

SEA ANEMONE A sedentary marine animal that resembles a flower. A COELENTERATE in POLYP form, the animal has a columnar body whose top is crowned by one or more circles of tentacles that surround its central mouth. Sea anemones are found everywhere from tide pools to the ocean's deepest trenches, with the giant ones growing up to one meter in diameter.

SEA CLIFF A 30-foot *ALVIN* look-alike built by the Navy during its undersea buildup in 1960s. Launched in 1968, the submersible can dive to depths of 6.1 kilometers, or 3.8 miles. In the 1990s, the Navy increased its sharing of *Sea Cliff* with civilian scientists and focused much of its work on deep environmental monitoring and repair.

SEA CUCUMBER An ECHINODERM having a long leathery body with tentacles around the front end near the mouth. They resemble squash or cucumbers and grow in length up to a foot. Also known as holothurians.

SEA DRAGON See *dragon fish.*

SEAFLOOR SPREADING The mechanism by which new seafloor CRUST is created at the volcanic RIFTS and spreads laterally as PLATES diverge.

SEA LILY An ancient type of deep animal that looks like a flower. It grows on a long stem and at its top has petallike arms for capturing particles of food from seawater. A stalked type of CRINOID, the sea lily thrived more than 100 million years ago during the CRETACEOUS Period and was thought largely extinct until discovered in profusion in the 1860s, making it one of the first of the deep's living fossils to come to light.

SEAMOUNT An individual peak extending above the seafloor.

SEA PEN A COELENTERATE cousin of CORALS and SEA ANEMONES that looks like a fat feather and lives on the bottom from shallow to deep water. About 300 known species. Many are bioluminescent.

SEA SCORPION An extinct ARTHROPOD with a long body and large pincers that lived in the sea during the Silurian Period more than 400 million years ago.

SEA SPIDER A spiderlike member of the phylum ARTHROPODA with very long legs, which lives in the deep sea and is related to spiders, scorpions, and mites on land. Deep-sea spiders have a conspicuous muscular proboscis, probably for sucking fluids out of SESSILE invertebrates. More than 100 SPECIES are known to crawl through the deep, and the largest individuals have leg spans of nearly two feet.

SEA SQUIRT A SESSILE marine animal that sometimes looks like a lumpy potato and lives in all seas to all depths. Some are quite delicate and beautiful. All actively pump water through their bodies to filter out suspended ORGANIC particles.

SEA STAR A more accurate name for starfish, which is not a fish but an ECHINODERM, usually in the form of a star, with five or more arms radiating from a central core.

SEA URCHIN An ECHINODERM that is somewhat globular in shape and has a shell covered with sharp spines.

SEAWOLF The world's second nuclear SUBMARINE, commissioned by the American Navy in 1957. It had an innovative sodium-cooled reactor that reached high temperatures but was ultimately judged unsafe. The submarine's unfueled but radioactive reactor shell was eventually dumped in the Atlantic at a spot some 150 miles east of Delaware Bay, at a depth of more than a mile.

SEDIMENT Particles of ORGANIC or INORGANIC origin that accumulate in loose form on the seabed.

SESSILE Permanently fixed or sedentary on the sea's bottom and not free to swim about.

SHINKAI 6500 A 31-foot Japanese SUBMERSIBLE that debuted in 1990 as the world's deepest piloted craft, able to dive to depths of 6.5 kilometers, or 4.0 miles.

SHRIMP A CRUSTACEAN important to commercial harvesting, with roughly 2,000 known species found everywhere, from coastal zones to deep trenches.

SIPHONOPHORE A type of COELENTERATE that forms midwater colonies.

SKATE A flat, cartilaginous fish with winglike PECTORAL fins that extend from the nose far back along the body, often like big triangles, the ends of which terminate in a slender tail. Related to the rays. Up to 26 feet long and usually dwelling on the bottom, skates have been found at depths ranging down to nearly 2 miles.

SLATJAW EEL A brownish deep EEL with a large mouth that lives down at least to depths of 3.6 kilometers, or 2.3 miles, and grows to lengths of a little more than 2 feet. Found widely in the North Atlantic.

SLIMEHEAD A family of deep predatory fishes whose members grow to lengths of about 2 feet, live to depths of about a kilometer, and include ORANGE ROUGHY, which is widely caught for human consumption. The family's common name derives from the fact that heads of these animals have a number of skin-covered mucus cavities, especially around the eyes.

SONAR An acronym for SOund Navigation And Ranging, a method of detecting and locating objects submerged in the sea by listening with underwater microphones for sound waves the objects reflect or produce. The name also applies to the apparatus that does such work. Sonars are widely used by SUBMARINES to track targets and by surface ships to map the seabed.

SOSUS An acronym for the SOund SUrveillance System of the American Navy, a global network of undersea microphones built at a cost of $16 billion during the COLD WAR to spy on sounds of Soviet ships and SUBMARINES in an effort to track their movements. In contrast to Navy microphones towed from ships, the SOSUS ones are moored on the seabed.

SPECIES The basic category of biological classification that ranks below genus and is undivided. Individuals of the same species closely resemble one another and are able to interbreed and produce viable offspring.

SPERM WHALE The largest toothed whale. It dives a mile or more deep to hunt prey and has no BALEEN.

SPREADING CENTER A synonym for a divergent PLATE boundary, a place where hot lava wells up and moves laterally to form new crust.

SQUID Any of several ten-armed CEPHALOPODS having a slender body and a pair of rounded or triangular CAUDAL fins and varying in length from inches to sixty feet and perhaps more. Two of the arms, called tentacles, are much longer than the others and are used to grab prey. The other arms then help carry prey to the central mouth.

STARFISH See *sea star.*

STOMIATID See *dragon fish.*

STRONTIUM-90 A radioactive ISOTOPE produced as a by-product of NUCLEAR FISSION in both reactors and atomic blasts, strontium-90 has a half-life of 28 years and a bad reputation among health physicists because it binds readily with human bones to replace calcium, causing cancer and other diseases.

SUBMARINE A vessel that can dive beneath the waves and navigate in the upper waters of the sea, often for military purposes. A large one can stay down for months and carry more than one hundred sailors.

SUBMERSIBLE A relatively small vessel, with a thick-walled personnel sphere and tiny windows, which can dive to great depths, often for scientific research. Usually carries no more than two or three people, has gear for short-term life support, and, unlike a SUBMARINE, requires a support ship to recharge its systems and house its crew. A submersible usually descends only during daylight hours, coming back to the surface before nightfall.

SULFIDE A compound formed with SULFUR that contains a more electropositive element, such as copper, zinc, or nickel. Such sulfides often occur in nature as ores that are mined commercially for the metals.

SULFUR A yellow mineral composed of the ELEMENT sulfur that is commonly found in association with hydrogen and HYDROCARBONS.

SWELL An ocean wave that is unbroken and has no foamy white crest.

SYMBIONT An organism living with a dissimilar one where the association is mutually beneficial. The state is known as symbiosis.

SYNTACTIC FOAM A brick-hard polymer made by dispersing tiny, hollow glass beads in a fluid polymer and then curing it. Its great buoyancy and strength make it an ideal flotation material for heavy gear that must withstand the crushing pressures of the deep.

TAG An acronym for Trans-Atlantic Geotraverse, an American expedition that in 1985 while towing dredges and other instruments over the seabed discovered a huge volcanic mound hidden in the recesses of the Mid-Atlantic Ridge. Subsequently named TAG, the mound is one of the largest active SULFIDE structures ever found on the seabed. Blistering hot, it swarms with eyeless shrimp and contains an unknown amount of gold.

TANAID A small type of marine ARTHROPOD.

TAQ See *Thermus aquaticus.*

TECTONICS The building and bending and destroying of the Earth's CRUST by such forces as EARTHQUAKES, folds, and FAULTS, and ultimately by the flow of heat from the Earth's interior.

THERMUS AQUATICUS A bacterium found in 1966 thriving in the hot springs of Yellowstone National Park in the United States at temperatures then thought high enough to kill all forms of life. Its DNA POLYMERASE ENZYME was eventually used to advance the development of the POLYMERASE CHAIN REACTION.

THRESHER A 278-foot American SUBMARINE that sank during sea trials in 1963 in Atlantic waters more than a mile and a half deep, killing 129 men. The Navy was unable to find the sunken hulk for many months. That humiliation prompted the service to embark on a vast buildup of undersea skills and forces, including whole fleets of deep vehicles, both piloted and robotic.

TIBURON An advanced scientific robot built with cold-war talent by the Monterey Bay Aquarium Research Institute. It debuted in 1996.

TITANIC In its day, the world's largest and most luxurious ship. On its inaugural voyage in 1912, carrying about 2,220 passengers and crew members (the exact number is unclear because of list discrepancies), the 882-foot-long liner headed westward across the Atlantic. On the bitterly cold night of Sunday, April 14, near midnight, it struck an iceberg off Newfoundland while moving at a speed of more than 20 knots. A little more than two and a half hours later, *Titanic* went down. Lost were more than 1,500 people.

TITANIUM A dark gray or silvery metal, lustrous, strong, hard, and corrosion-resistant. About 40 percent lighter than steel, titanium is often used to make the personnel spheres of deep SUBMERSIBLES.

TOPOGRAPHY The contour of a surface, in oceanography usually of the seabed.

TRANSPONDER A small electronic device on a vehicle or moored to the deep seabed that listens for sounds at special frequencies. When it hears one, it sends out audible chirps in response, allowing the calculation of distance between sender and receiver and the creation of networks of transponders for SUBMERSIBLE navigation through dark waters.

TRAWL A sturdy bag or net that is towed on a long line, often across the bottom, to trap marine animals. It is used by both commercial fishermen and marine biologists.

TRENCH A long, narrow, and deep depression on the ocean floor that has relatively steep sides.

TRIESTE A 50-foot BATHYSCAPH made by the Swiss inventor Auguste Piccard and named after the city in Italy that was one of its patrons. The craft debuted in 1953 and was purchased in 1958 by the American Navy, its first piloted craft for deep exploration.

TRIESTE II A pair of BATHYSCAPHS that the United States Navy built during its crash program to develop deep forces. The first, 75 feet long, debuted in 1964 and left service in 1966. The second, 79 feet long, dove between 1966 and 1982. It bore little resemblance to its homely predecessors, looking more like a potent SUBMARINE. Giant propellers nudged the behemoth forward through the darkness below.

TRILOBITE An extinct class of ARTHROPODS that dominated the seas during the CAMBRIAN Period. One of the earliest-known fossils. Its body was flattened and oval, varying in length from an inch or less to up to two feet.

TUBE WORMS A family of deep-sea worms that possess no mouth, no gut, no anus, and are unable to feed by normal means. While some species are tiny, others rapidly grow quite large, their tubes up to nine or ten feet long. Many tube worms bear a bright red plume that extends from a top of the rigid white tube, which is made of chitin, a horny material. The worms can quickly withdraw the plumes into the tubes to dodge predators. The delicate plume works like a gill or a lung to extract environmental gases. Found extensively at hot volcanic VENTS and COLD SEEPS, tube worms thrive in a SYMBIOTIC relationship with bacteria that live in their tissues, the microbes using HYDROGEN SULFIDE transported by the blood of tube worms as a source of energy, much as plants use sunlight.

TURTLE A 26-foot *ALVIN* look-alike built by the Navy during its undersea buildup in the 1960s. Launched in 1968, the SUBMERSIBLE can dive to depths of 3.0 kilometers, or 1.9 miles. In the 1990s, the Navy began wide sharing of *Turtle* with civilian scientists.

URANIUM A rare natural ELEMENT widely mined and used as fuel for nuclear reactors and weapons, and as a nuclear feedstock in making PLUTONIUM.

USGS United States Geological Survey. An arm of the Interior Department that makes maps for the United States and conducts much research into the nation's mineral deposits, on both land and sea.

VAMPYROTEUTHIS The so-called vampire squid. A rare living fossil from a time before CEPHALOPODS went down their separate evolutionary ways as SQUID and OCTOPI. The animal has ten limbs, like a squid, but only eight of them are meaty, like those of an octopus. The remaining two are slender threads that have an unknown function, probably sensory. The creature is apparently a resident of the OXYGEN MINIMUM.

VENT An opening on land or seabed that emits volcanic materials, including hot gas, lava, and water. Along volcanic RIFTS, the vents are oftentimes hot springs, sometimes gushing and churning violently.

VENTANA An 8-foot robot run by the MONTEREY BAY AQUARIUM RESEARCH INSTITUTE for midwater and bottom scientific research, initially beneath the waves of Monterey Bay. Equipped with powerful cameras and motors and great agility, linked to the surface with FIBER-OPTIC lines, it debuted in 1988 and proceeded to help researchers find new creatures at a rate of about a dozen species a year and to illuminate how hundreds of others behave and hunt, live and die.

VENTRAL Pertaining to the belly or abdominal area of an animal.

VERTEBRATE An animal with a backbone and usually a well-developed brain and a skeleton of bone or cartilage. Vertebrates include fish, amphibians, reptiles, birds, and mammals.

VISCOSITY A property of fluids, based on internal friction, that manifests as resistance to flow. Honey is very viscous.

VOLCANO A vent in the Earth's surface through which melted rock and associated gases and fluids erupt, as well as the structure formed by ejected material.

WESTERLIES Steady winds found in both northern and southern hemispheres between 30 degrees and 60 degrees of LATITUDE, moving west to east, the direction of the Earth's rotation. At the planet's surface, the speed of the westerlies is about 6 miles per hour. But higher up, at jet-stream heights, the winds can move at speeds of hundreds of miles per hour.

WHITECAP A wave in the open ocean that breaks because of blowing wind, topping it with whitish foam.

WOODS HOLE Shorthand for Woods Hole Oceanographic Institution, a private center founded on Cape Cod in 1930 to investigate all aspects of oceanography. With its acquisition of *ALVIN* in 1964, Woods Hole became the world leader in piloted deep exploration, making hundreds of discoveries over the decades.

ZOOPLANKTON Animal forms of plankton.

Notes

One. LAIR

1. Robert S. Dietz, "The Underwater Landscape," in C. P. Idyll, ed., *Exploring the Ocean World: A History of Oceanography* (Crowell, New York, 1969), p. 22.

2. C. P. Idyll, *Abyss: The Deep Sea and the Creatures That Live in It* (Crowell, New York, 1963), p. 229. Cheryl Lyn Dybas, "The Oarfish," *Oceanus,* Vol. 36, No. 1, Spring 1993, pp. 98–99.

3. Idyll, *Abyss,* pp. 196–203, 277–280. N. B. Marshall, *Aspects of Deep Sea Biology* (Philosophical Library, New York, 1954), p. 13. W. B. Scott and M. G. Scott, *Atlantic Fishes of Canada* (University of Toronto, Toronto, 1988), pp. 242–253.

4. Idyll, *Abyss,* pp. 222, 224–225. Daniel Cohen, *A Modern Look at Monsters* (Dodd Mead, New York, 1970), pp. 26–31. Daniel Cohen, *Encyclopedia of Monsters* (Dodd Mead, New York, 1982), pp. 163–169. Thomas Helm, *Dangerous Sea Creatures: A Complete Guide to Hazardous Marine Life* (Funk & Wagnalls, New York, 1976), pp. 219–220. Bernhard Grzimek, *Grzimek's Animal Life Encyclopedia: Mollusks* (Van Nostrand, New York, 1974), pp. 210–211. Ralph Buchsbaum et al., *Animals Without Backbones* (University of Chicago, Chicago, 1987), p. 285. Clyde F. E. Roper and Kenneth J. Boss, "The Giant Squid," *Scientific American,* Vol. 246, No. 4, April 1982, pp. 96–105. Cleveland P. Hickman, *Biology of the Invertebrates* (Mosby, St. Louis, 1973), pp. 394–407.

5. Margaret Deacon, *Scientists and the Sea, 1650–1900: A Study of Marine Science* (Academic Press, New York, 1971), pp. 281–286. Dean King et al., *A Sea of Words* (Henry Holt, New York, 1995), pp. 229, 343.

6. Susan Schlee, *The Edge of an Unfamiliar World: A History of Oceanography* (Dutton, New York, 1973), p. 85. For the Ross quote, Rachel L. Carson, *The Sea Around Us* (Oxford, New York, 1951), p. 39. For a concise overview of this history, see Dietz, "The Underwater Landscape," p. 22.

7. Schlee, *The Edge,* pp. 82–89. James Hamilton-Paterson, *The Great Deep: The Sea and Its Thresholds* (Random House, New York, 1992), pp. 167–173. For the Forbes quote, see Carson, *The Sea,* pp. 38–39.

8. For the telegraph and the Atlantic cable, see Bern Dibner, "Communications," in Melvin Kranzberg and Carroll W. Pursell, Jr., *Technology in Western Civilization,* Volume I

(Oxford, New York, 1967), pp. 454–460. For Maury, Field, and the Mediterranean cable, see Schlee, *The Edge,* pp. 55–58, 88–89. For specific animals on the Mediterranean cable, see Marshall, *Aspects,* p. 17. For an analysis of how cable laying aided ocean-bottom exploration, see Chandra Mukerji, *A Fragile Power: Scientists and the State* (Princeton University, Princeton, N.J., 1989), pp. 22–30.

9. Schlee, *The Edge,* pp. 90–92. Charles Darwin, *The Origin of Species* and *The Descent of Man* (Modern Library, New York, 1948), pp. 256, 261, 271. On the antiquity of brachiopods, see Buchsbaum, *Animals Without Backbones,* pp. 243–247.

10. Schlee, *The Edge,* pp. 92–95.

11. Ibid., pp. 99–103. See also Deacon, *Scientists and the Sea,* pp. 306–310.

12. Schlee, *The Edge,* p. 103. See also Deacon, *Scientists and the Sea,* pp. 312–328. Idyll, *Exploring,* pp. 11–12. C. Wyville Thomson, *The Depths of the Sea* (Macmillan, London, 1874), p. 79.

13. Walter James Miller and Frederick Paul Walter, trans., *Jules Verne's Twenty Thousand Leagues Under the Sea: The Definitive, Unabridged Edition Based on the Original French Texts* (Naval Institute Press, Annapolis, 1993), pp. vii–xxii.

Miller and Walter argue that hundreds of technical errors often attributed to Verne are in fact the work of translators. Their case is persuasive, and Verne clearly was prophetic in seeing so clearly how the machinery of underwater travel would evolve in future decades. But in biology, if not technology, he was no visionary.

14. Ibid., pp. 95–100, 281–283.

15. Marshall, *Aspects,* pp. 18–19. For an analysis of how *Challenger* and much ocean exploration of the nineteenth century were meant to aid cable laying, see Mukerji, *A Fragile Power,* pp. 22–30.

16. Schlee, *The Edge,* pp. 107–108.

17. Ibid., pp. 111–125, 147–152. Deacon, *Scientists and the Sea,* pp. 351.

18. Schlee, *The Edge,* pp. 116–118.

19. Ibid., pp. 143–149, 152–153. Harold V. Thurman, *Introductory Oceanography* (Macmillan, New York, 1994), p. 7.

20. Schlee, *The Edge,* pp. 122–125. For a modern viewpoint on the antiquity of some deep life, see J. D. Gage and P. A. Tyler, *Deep-Sea Biology* (Cambridge University Press, New York and Cambridge, 1991), p. 366.

21. Schlee, *The Edge,* pp. 143–150. Thurman, *Introductory Oceanography,* pp. 115–116. Gage and Tyler, *Deep-Sea Biology,* pp. 263–282.

22. Schlee, *The Edge,* p. 112.

23. William Beebe, *Half Mile Down* (Harcourt Brace, New York, 1934), pp. 87–88. Marshall, *Aspects,* p. 157.

24. Beebe, *Half Mile,* pp. 93, 172, 231–250.

25. Ibid., pp. 152–154.

26. Ibid., pp. 111, 171.

27. Ibid., pp. 172–173.

28. Ibid., p. 207.

29. Ibid., pp. 219–220.

30. Idyll, *Abyss,* pp. 240–241. Richard Ellis, *Monsters of the Sea* (Knopf, New York, 1994), pp. 11–13.

Judith E. Winston cited the coelacanth in suggesting that big undiscovered animals may yet ply the deep. "It is not unrealistic," she wrote, "to believe that other animals, even very large animals, could still be hidden there." Judith E. Winston, "Systematics and Marine Conservation," in Niles Eldredge, ed., *Systematics, Ecology and the Biodiversity Crisis* (Columbia University Press, New York, 1992), p. 148.

31. Anton F. Bruun et al., *The Galathea Deep Sea Expedition: 1950–1952* (Macmillan, New York, 1956), pp. 22, 112–118, 160, 173–178, 182–187. Idyll, *Abyss,* pp. 241–244. For more on the mollusk, see A. S. Romer, "Darwin and the Fossil Record," in Philip

Appleman, ed., *Darwin: A Norton Critical Edition* (Norton, New York, 1970), pp. 374–375. J. Frederick Grassle and Nancy J. Maciolek, "Deep-Sea Species Richness: Regional and Local Diversity Estimates from Quantitative Bottom Samples," *American Naturalist,* Vol. 139, February 1992, pp. 313–341. Gary C. B. Poore and George D. F. Wilson, "Marine Species Richness," *Nature,* Vol. 361, February 18, 1993, pp. 597–598. P. John D. Lambshead, "Recent Developments in Marine Benthic Biodiversity Research," *Oceanus,* Vol. 19, No. 6, 1993, pp. 5–24. The National Research Council, *Understanding Marine Biodiversity* (National Academy Press, Washington, 1995), pp. 44–45. William J. Broad, "The World's Deep, Cold Sea Floors Harbor a Riotous Diversity of Life," *The New York Times,* October 17, 1995, p. C1.

Discovered among the invertebrate swarms was a whole new phylum, one of the most basic divisions of the animal kingdom. For the discovery of loriciferans, tiny animals that burrow into the seabed, see Edward O. Wilson, *The Diversity of Life* (Norton, New York, 1992), pp. 131–132.

32. For an early contrasting of land and sea as habitats, see Bruun, *The Galathea,* pp. 11–12, 191.

33. For a review of the percentages and their implications, see G. Richard Harbison, "The Gelatinous Inhabitants of the Ocean Interior," *Oceanus,* Vol. 35, No. 3, Fall 1992, p. 19. See also James J. Childress, "Oceanic Biology: Lost in Space?" in P. Brewer, ed., *Oceanography: The Present and Future* (Springer-Verlag, New York, 1983), pp. 127–135. For the bottom of the photic zone as the edge of the deep: interviews, Bruce H. Robison of the Monterey Bay Aquarium Research Institute and Sylvia A. Earle of Deep Ocean Exploration and Research, November 29, 1996.

34. For the 1991 estimate, see J. Thiede and K. J. Hsü, "The Future of Ocean Resources," in J. Thiede and K. J. Hsü, eds., *Use and Misuse of the Seafloor* (Wiley, New York, 1992), p. 413.

35. Schlee, *The Edge,* pp. 132–134. Eric L. Mills and Jacqueline Carpine-Lancre, "The Oceanographic Museum of Monaco," in Elisabeth Mann Borgese, ed., *Ocean Frontiers: Explorations by Oceanographers on Five Continents* (Abrams, New York, 1992), pp. 121–135. Interview, Steve Johnson, Wildlife Conservation Society (formerly the New York Zoological Society), May 10, 1996. Beebe, *Half Mile,* pp. ix, 227. Roger Revelle, "The Scripps Institution of Oceanography," and Arthur G. Gaines, Jr., "The Woods Hole Oceanographic Institution," both in Borgese, *Ocean Frontiers,* pp. 16–53, 56–93, respectively. Winslow Carlton, "Bankers' Row," and William MacLeish, "Woods Hole Oceanographic Institution," both in Mary Lou Smith, ed., *Woods Hole Reflections* (Woods Hole Historical Collection, Woods Hole, Mass., 1983), pp. 140–147, 186–197, respectively. Ragnar Sparck, "Background and Origin of the Expedition," in Bruun, *The Galathea,* pp. 15–17. For a sociological analysis of the ups and downs of state interest in undersea exploration, see Mukerji, *A Fragile,* pp. 22–61.

TWO. BATTLE ZONE

1. Jacques Piccard and Robert S. Dietz, *Seven Miles Down: The Story of the Bathyscaph Trieste* (Putnam, New York, 1961), pp. 30–76. For more on Piccard's history, see Isaac Asimov, *Asimov's Biographical Encyclopedia of Science and Technology* (Doubleday, New York, 1982), pp. 694–695.

2. Piccard and Dietz, *Seven Miles,* pp. 30–76.

3. Ibid., pp. 72–73, 234–235.

4. Ibid., pp. 86–87, 237.

5. Ibid., pp. 77–109. For more on *Trieste*'s dives that summer, see Victoria A. Kaharl, *Water Baby: The Story of Alvin* (Oxford University Press, New York, 1990), p. 15.

6. Piccard and Dietz, *Seven Miles,* pp. 87–89, 236–238. For a brief history of the

SOSUS system, see Jeffrey T. Richelson, *The U.S. Intelligence Community* (Ballinger, Cambridge, Mass., 1985), pp. 147–149.

7. Piccard and Dietz, *Seven Miles,* p. 92.

8. Ibid., pp. 115–116, 123–132, 135, 140–158, 228–229.

9. Ibid., pp. 160–180.

10. Don Walsh, "Looking Backwards at the Future," *Naval Institute Proceedings,* January 1985, pp. 103–104. Walsh, "Thirty Thousand Feet and Thirty Years Later: Some Thoughts on the Deepest Presence Concept," *Marine Technology Society Journal,* June 1990, Vol. 24, No. 2, pp. 7–8. For Rechnitzer quote, see Kaharl, *Water Baby,* p. 16.

It should be noted that the fighting Navy was interested in oceanography in general, and dramatically stepped up its funding of the field in the years immediately after World War Two. See H. W. Menard, *The Ocean of Truth: A Personal History of Global Tectonics* (Princeton University Press, Princeton, 1986), pp. 37–43.

11. Kaharl, *Water Baby,* pp. 15–48.

12. Norman Polmar, *Death of the Thresher* (Chilton, New York, 1964), pp. 1–5. Robert W. Love, Jr., *History of the U.S. Navy,* Volume Two, 1942–1991 (Stackpole, Harrisburg, 1992), pp. 416, 480–481. *The New York Times Index* (New York Times, New York, 1963), pp. 707–708. Tom Clancy, *Submarine: A Guided Tour Inside a Nuclear Warship* (Berkley, New York, 1993), pp. 17–19. For silencing and submarine depths, see Patrick Tyler, *Running Critical* (Harper & Row, New York, 1986), pp. 57–58. On the link between *Thresher* and missile subs, interview, John P. Craven, formerly chief scientist for the Navy's Polaris program, July 2, 1993.

13. Polmar, *Death,* pp. 35–36, 55, 89.

14. Ibid., pp. 60–61, 91, 141. James H. Wakelin, Jr., *"Thresher:* Lesson and Challenge," *National Geographic,* June 1964, pp. 759–763. For the mistaken photo, see "Back to the Hunt," *Newsweek,* June 10, 1963, pp. 31–32. Donald L. Keach, "Down to *Thresher* by Bathyscaph," *National Geographic,* June 1964, pp. 764–777.

15. Polmar, *Death,* pp. 125. For a general description of the Stephan panel, also see Norman Polmar and Thomas B. Allen, *Rickover: Controversy and Genius* (Simon & Schuster, New York, 1982), p. 435, and for details of its work see Edwin A. Link, "Tomorrow on the Deep Frontier," *National Geographic,* June 1964, pp. 778–801. Paul H. Nitze, "Harnessing the Ocean Depths," *Vital Speeches,* October 1, 1964, pp. 749–751.

16. Kaharl, *Water Baby,* pp. 45–48.

17. Wakelin, *"Thresher:* Lesson and Challenge," p. 763.

18. For overviews of parts of the undersea fleet, see Frank Busby, *Undersea Vehicles Directory—1990–91* (Busby Associates, Arlington, Va., 1990). Also see various annual editions of *Jane's Fighting Ships* (McGraw-Hill, New York, and Jane's Information Group, Alexandria, Va.), especially the 1968–69 edition, pp. 471–472; the 1972–73 edition, pp. 558–560; and the 1993–94 edition, pp. 822–823. For *NR-1* details, see Ivar G. Babb et al., "Dual Use of a Nuclear Powered Research Submersible: The U.S. Navy's *NR-1,"* *Marine Technology Society Journal,* Vol. 27, No. 4, Winter 1993–94, pp. 39–48.

19. For robots, see "Naval Ocean Systems Center Underwater Vehicle History," Naval Ocean Systems Center, San Diego, Technical Document 1530, April 1989, pp. 1–31. For early laser work, see A. W. Palowitch, "Lasers in Underwater Imaging," *Underwater News & Technology,* Vol. 1, No. 1, July-August 1993, pp. 4–6. For "pingers," see Link, "Tomorrow on the Deep Frontier," p. 787. For multibeam sonar, see Robert Tyce et al., "NECOR Sea Beam Data Collection and Processing Development," *Marine Technology Society Journal,* Vol. 21, No. 3, June 1987, pp. 80–92. For bottom mapping, see "Navy Plumbs the Ocean's Depths," *Business Week,* July 6, 1968, pp. 58–59, and Navy's first Court of Inquiry report on the *Scorpion,* which was declassified in October 1993 and a copy obtained from Navy public affairs, pp. 1064–1065. For satellites, see Love, *History of the U.S. Navy,* p. 645, and J. E. D. Williams, *From Sails to Satellites: The Origin and Development of Navigational Science* (Oxford, New York, 1992), pp. 238–241.

20. For an analysis of how scientists lobbied on behalf of a deep naval push, see Robert C. Herold and Shane E. Mahoney, "Military Hardware Procurement," *Comparative Politics,* Vol. 6, No. 4, July 1974, pp. 571–599. Herold was a member of the Deep Submergence Systems Project, his observations those of a participant. Link, "Tomorrow on the Deep Frontier," p. 800.

21. Interviews, former Central Intelligence Agency official, May 18, 1994, May 19, 1994; Federal Intelligence expert, May 25, 1994; Robert C. Herold, May 26, 1994. Herold worked for the Deep Submergence Systems Project between 1966 and 1969. He also described some of the intelligence links in Herold and Mahoney, "Military Hardware Procurement," pp. 581–588.

A key broker of intelligence support was John S. Foster, Jr., a former director of the government's Livermore nuclear-weapons laboratory, who in 1965 became the Pentagon's top scientist as Undersecretary for Defense Research and Engineering. At Livermore, Foster had worked hard to help Federal intelligence analysts assess the state of Soviet nuclear weaponry. At the Pentagon, he showed great interest in aiding undersea endeavors that promised to further such ends.

22. Interview, John P. Craven, July 3, 1993. For Craven's pre-*Thresher* frame of mind, see Herold and Mahoney, "Military Hardware Procurement," pp. 583–584. For Craven and Project Seabed, see Robert C. Herold, *The Politics of Decision-Making in the Defense Establishment: A Case Study* (Ph.D. Thesis, George Washington University, September 1969), pp. 106–108, 119, 127. John P. Craven, "Sea Power and the Sea Bed," U.S. Naval Institute *Proceedings,* April 1966, pp. 36–51.

Herold suggests that Craven was less of an adventurer than he seemed, that his writings and persuasive efforts were meant at least partly to create symbols to attract military support and financial backing for his work, no matter that its actual outcome might simply be a better understanding of the planet's most exotic environment. Craven was, after all, a technologist, not a member of the uniformed military. It seems plausible that, at least on one level, he was a member of the scientific elite trying to exploit the *Thresher* disaster to his own ends. For Craven's manipulation of symbols, see Herold, *The Politics of Decision-Making,* p. 111.

23. John P. Craven, "The Geopolitical Significance of a Deepest Ocean Presence," *Marine Technology Society Journal,* June 1990, Vol. 24, No. 2, pp. 13–15. Walsh, "Thirty Thousand Feet," p. 8. Sylvia A. Earle, "Ocean Everest—An Idea Whose Time Has Come," *Marine Technology Society Journal,* June 1990, Vol. 24, No. 2, p. 10. These authors describe the 3 percent in conflicting ways. If the Earth's surface is 196.9 million square miles, then the oceans, 71 percent of that surface, comprise 139.8 million square miles, and 3 percent of that is 4.2 million square miles. The area of Europe, in contrast, is considered to be 3.8 million square miles.

24. For an example of an early drawing, see Daniel Behrman, *The New World of the Oceans: Men and Oceanography* (Little, Brown, Boston, 1969), p. 345.

25. Interview, James Alfred Locke Miller, Jr., former *Mizar* crew member, March 15, 1993.

26. For the Palomares search and recovery, see Flora Lewis, *One of Our H-Bombs Is Missing* (McGraw-Hill, New York, 1967); Christopher Morris, *The Day They Lost the H-Bomb* (Coward-McCann, New York, 1966); and Kaharl, *Water Baby,* pp. 65–81. For bomb description, see Chuck Hansen, *U.S. Nuclear Weapons: The Secret History* (Orion, New York, 1988), pp. 150–154, 224, 226. For President Johnson's order, see Craven, "Geopolitical Significance," p. 13. For the "grave danger" quote, see Lewis, p. 169; for map, pp. 134–135.

27. Lewis, *One of Our H-Bombs,* pp. 159, 162, 182–208. Kaharl, *Water Baby,* pp. 65–71.

28. Lewis, *One of Our H-Bombs,* pp. 197–210. Kaharl, *Water Baby,* pp. 71–79.

29. Lewis, *One of Our H-Bombs,* pp. 211–225. Kaharl, *Water Baby,* pp. 78–79.

30. Interview, J. Bradford Mooney, Jr., September 26, 1994. Lewis, *One of Our H-Bombs,* pp. 225–229. Morris, pp. 168–183. Kaharl, *Water Baby,* pp. 78–79.

31. For a description of the Soviet launch sites, see Nicholas L. Johnson, *Soviet Space Programs: 1980–1985* (American Astronautical Society, San Diego, 1987), pp. 27–33. For Soviet RV design, see Robert P. Berman and John C. Baker, *Soviet Strategic Forces* (Brookings Institution, Washington, D.C., 1982), pp. 89–90. For heavy RVs in testing, see Thomas B. Cochran et al., *Nuclear Weapons Databook: Volume Two, U.S. Nuclear Warheads Production* (Ballinger, Cambridge, Mass., 1987), p. 41. For Soviet impact areas, see William J. Arkin and Richard W. Fieldhouse, *Nuclear Battlefields* (Ballinger, Cambridge, Mass., 1985), p. 128. For Kamchatka impacts, interview, James E. Oberg, May 31, 1994. Oberg is author of *Red Star in Orbit* (Random House, New York, 1981). For Pacific RV splashdowns, see Nicholas L. Johnson, *The Soviet Year in Space: 1986* (Teledyne Brown, Colorado Springs, 1987), p. 10. For longer-range psychology, see "The Soviet Military Technological Challenge," *The Center for Strategic Studies,* Georgetown University, Washington, Special Report No. 6, September 1967, p. 53.

32. Interview, James E. Oberg, May 31, 1994. For the pressure to get additional intelligence during this period, see Victor Marchetti and John D. Marks, *The CIA and the Cult of Intelligence* (Dell, New York, 1975), pp. 299–300. For the intelligence community's rejection of a surface approach to RV retrieval, see Willard Bascom, *The Crest of the Wave: Adventures in Oceanography* (Doubleday/Anchor, New York, 1990), pp. 241–242.

33. Interview, former Defense Department expert, May 26, 1994. Interview, Federal intelligence expert, May 31, 1994.

34. For *Halibut* overview, see William J. Broad, "Navy Has Long Had Secret Subs for Deep-Sea Spying, Experts Say," *The New York Times,* February 7, 1994, p. A1. For quote and an early article on *Halibut,* see Christopher Drew et al., "A Risky Game of Cloak-and-Dagger Under the Sea," *Chicago Tribune,* January 7, 1991, p. 1. For general description of the sub, see *Jane's Fighting Ships, 1972–73* (McGraw-Hill, New York, 1972), pp. 420–421. For color photographs, letter to author, John P. Craven, May 25, 1994.

For the most detailed look at the sub's gear and preparations, see Roger C. Dunham, *Spy Sub: A Top Secret Mission to the Bottom of the Pacific* (Naval Institute Press, Annapolis, 1996). For the well, see pp. 22–23, and for cable problems, pp. 207–209. Dunham was a member of the crew. In order to pass a Defense Department classification review, he altered such things as the spy sub's name and target names and locations. Interview, Roger C. Dunham, May 17, 1996.

35. Interview, John P. Craven, July 3, 1993. For *Halibut* carrying DSRV simulator, see *Jane's Fighting Ships, 1976–77* (McGraw-Hill, New York, 1976), p. 551.

36. Interviews, former Defense Department expert, 1994.

37. Love, *History of the U.S. Navy,* pp. 562–568. See also Jeffrey Richelson, *American Espionage and the Soviet Target* (Morrow, New York, 1987), pp. 158–159. For the intelligence context of the operation, see Angelo Codevilla, *Informing Statecraft: Intelligence for a New Century* (Free Press, New York, 1992), pp. 109–110.

38. For official U.S. information on the Soviet sub, see press release "Glomar Explorer: Recovery and Burial of Soviet Sailors," Central Intelligence Agency, Washington, D.C., November 12, 1993. For a detailed but somewhat muddled account of the sub, see Roy Varner and Wayne Collier, *A Matter of Risk: The Inside Story of the CIA's Hughes Glomar Explorer Mission to Raise a Russian Submarine* (Random House, New York, 1978). For slim references, see Dunham, *Spy Sub,* pp. 196–197, 210. The Soviet sub is widely reported to have been a type known as Golf and to have carried two or three missiles.

The *Glomar Explorer* was a giant vessel built by CIA in the guise of a deep-sea mining ship for the industrialist Howard R. Hughes. At the behest of the Nixon Administration, it tried to physically recover a section of the Soviet sub years after it had been reconnoitered by the *Halibut.* See pp. 78–80.

39. After the collapse of the Soviet Union, the Russian press dug into the sinking episode. For English translations of a three-part series in *Izvestia,* see Joint Publications Research Service, JPRS-UMA, July 22, 1992, pp. 8–9, and July 29, 1992, pp. 12–15. These

publications are available from the Department of Commerce, National Technical Information Service, Springfield, Virginia.

40. Interview, Federal intelligence expert, February 4, 1994. The expert said the United States at the time had captured radio signals from test flights of Soviet missiles but had next to nothing to aid their interpretation. Simply monitoring a flight, which was done routinely by intelligence agencies, gave no clue as to how close the missile had come to its intended target. "Telemetry is meaningless without data points," the expert said of the intercepted signals. It should also be noted that, starting in the mid-1970s, the accuracy of Soviet missiles improved to such an extent that American land-based missiles were judged to have become vulnerable to a preemptive surprise attack. In response, the United States embarked on a long, costly, and ultimately futile search for new ways to base missiles securely on land, including such schemes as the Carter Administration's shell-game plan to rotate real and fake missiles through bunkers spread across the western United States. Lastly, it should be noted that the Navy was the beneficiary of such technical trends, politically at least, since its missile-carrying submarines were mobile and hard to target and ultimately were judged to be a more secure force than bombers and land-based missiles for the purpose of nuclear deterrence. That is still the case today.

For missile accuracy as a strategic issue, see Codevilla, *Informing Statecraft,* pp. 122–123.

41. Varner and Collier, *A Matter of Risk,* p. 14. While valuable in some respects, this book contains hints of CIA disinformation, such as the suggestion that the Soviet sub was discovered by *Mizar,* p. 18. At the time of the book's writing, *Halibut*'s powers of deep reconnaissance were a top national secret protected by a body of lies. For a technical description of the underwater listening gear that heard the sub's death throes, see Richelson, *American Espionage,* p. 169.

42. Interviews, *Halibut* crew member, December 8, 1994, and April 4, 1995. For a mission overview, see Dunham, *Spy Sub,* especially pp. 10–11, 121–213. See also letter, John P. Craven to Hawaii Senator Daniel K. Akaka, November 24, 1993. The letter was follow-up testimony to a Senate hearing. For the sub's condition, interview, Federal salvage contractor, February 4, 1994. This person worked on the *Glomar Explorer* and had intimate knowledge of the sub's condition.

43. For service award, letter, Craven to Akaka. For presidential award, copy of citation document from files of USS *Bowfin* Submarine Museum, Honolulu, Hawaii. For Senate quote, letter, Craven to Akaka. Interview, John P. Craven, June 16, 1993.

44. Interview, intelligence expert, February 4, 1994. For an overview of one of *Halibut*'s successors, the *Parche,* see Ed Offley, "Secret Navy Sub Finds New Home at Bangor Base," *Seattle Post-Intelligencer,* November 23, 1994, p. A1. For a look at part of the spy fleet, see Norman Polmar, "How Many Spy Subs . . . ?" U.S. Naval Institute *Proceedings,* December 1996, pp. 87–88. For a review of spy-satellite limits and the struggle to overcome them, see William J. Broad, "U.S. Designs Spy Satellites to Be More Secret Than Ever," *The New York Times,* November 3, 1987, p. C1.

45. For Nixon's charge, see Love, *History of the U.S. Navy,* p. 575.

46. For *NR-1* recovery, Robert D. Ballard, "NR-1: The Navy's Inner-Space Shuttle," *National Geographic,* April 1985, pp. 450–456. For Dragon Shield, interviews, former Defense Department expert, 1994.

47. "Presumed Lost," *Newsweek,* June 17, 1968, p. 74. Clancy, *Submarine,* pp. 21–22. For bomb description, see Hansen, *U.S. Nuclear Weapons,* pp. 207–208.

48. Interview, John P. Craven, July 3, 1993. Navy's first Court of Inquiry report on the *Scorpion,* pp. 1060–1061. Navy's second Court of Inquiry report on the *Scorpion,* which was declassified in October 1993 and a copy obtained from Navy public affairs, pp. 240, 242.

49. Henry R. Richardson and Lawrence D. Stone, "Operations Analysis During the Underwater Search for Scorpion," *Naval Research Logistics Quarterly,* Vol. 18, No. 2, June 1971, pp. 141–157. Navy's first Court of Inquiry report on the *Scorpion,* p. 1045.

50. Interview, John P. Craven, July 3, 1993. Navy's first Court of Inquiry report on the

Scorpion, pp. 1079–1083. Navy's second Court of Inquiry report on the *Scorpion,* pp. 240–257. Also see Richardson and Stone, "Operations Analysis," pp. 142–143.

51. Letters to author, John P. Craven, April 12, 1994; April 21, 1994; May 14, 1994; May 16, 1994.

52. Letter, Craven to Akaka. Interview, John P. Craven, June 16, 1993.

53. For an overview of press coverage, see Varner and Collier, *A Matter of Risk,* pp. 200–221. See also Seymour M. Hersh, "C.I.A. Salvage Ship Brought Up Part of Soviet Sub Lost in 1968, Failed to Raise Atom Missiles," *The New York Times,* March 19, 1975, p. 1, and Seymour M. Hersh, "Human Error Is Cited in '74 Glomar Failure," *The New York Times,* December 9, 1976, p. 1. See also David M. Alpern, "CIA's Mission Impossible," *Newsweek,* March 31, 1975, pp. 24–32.

54. Interview, Federal salvage contractor, February 4, 1994.

55. Robert M. Gates, *From the Shadows: The Ultimate Insider's Story of Five Presidents and How They Won the Cold War* (Simon & Schuster, New York, 1996), pp. 553–554. Press release, "Glomar Explorer," Central Intelligence Agency, Washington, November 12, 1993. John-Thor Dahlburg, "CIA's Raising of Soviet Sub Told," *The Los Angeles Times,* October 17, 1992, p. A2. William J. Broad, "Russia Says U.S. Got Sub's Atom Arms," *The New York Times,* June 20, 1993, p. A14.

56. For an example from the space program of the Nixon Administration's political use of contracts, see John M. Logsdon, "The Space Shuttle Program: A Policy Failure?" *Science,* Vol. 232, May 30, 1986, pp. 1099–1105.

57. For DSRV descriptions, see Scott Carpenter, "Escape from the Deep," *Popular Science,* September 1969, pp. 79–81, 214; Charles LaFond, "Hope for Disabled Undersea Vessels," *Science News,* September 12, 1970, pp. 231–233. For *NR-1* towing, see Norman Polmar, "The Deep Submergence Vehicle Fleet," U.S. Naval Institute *Proceedings,* June 1986, pp. 119–120.

58. For the intelligence tie, Herold and Mahoney, "Military Hardware Procurement," pp. 584-588. For the final cost, see *Jane's Fighting Ships, 1975–76* (McGraw-Hill, New York, 1975), pp. 531–532.

59. Interviews, Lieut. Gen. Daniel O. Graham, former director of the Defense Intelligence Agency and former deputy director of the CIA, January 27, 1994, and Angelo M. Codevilla, former senior staffer of the U.S. Senate Select Committee on Intelligence, May 25, 1994. Also see Bob Woodward, *Veil: The Secret Wars of the CIA, 1981–1987* (Simon & Schuster, New York, 1987), pp. 448–449.

60. Interview, Angelo M. Codevilla, December 13, 1993. Woodward, *Veil,* p. 449–450.

61. Ibid.

62. The various Soviet sinkings are listed in a Defense Intelligence Agency report obtained by Greenpeace in Washington, D.C., under the Freedom of Information Act, and in Joshua Handler et al., "Naval Safety 1989: The Year of the Accident" (Greenpeace, Washington, D.C., April 1990), pp. 24–25, and in a detailed bibliography of Russian sub accidents compiled by Handler, which ran to forty-five pages in February 1993. For the *Yankee*'s condition, see William J. Broad, "Sunken Soviet Sub Leaks Radioactivity in Atlantic," *The New York Times,* February 8, 1994, p. A13. For the Navy's center near Washington for processing deep Soviet gear, interview, former CIA official, January 27, 1994.

63. Nicholas Wade, "Do Moles Matter?" *The New York Times Magazine,* March 20, 1994, p. 22.

64. Interview, former Federal salvage specialist, June 6, 1994.

65. Interview, John P. Craven, July 3, 1993.

66. For the policy declaration, see Sean O'Keefe et al., ". . . From the Sea," U.S. Naval Institute *Proceedings,* November 1992, pp. 93–96. O'Keefe at the time was secretary of the Navy. For a reaffirmation and expansion of the policy, see John H. Dalton et al., "Forward . . . from the Sea," U.S. Naval Institute *Proceedings,* December 1994, pp. 46–49. Dalton is secretary of the Navy.

The campaign for the shift undoubtedly gained strength in the wake of the Gulf War, during which, early in 1991, the United States Navy found itself facing a series of unexpected troubles while fighting Iraq from within the confines of the Persian Gulf, whose waters have an average depth of only 328 feet.

67. For a review of the industrial boom of the 1960s, see "Oceanography: Work Beneath the Waves," *Time,* January 19, 1968, pp. 68–75. For a sketch of how commercial robots developed in the aftermath of the oil wars, see Tony Hayward, "Introduction," *ROV Review* (Windate Enterprises, Spring Valley, Calif., 1988), pp. 3–4.

It is worth noting, too, that the undersea industry, like the military, used the revolution in computer chips to make its gear increasingly dexterous and smart. By the 1980s, the field of deep commercial technology in many respects was ripe for breakthroughs, strengthening the repercussions of the military spinoffs.

68. Robert D. Ballard with Rick Archbold, *The Discovery of the Titanic* (Warner, New York, 1989), pp. 148–153, 162, 208–209, 213–220. William J. Broad, "Titanic Wreck Was Surprise Yield of Underwater Tests for Military," *The New York Times,* September 8, 1985, p. A1. United Press International, "Robot That Spotted Titanic Also Found Sunken Submarine," *The New York Times,* September 17, 1985, p. A23. Broad, "Items on Titanic Brought to U.S.," *The New York Times,* July 7, 1993, p. A11. Gillian Hutchinson, *The Wreck of the Titanic* (National Maritime Museum, Greenwich, England, 1994).

69. Robert D. Ballard with Malcolm McConnell, *Explorations: My Quest for Adventure and Discovery Under the Sea* (Hyperion, New York, 1995), pp. 321–346. Ballard with Rick Archbold, *The Discovery of the Bismarck* (Warner, New York, 1990). Ballard with Rick Archbold, *The Lost Ships of Guadalcanal* (Warner, New York, 1993). Ballard, "Riddle of the Lusitania," *National Geographic,* April 1994, pp. 68–85.

Interestingly, Ballard's commercial push began quite early, even as the *Titanic* discovery was still fresh. His private company sought to sell civilian versions of the military gear that had found the sunken liner, although it is unclear how successful his efforts were at the time, given the security constraints of the cold war. See William J. Broad, "Finder of Titanic Aims to Capitalize," *The New York Times,* September 11, 1985, p. D26.

70. Interview, Robert D. Ballard, January 20, 1995. William J. Broad, "Secret Sub to Scan Sea Floor for Roman Wrecks," *The New York Times,* February 7, 1995, p. C1. For *NR-1* details and general sharing, see Babb et al., "Dual Use," pp. 39–48. For *NR-1,* see Ballard, "NR-1: The Navy's Inner-Space Shuttle," pp. 450–459. For how the *Jason* robot helped pave the way for Ballard's 1995 Mediterranean foray, see Anna Marguerite McCann and Joann Freed, "Deep Water Archaeology: A Late-Roman Ship from Carthage and an Ancient Trade Route Near Skerki Bank off Northwest Sicily," *The Journal of Roman Archaeology,* Supplementary Series No. 13 (Ann Arbor, Mich., 1994). See also Ballard and Rick Archbold, *The Lost Wreck of the Isis* (Madison, Toronto, 1990).

71. Lawrence D. Stone, "Search for the SS Central America: Mathematical Treasure Hunting," *Interfaces,* January-February 1992, pp. 32–54. Tim Noonan, "The Greatest Treasure Ever Found," *Life,* March 1992, pp. 32–42. Interviews, Ted Brockett and John D. Moore, president and vice president of Sound Ocean Systems, Inc., which developed the robot, January 18, 1993.

72. For the *Mir*'s filming of the *Titanic,* see Joseph MacInnis, *Titanic in a New Light* (Thomasson-Grant, Charlottesville, Va., 1992), as well as William J. Broad, "Deepest Wrecks Now Visible to Undersea Cameras," *The New York Times,* February 2, 1993, p. C1. For *Mir*'s gold hunt, see Valerie Moore, "Mysteries of the Deep," *Sky* magazine of Delta Airlines, Vol. 22, No. 8, August 1993, pp. 30–40. For an overview of Moscow's deep gear bonanza, see Deam Given, "Underwater Technology in the USSR," *Oceanus,* Vol. 34, No. 1, Spring 1993, pp. 67–73. Also see Anatoly M. Sagalevitch, " 'Mir' Submersibles in Science and Underwater Technology," *MTS 94 Conference Proceedings* (Marine Technology Society, Washington, 1994), pp. 241–244. For comprehensive overviews of Russian gear, see World Technology Evaluation Center, "WTEC Panel Report on Research Submersibles and Undersea Technol-

ogies," Loyola College, Baltimore, June 1994, and World Technology Evaluation Center, "WTEC Panel Report on Submersibles and Marine Technologies in Russia's Far East and Siberia," Loyola College, Baltimore, August 1996. For U.S. trends, see Gregory Stone, "Dual Use of Military Assets for Environmental Research," and Robert B. Oswald, "Dual Use Technology: Using Department of Defense Technology to Support Marine Research," both in *Marine Technology Society Journal,* Vol. 27, No. 4, Winter 1993–94, pp. 3 and 49–51, respectively. See also Edward C. Whitman, "Defense Conversion in Marine Technology—Past, Present, and Potential Possibilities," *MTS 94 Conference Proceedings* (Marine Technology Society, Washington, 1994), pp. 121–126. For a popular version of this article, see Edward C. Whitman, "Defense Conversion in Marine Technology," *Sea Technology,* November 1994, pp. 21–25. See also National Research Council, *Undersea Vehicles and National Needs* (National Academy Press, Washington, 1996), pp. 2, 15–16.

The Clinton Administration, which took office in January 1993, pushed hard for the public release of all kinds of military technologies from the cold war, including deep ones. The action was led by Vice President Al Gore, a liberal former Senator with broad experience in military affairs. Some of that Democratic momentum was lost when control of both houses of Congress was won by the Republicans, who took over in January 1995 and generally frowned on dual-use military programs. After that shift, some aspects of the conversion effort moved at a slower pace.

73. Interview, N. Eugene Smith, NOAA's liaison to the Navy, October 2, 1996. N. Eugene Smith, "DSV Sea Cliff—1992 Hawaiian Science Cruise," *Underwater Intervention '93 Conference Proceedings* (Marine Technology Society, Washington, 1993), p. 16. N. Eugene Smith, "Civilian Scientists Use the Navy's Deep Submergence Vehicles," *Marine Technology Society Journal,* Vol. 27, No. 4, Winter 1993, pp. 64–69. "Naval Ocean Systems Center Underwater Vehicle History," U.S. Navy, Naval Ocean Systems Center, San Diego, undated, pp. 1–2, 27. Letter to author, Randy Koski, U.S. Geological Survey, a member of the Oregon team, September 15, 1994. Interview, Randy Koski, September 20, 1994. For *Jason* sharing, see National Research Council, *Undersea Vehicles,* p. 15.

74. Marcus Langseth et al., "SCICEX-93: Arctic Cruise of the U.S. Navy Nuclear Powered Submarine USS Pargo," *Marine Technology Society Journal,* Vol. 27, No. 4, Winter 1993–94, pp. 4–12. William J. Broad, "U.S. Navy's Attack Subs to Be Lent for Study of Arctic Icecap," *The New York Times,* February 21, 1995, p. C1.

75. William J. Broad, "Long-Secret Navy Devices Allow Monitoring of Ocean Eruption," *The New York Times,* August 20, 1993, p. A1. Broad, "Navy Listening System Opening World of Whales," *The New York Times,* August 23, 1993, p. A12. See also Ivan Amato, "A Sub Surveillance Network Becomes a Window on Whales," *Science,* Vol. 261, July 30, 1993, pp. 549–550. Interview, Amos S. Eno, June 26, 1996. Broad, "Anti-Sub Seabed Grid Thrown Open to Research Uses," *The New York Times,* July 2, 1996, p. C1.

76. Alan Gordon, "Underwater Laser Line Scan Technology," Underwater Intervention '93 Conference Proceedings (Marine Technology Society, Washington, 1993), pp. 164–170. James Leatham and Bryan W. Coles, "Use of Laser Sensors for Search & Survey," Underwater Intervention '93 Conference Proceedings (Marine Technology Society, Washington, 1993), pp. 171–186. For a popular account, see William J. Broad, "Deepest Wrecks Now Visible to Undersea Cameras," *The New York Times,* February 2, 1993, p. C1.

77. "Advanced Unmanned Search System," Naval Command, Control and Ocean Surveillance Center, RDT&E Division, San Diego, Document No. 2348, December 1992. For a similar description, see James Walton et al., "Advanced Unmanned Search System," Underwater Intervention '93 Conference Proceedings (Marine Technology Society, Washington, 1993), pp. 243–249.

78. Ian R. McDonald et al., "Deep-Ocean Use of the SM2000 Laser Line Scanner on Submarine NR-1 Demonstrates System Potential for Industry and Basic Science," *Conference Proceedings, Oceans '95,* Marine Technology Society and Institute of Electrical and Electronic Engineers, October 1995, Vol. 1, pp. 555–565.

79. Interview, David L. Mearns, business manager, Oceaneering Technologies, March 29, 1995. Letter to author, George H. Seltzer, program manager, Oceaneering Technologies, September 16, 1994. "Magellan 725," product brochure, Oceaneering Technologies (Upper Marlboro, Md., undated). Alan Cowell, "Italian Obsession: Was Airliner Shot Down?" *The New York Times,* February 10, 1992, p. A7. Craig Mullen, "Eastport International Solves the M/V Lucona Sinking Mystery," *Waves,* September-October 1991, pp. 18–19. Eastport was a previous name of Oceaneering Technologies.

80. William J. Broad, "Map Makes Ocean Floors as Knowable as Venus," *The New York Times,* October 24, 1995, p. C1, and "Correction," October 25, 1995, p. A2. It turns out that the surface of Venus was mapped by the *Magellan* spacecraft to reveal objects as small as hundreds of feet across, while the seafloor map can make out only features that are miles in diameter. Mark Carlowicz, "New Map of Seafloor Mirrors Surface," *EOS,* Vol. 76, No. 44, October 31, 1995, p. 441. For the genesis of gravity mapping, see Stephen S. Hall, *Mapping the Next Millennium* (Vintage, New York, 1993), pp. 71–87.

81. William J. Broad, "Navy Is Releasing Treasure of Secret Data on World's Oceans," *The New York Times,* November 28, 1995, p. C1. For a detailed description of the data, see Kenneth E. Hawker, Jr., ed., "Scientific Utility of Naval Environmental Data: A MEDEA Special Task Force Report," MEDEA, McLean, Va., June 1995. The MEDEA office is run by the Mitre Corporation but the group itself is a Federal advisory panel that provides scientific counsel to the American intelligence community, including the Central Intelligence Agency.

82. Interview, Graham Hawkes, July 2, 1993. See also William J. Broad, "Graham Hawkes: Racing to the Bottom of the Deep, Black Sea," *The New York Times,* August 3, 1993, p. C1.

83. Interview, Graham Hawkes, July 1, 1993.

84. For a description of *Deep Flight Two,* see Graham Hawkes and Philip J. Ballou, "The Ocean Everest Concept: A Versatile Manned Submersible for Full Ocean Depth," *Marine Technology Society Journal,* June 1990, Vol. 24, No. 2, pp. 79–86. See also Gregory T. Pope, "Deep Flight," *Popular Mechanics,* April 1990, pp. 70–72. For the Navy's work in ceramic casings and its view of Hawkes, see Ramon R. Kurkchubasche, "Underwater Applications for Structural Ceramics," *Underwater Intervention '94 Conference Proceedings* (Marine Technology Society, Washington, 1994), pp. 357–363.

85. Interview, Graham Hawkes, July 2, 1993.

Three. GARDEN OF EDEN

1. Broad, "Long-Secret Navy Devices."

2. Cindy Lee Van Dover, "Do 'Eyeless' Shrimp See the Light of Glowing Deep-Sea Vents?" *Oceanus,* Winter 1988–89, Vol. 31, No. 4, pp. 47–52. For her career in general, see Cindy Lee Van Dover, *The Octopus's Garden: Hydrothermal Vents and Other Mysteries of the Deep Sea* (Addison, New York, 1996). For her description of the Juan de Fuca voyage, see pp. 115–126; for vent glow and not-so-blind shrimp, pp. 127–138.

Cindy's genius was to turn off all *Alvin*'s lights during one dive as the sub sat motionless next to a chimney spewing extraordinarily hot water. An electronic still camera hundreds of times more sensitive than regular film or the human eye peered into the pitch blackness, detecting an eerie glow. With cool logic, she had predicted it must be so because "eyeless" shrimp that swarm by the thousands around some Atlantic chimneys have rudimentary eyes on their backs. In discovering the eyes and glow, she bolstered the remarkable idea that some fauna seek out faint volcanic lights as a kind of natural advertisement for deep food.

3. Interviews at sea were conducted from October 9 to 25, 1993. I've taken the liberty of not noting the dates of the individual interviews during the voyage. Names of the interviewees

should be obvious in the text. For a voyage overview filed when I returned to New York, see William J. Broad, "A Voyage into the Abyss: Gloom, Gold and Godzilla," *The New York Times,* November 2, 1993, p. C1.

4. For details on Ten North, see R. A. Haymon et al., "Volcanic Eruption of the Mid-ocean Ridge along the East Pacific Rise," *Earth and Planetary Science Letters,* 1993, Vol. 119, pp. 85–101. See also Richard A. Lutz and Rachel M. Haymon, "Rebirth of a Deep-Sea Vent," *National Geographic,* November 1994, pp. 114–126, and William J. Broad, "In Ocean Depths, Scientists Find Fast Growth," *The New York Times,* October 20, 1994, p. B13.

5. For Godzilla discovery, see Veronique Robigou and John R. Delaney, "Large Massive Sulfide Deposits in a Newly Discovered Active Hydrothermal System, the High-Rise Field, Endeavour Segment, Juan de Fuca Ridge," *Geophysical Research Letters,* September 3, 1993, Vol. 20, No. 17, pp. 1887–1890.

6. Interview, John R. Delaney, September 12, 1993.

7. John R. Delaney, "An Interdisciplinary Response to the Earthquake Swarms and Submarine Eruption, CoAxial Segment of the Juan de Fuca Ridge," proposal to the National Science Foundation, August 20, 1993, p. 7.

8. The geology material I read on the voyage included Peter A. Rona, "The Dynamic Abyss," in Joseph MacInnis, ed., *Saving the Oceans* (Key Porter, Toronto, 1992), pp. 99–116, and Frank Press and Raymond Siever, *Understanding Earth* (Freeman, New York, 1994). Another book I gave close attention to throughout the voyage was Kaharl, *Water Baby.*

9. Kaharl, *Water Baby,* pp. 153–160. Also see Robert D. Ballard, *Exploring Our Living Planet* (National Geographic Society, Washington, 1988), pp. 110–121, and Ballard with Malcolm McConnell, *Explorations: My Quest for Adventure and Discovery Under the Sea* (Hyperion, New York, 1995), pp. 103–130.

10. Kaharl, *Water Baby,* p. 173.

11. Ibid., pp. 173–174. John B. Corliss et al., "Submarine Thermal Springs on the Galápagos Rift," *Science,* March 16, 1979, pp. 1073–1082. John M. Edmond and Karen Von Damm, "Hot Springs on the Ocean Floor," *Scientific American,* April 1983, p. 78. See also Venena Tunnicliffe, "Hydrothermal-Vent Communities of the Deep Sea," *American Scientist,* July-August 1992, pp. 336–349. For a comprehensive overview, see Susan E. Humphris et al., *Seafloor Hydrothermal Systems: Physical, Chemical, Biological and Geological Interactions* (American Geophysical Union, Washington, 1995).

12. Quoted in Kaharl, *Water Baby,* p. 175.

13. Ibid., pp. 199–201.

14. Ibid., p. 202. See also John M. Edmond and Karen Von Damm, "Hydrothermal Activity in the Deep Sea," *Oceanus,* spring 1992, p. 76, and Edmond and Von Damm, "Hot Springs."

15. For a vent-discovery overview, see Thurman, *Introductory Oceanography,* p. 445–447. Also see Peter A. Rona, "Deep-Sea Geysers of the Atlantic," *National Geographic,* October 1992, pp. 105–109. For the flow of seawater through the volcanic rifts, see Press and Siever, *Understanding Earth,* p. 394.

16. Kaharl, *Water Baby,* p. 202.

17. Joseph Cone, *Fire Under the Sea: The Discovery of the Most Extraordinary Environment on Earth—Volcanic Hot Springs on the Ocean Floor* (Morrow, New York, 1991), pp. 69–90, 167–205. J. B. Corliss, J. A. Baross, S. E. Hoffman, "An Hypothesis Concerning the Relationship Between Submarine Hot Springs and the Origin of Life on Earth," *Oceanologica Acta,* supplement, 1981, pp. 59–69.

18. William J. Broad, "Clues to Fiery Origin of Life Sought in Hothouse Microbes," *The New York Times,* May 9, 1995, p. C1. M. Mitchell Waldrop, "Goodbye to the Warm Little Pond?" *Science,* November 23, 1990, pp. 1078–1080. Waldrop, "The Golden Crystal of Life," *Science,* November 23, 1990, p. 1080. For a detailed overview of the thesis by a

Nobel laureate, see Christian de Duve, *Vital Dust: Life as a Cosmic Imperative* (Basic Books, New York, 1995), pp. 2–32, 100–101.

19. For a description of some of the device's major accomplishments, see Edward T. Baker, "Megaplumes," *Oceanus,* Winter 1991–92, p. 84.

20. Inspired by snowstorms blowing in reverse and undulating thickets of deep life, as well as a potent cup of latte made for me by Gary, from go-go Seattle, the nation's coffee capital, I wrote up a story that day and filed it by satellite with the help of the ship's radio expert. William J. Broad, "Life Springs Up in Ocean's Volcanic Vents, Deep Divers Find," *The New York Times*, October 19, 1993, p. C4.

21. Thomas Gold, "The Deep, Hot Biosphere," *The Proceedings of the National Academy of Sciences,* Vol. 89, July 1992, pp. 6045–6049. Also see Jody W. Deming and John A. Baross, "Deep-sea Smokers: Windows to a Subsurface Biosphere?" *Geochimica et Cosmochimica Acta,* 1993, Vol. 57, pp. 3219–3230.

22. The dive occurred on October 24, 1993, and was documented in notes, personal audio recordings, and audio and video recordings made automatically by *Alvin*'s gear. Over the submersible's lifetime, it was the 2,681st dive.

Four. LOST WORLDS

1. T. Bentley Duncan, *Atlantic Islands: Madeira, the Azores and the Cape Verdes in Seventeenth-Century Commerce and Navigation* (University of Chicago, 1972), pp. 113–136. Charles R. Boxer, *Four Centuries of Portuguese Expansion, 1415–1825* (University of California, Berkeley, 1969), pp. 1–21. A. H. de Oliveira Marques, *History of Portugal* (Casa da Moeda, Lisbon, 1991), pp. 33–41, 52–59. Portugal Regional Tourism Board, "Terceira," Angra do Heroismo, Azores, undated. For the role of the wind, see Felipe Fernández-Armesto, ed., *The Times Atlas of World Exploration* (HarperCollins, New York, 1991), pp. 40–49. For an overview on the opening of the Atlantic, see R. R. Palmer and Joel Colton, *A History of the Modern World* (Knopf, New York, 1966), pp. 90–93, 109–111.

2. Winston Graham, *The Spanish Armadas* (Doubleday, Garden City, N.Y., 1972), pp. 184–199.

3. Margaret Rule, *The Mary Rose: The Excavation and Raising of Henry VIII's Flagship* (Conway Maritime Press, London, 1983), pp. 103–201.

4. Interviews, Robert F. Marx, January 19, April 16, 1995. Marx, "Phoenician Explorations, Ltd., Azorean Underwater Archaeological Project, Plan of Operations," September 25, 1994, pp. 1–2. Robert F. Marx and Jenifer Marx, *The Search for Sunken Treasure: Exploring the World's Great Shipwrecks* (Key Porter, Toronto, 1993). Chapter Eleven of this book gives Marx's general view of the promise of deep recovery, pp. 172–185.

5. Interviews, Robert F. Marx, May 30, June 22, September 12, 1995. Nancy Gibbs, "The Race for Sunken Treasure: The Ocean Gold Rush," *Time,* international edition, October 25, 1993, pp. 48–55. Marx, "Phoenician Explorations."

6. Interview, Robert F. Marx, May 30, 1995. See also William J. Broad, "Watery Grave of the Azores to Yield Shipwrecked Riches," *The New York Times,* June 6, 1995, p. C1.

7. Interviews, Robert F. Marx, January 19, September 8, September 10, 1995.

8. One article called Marx "a person with little scientific rigor" who made an art of "procuring the good will of politicians and governments." Daniel Adrião, "Gomes Da Silva Becomes Part of the Treasure Hunt," *Seminario* (Lisbon), August 27, 1994, pp. 50–51. Rui Gomes da Silva was Marx's lawyer and a rising young legislator, who coauthored the new law. *O Independente* called the law a "shipwreck of principles" that destroyed "a healthy blindness that does not distinguish between a rotten piece of wood and a gold cross with encrusted rubies." Pedro Marta Santos, "Fruits of the Sea," Vida supplement of *O Independente* (Lisbon), April 28, 1995, pp. 14–16.

9. Interview, Francisco J. S. Alves, September 14, 1995.

I've taken the liberty of not noting individually all the other interviews conducted on this day, including those before and after the dive in Angra Bay. The names of the interviewees should be obvious in the text.

10. *All Lisbon and Its Surroundings* (Editorial Escudo de Oro, Lisbon, 1994), pp. 50–51. Broad, "Watery Grave." "Presentation," Archaeonautica Studies Center, Lisbon, undated one-page description of the group and its activities. See also Archaeonautica Correspondence, Archaeonautica Studies Center, Lisbon, No. 2, 1st semester of 1995.

11. Interview, Francisco J. S. Alves, September 15, 1995. I've taken the liberty of not noting individually the other interviews conducted on this day.

12. Interview, Francisco Ernesto de Oliveira Martins, September 16, 1995.

13. Interview, António Mega Ferreira, September 12, 1995. For the fair and its origins see "Expo '98, the Last Exposition of the 20th Century," The Commission for the Lisbon World Exposition, Lisbon, 1995, pp. 3–6.

14. For his action-oriented life, see Robert F. Marx, *Still More Adventures* (Mason, New York, 1976). For some of his archaeological adventures, see Robert F. Marx, *The Underwater Dig: Introduction to Marine Archaeology* (Gulf, Houston, 1990).

15. Interview, Robert F. Marx, September 18, 1995.

Despite Marx's optimism, the law and the ship-recovery plan were shortly thereafter put on hold as the governing Social Democrats, plagued by an economic slump and corruption scandals, lost the Portuguese elections of October 1995. Winning the elections were the Socialists, who had been out of power for ten years and were generally opposed to the free-market recovery of shipwrecks.

16. Interviews and observations were conducted from August 8 to August 19, 1996, the central leg of a monthlong expedition. I've taken the liberty of not noting the individual dates of interviews during the voyage. Names of the interviewees should be obvious in the text. For an expedition overview, see William J. Broad, "Titanic Relics Open Era in Deep-Sea Commerce," *The New York Times,* August 25, 1996, p. A24. For background on the operation, see William J. Broad, "Items on Titanic Brought to U.S.," *The New York Times,* July 7, 1993, p. A11, and Broad, "Deep-Sea Scavenging Gets Easier," *The New York Times,* July 18, 1993, section 4, p. 6.

17. For an overview of the restoration process, see Charles A. Haas, ed., *Voyage,* No. 22, Winter 1996, pp. 51–83. This journal is published by Titanic International, Inc., based in Freehold, New Jersey. The issue profiles the work of *Titanic* conservators working in the French village of Semur-en-Auxois.

18. Ken Ringle, "New Depths for Titanic Promoter? Cruise to Ship Site Reopens Controversy," *The Washington Post,* August 6, 1996, p. B1.

19. For a review of hunts failed and successful, see John P. Eaton and Charles A. Haas, *Titanic: Triumph and Tragedy* (Norton, New York, 1995), pp. 301–316. For the finding, see Ballard and Archbold, *Discovery of the Titanic,* pp. 59–84. For the military context, see Broad, "Titanic Wreck Was Surprise Yield."

20. Ballard and Archbold, *Discovery of the Titanic,* pp. 137–193, 222–226.

21. Eaton and Haas, *Titanic: Triumph and Tragedy,* pp. 308–309. Ballard and Archbold, *Discovery of the Titanic,* pp. 209–210, 213–221. "The Salvage Controversy," *USA Today* the magazine, March 1995, p. 68. Eva Hart, the last *Titanic* survivor said to have a clear memory of the disaster, who died in early 1996, during her final years denounced salvage efforts as grave robbing. See Robert McG. Thomas, Jr., "Eva Hart, 91, a Last Survivor with Memory of Titanic, Dies," *The New York Times,* February 16, 1996, p. A30.

22. Interviews, George Tulloch, March 7, December 8, 1996. Ringle, "New Depths for Titanic Promoter."

23. Eaton and Haas, *Titanic: Triumph and Tragedy,* pp. 310–312. Note that critics of the recovery work charge that some artifacts were also removed from the ship itself, not just

the debris field. See Ballard and Archbold, *The Discovery of the Titanic,* pp. 213–221, especially p. 214, and Don Lynch and Ken Marschall, *Titanic: An Illustrated History* (Hyperion, New York, 1992), pp. 208–209.

24. For the *Mir*'s filming of the *Titanic,* see MacInnis, *Titanic in a New Light,* and Broad, "Deepest Wrecks."

25. Eaton and Haas, *Titanic: Triumph and Tragedy,* p. 313. William Thomas, "Memphis Salvagers Plan Trip to Titanic," *The Commercial Appeal* (Memphis), August 15, 1992, p. A1. United Press International, "Texan Loses Bid to Salvage Titanic," October 3, 1992.

26. Eaton and Haas, *Titanic: Triumph and Tragedy,* pp. 313–316. William J. Broad, "Items on Titanic Brought to U.S.," *The New York Times,* July 7, 1993, p. A11. For a detailed description of the 1993 expedition, see Charles A. Haas, "Destination, North Atlantic," "Dives 1 and 2: Precision in Preparation," "A Voyage for Knowledge," *Voyage,* No. 18, Autumn 1994, pp. 40–65, and John P. Eaton, "Descent to History," *Voyage,* No. 18, Autumn 1994, pp. 66–77. For the British exhibition, see Gillian Hutchinson, *The Wreck of the Titanic* (National Maritime Museum, Greenwich, England, 1994).

27. Jamie Portman, "Cameron Can't Avoid the Big Picture: Director Embarks On Another Blockbuster—This Time About Titanic," *The Gazette* (Montreal), March 5, 1996, p. C9. John Brodie and Andrew Hindes, "Fox, Par Board 'Titanic,' " *Variety,* July 22, 1996, p. 17.

28. For Garzke's analytical work, see William J. Broad, "New Idea on Titanic Sinking Faults Steel as Main Culprit," *The New York Times,* September 16, 1993, p. A14.

29. Interview, George Tulloch, July 28, 1996.

30. Interview, Carl Boyd, July 11, 1995. Nigel Pickford, *The Atlas of Shipwrecks and Treasure: The History, Location, and Treasures of Ships Lost at Sea* (Dorling Kindersley, New York, 1994), pp. 116–117. Paul R. Tidwell, "Operation Rising Sun," set of reproductions of archival documents, undated. William J. Broad, "Lost Japanese Sub with 2 Tons of Axis Gold Found on Floor of Atlantic," *The New York Times,* July 18, 1995, p. C1.

31. Interview, Jesse D. Taylor, July 12, 1995. Broad, "Lost Japanese Sub."

32. Interview, Paul R. Tidwell, July 11, 1995.

33. Interviews, Tidwell, July 6, 11, 13, 14, 1995. Résumé, Tidwell. United Press International, "In Search of Sunken Treasure—Copper," March 28, 1982.

34. Interviews, Tidwell, July 11, 13, 1995.

35. Pickford, *The Atlas of Shipwrecks,* pp. 116–117.

36. Interview, Tidwell, July 13, 1995.

37. Interview, Tidwell, July 13, 1995. See also David M. Graham, "Seven-Vessel Fleet Marks Russian CGGE," *Sea Technology,* June 1994, pp. 67–69.

38. Letter and manuscript to author, Andrew Dougherty, July 20, 1994. Dougherty is the son of Tidwell's lawyer and a member of the expedition, who later wrote an overview of the voyage.

39. Interviews, Tidwell, July 11, 1995, and David W. Jourdan, July 12, 1995.

40. Ibid. Mark Guidera, "High-tech Hunters Close In on Lost Sub," *The Sun* (Baltimore), July 23, 1995, p. 1B.

41. Interview, Tidwell, July 13, 1995.

42. Dougherty manuscript. Interviews, Tidwell, July 6, 11, 13, 14, 1995. Interviews, Jourdan, July 12, 13, 1995. Jourdan, "Searching for the I-52," *Meridian Passages,* Vol. 2, No. 2, Spring 1995, pp. 1–3. Tidwell et al., "The Search and Discovery of the I-52," *Conference Proceedings, Oceans '95, Marine Technology Society and Institute of Electrical and Electronic Engineers, October 1995,* Vol. 3, p. 1516. These sources are also the basis for the voyage narrative in the next several paragraphs.

43. Interview, Tidwell, July 6, 1995.

44. Interview, Carl Boyd, July 11, 1995.

45. Interview, Jourdan, July 13, 1995.

46. Interview, David Wyatt, July 13, 1995.

47. Interview, Tidwell, July 13, 1995.

One by-product of such activity is going to be the writing and rewriting of international law, especially relating to newer ships. The antagonists of the cold war did their work surreptitiously. But the public opening of the deep is going to beget new frictions, not the least being the kind of feud that eventually erupted between Tidwell and Tokyo, the putative owner of the *I-52*'s gold. After the discovery was announced, the Japanese government said the submarine and its cargo of $25 million in gold bars remained its property. Tidwell, while cordial, vowed to press ahead with the recovery and prepared for a return voyage.

See Vanora Bennett, "Japan, Americans Argue Over Undersea Treasure Trove," Reuters, July 27, 1995.

Five. CANYON

1. National Oceanic and Atmospheric Administration, "Monterey Bay, National Marine Sanctuary," U.S. Department of Commerce, NOAA, Washington, undated.

2. Rick Gore, "Between Monterey Tides," *National Geographic,* February 1990, pp. 2–43. "The Monterey Bay Connection: a Prospectus for Scientific Excellence in the 1990's," U.S. Department of Commerce, NOAA, June 1992. Gary B. Griggs, "Monterey Bay, National Center for Marine Science," *Sea Technology,* May 1995, pp. 43–53.

3. For a MBARI overview, see William J. Broad, "Fantastic World Found in Sea's Middle Depths," *The New York Times,* July 26, 1994, p. C1.

4. Interview, David Packard, July 17, 1994. Judith L. Connor and Nora L. Deans, "David Packard and Julie Packard: Monterey Bay Profiles in Depth," *Oceanus,* Summer 1993, pp. 74–79. David Packard, *The HP Way: How Bill Hewlett and I Built Our Company* (HarperCollins, New York, 1995).

5. Interview, Peter G. Brewer, July 11, 1994. Unless otherwise noted, all interviews occurred at MBARI between July 11 and July 15, 1994. I've taken the liberty of not always noting the individual dates of these interviews. The names of the interviewees should be obvious in the text.

6. Sagalevitch, " 'Mir' Submersibles," p. 242. Bruce Robison et al., "Cold Seep Research Program," annual research plan, Monterey Bay Aquarium Research Institute, 1993. Gage and Tyler, *Deep-Sea Biology,* pp. 140, 369–371. Thurman, *Introductory Oceanography,* pp. 447–451.

7. For more on the SWATH ship, see James B. Newman and Bruce H. Robison, "Development of a Dedicated ROV for Ocean Science," *MTS Journal* (Marine Technology Society), Vol. 26. No. 4, pp. 46–53. D. M. A. Hollaway, Roy D. Haul, and Robertson P. Dinsmore, "Waves and the Future: Swath for Ocean Science," *MTS 94 Conference Proceedings* (Marine Technology Society, Washington, 1994), pp. 548–551. Daniel Rolland and Thomas Demas, "Oceanographic Research Ships of the Twenty-First Century," *MTS 94 Conference Proceedings* (Marine Technology Society, Washington, 1994), pp. 556–562. For the Navy's surveillance use, see *Jane's Fighting Ships, 1993–94* (Jane's Information Group, Alexandria, Va., 1993), p. 815. For SWATH's overall naval history, see William J. Broad, "Calming Stormy Seas with New Kinds of Ship Hulls," *The New York Times,* August 15, 1995, p. C1.

8. I was unable to talk to the project's head, Khosrow Lashkari, during my MBARI visit but eventually spoke with him on June 24, 1996, and learned more of the observatory's work. See William J. Broad, "Anti-Sub Seabed Grid Thrown Open to Research Uses," *The New York Times,* July 2, 1996, p. C1.

9. For more on the aquarium, see "Monterey Bay Aquarium," Monterey Bay Aquarium Foundation, 1992. Also see Judith Connor and Charles Baxter, "Kelp Forests," Monterey Bay Aquarium Foundation, 1989.

10. Interview, Bruce H. Robison, July 12, 1994. For an overview of his research, see Robison, "Light in the Ocean's Midwaters," *Scientific American,* July 1995, pp. 60–64.

11. For details of this history, see Bruce Robison, "Midwater Biological Research with the WASP ADS," *Marine Technology Society Journal,* Vol. 17, No. 3, Fall 1983, pp. 21–27. Also see Graham Hawkes, "The Future of Atmospheric Diving Systems and Associated Manipulator Technology with Special Reference to the New Omads, Deep Rover," *Marine Technology Society Journal,* Vol. 17, No. 3, Fall 1983, pp. 51–60.

12. E. A. Widder et al., "Bioluminescence in the Monterey Submarine Canyon: Image Analysis of Video Recordings from a Midwater Submersible," *Marine Biology,* Vol. 100, 1989, pp. 541–551.

13. Interview, Bruce H. Robison, July 14, 1994.

14. For more on the minimum, see N. B. Marshall, *Aspects of Deep Sea Biology* (Philosophical Library, New York, 1954), pp. 174–177, and Thurman, *Introductory Oceanography,* pp. 200–219.

15. Cheryl Lyn Dybas, "Beautiful, Ethereal Larvaceans Play a Central Role in Ocean Ecology," *Oceanus,* Vol. 36, No. 2, Summer 1993, pp. 84–86.

16. John Travis, "Invader Threatens Black, Azov Seas," *Science,* November 26, 1993, pp. 1366–1367. Richard C. Brusca and Gary J. Brusca, *Invertebrates* (Sinauer, Sunderland, Mass., 1990), pp. 263–277. Cheryl Lyn Dybas, "Comb Jelly: A Jewel of a Creature," *Sea Frontiers,* December 1994, pp. 12–13, 55.

17. Marshall, *Aspects,* pp. 218–231.

18. Thurman, *Introductory Oceanography,* p. 345.

19. Ralph Buchsbaum, Mildred Buchsbaum, John Pearse, Vicki Pearse, *Animals Without Backbones* (University of Chicago, Chicago, 1987), pp. 114–116.

20. Brusca and Brusca, *Invertebrates,* pp. 211–257.

21. For more on the deep-scattering layer, see Thurman, *Introductory Oceanography,* pp. 345, 347.

22. For more on the uniqueness of such adaptations, see G. Richard Harbison, "The Gelatinous Inhabitants of the Ocean Interior," *Oceanus,* Fall 1992, pp. 18–23.

23. Ibid., p. 22.

24. For more on *Architeuthis* and the behavioral secrets of deep squids, see William J. Broad, "Squids Emerge as Smart, Elusive Hunters of Mid-Sea," *The New York Times,* August 30, 1994, p. C1.

25. Interview, Steve Etchemendy, July 12, 1994. For more on the new robot, see Newman and Robison, "Development of a Dedicated ROV," pp. 46–53.

26. Interview, David Packard, July 17, 1994. See also Connor and Deans, "David Packard," p. 75, and "Packard, David," *Current Biography,* 1969, pp. 318–321.

27. Interview, David Packard, July 17, 1994. See also Connor and Deans, "David Packard," p. 78.

28. Thurman, *Introductory Oceanography,* pp. 37–39, 379–382; Press and Siever, *Earth,* pp. 349–351; Elliott A. Norse, ed., *Global Marine Biological Diversity* (Island Press, Washington, 1993), pp. 27–29; Gregg Easterbrook, *A Moment in Time: The Coming Age of Environmental Optimism* (Viking, New York, 1995), pp. 290–292, 295–298; and William K. Stevens, "Emissions Must Be Cut to Avert Shift in Climate, Panel Says," *The New York Times,* September 20, 1994, p. C4.

29. Michael Parfit, "Diminishing Returns," *National Geographic,* November 1995, pp. 2–37. Kent Jeffreys, "Rescuing the Oceans," in Ronald Bailey, ed., *The True State of the Planet* (Free Press, New York, 1995), pp. 296–299, 305–309. Lester Brown and Hal Kane, *Full House: Reassessing the Earth's Population Carrying Capacity* (Norton, New York, 1994), pp. 80–81. William K. Stevens, "Feeding a Booming Population Without Destroying the Planet," *The New York Times,* April 5, 1994, p. C1. Gage and Tyler, *Deep-Sea Biology,* pp. 394–396.

30. Packard died in 1996, leaving many of his billions to charity. See Lawrence M.

Fisher, "David Packard, 83, Pioneer of Silicon Valley, Is Dead," *The New York Times,* March 27, 1996, p. D20. See also Carey Goldberg, "With Fortune Built, Packard Heirs Look to Build a Legacy," *The New York Times,* May 6, 1996, p. A1.

SIX. FIELDS OF GOLD

1. Morris Goran, "Haber, Fritz," *Dictionary of Scientific Biography,* Vol. Five (Scribner's, New York, 1972), pp. 620–623. Isaac Asimov, "Haber, Fritz," *Asimov's Biographical Encyclopedia of Science and Technology* (Doubleday, Garden City, N.Y., 1982), pp. 624–625. For a review of the extraction project, see C. P. Idyll, ed., *Exploring the Ocean World* (Crowell, New York, 1969), pp. 18–19, 170–171. For a *Meteor* overview, see Thurman, *Introductory Oceanography,* pp. 10, 21.

2. Willard Bascom, *The Crest of the Wave: Adventures in Oceanography* (Doubleday/Anchor, New York, 1990), pp. 194–200.

3. Ibid., pp. 200–213. Kenneth J. Hsü, *Challenger at Sea: A Ship That Revolutionized Earth Science* (Princeton University Press, Princeton, N.J., 1992), pp. 3–20. For the political demise of the deep-drilling venture, see Daniel S. Greenberg, *The Politics of Pure Science* (New American Library, New York, 1967), pp. 170–206.

4. Bascom, *The Crest,* pp. 219–240. Willard Bascom, "Mining in the Sea," in Lewis M. Alexander, ed., *The Law of the Sea: Offshore Boundaries and Zones* (Ohio State University Press, Columbus, Ohio, 1967), pp. 160–171. David W. Pasho, "Canada and Ocean Mining," *Marine Technology Society Journal,* Vol. 19, No. 4, 1985, pp. 26–30.

5. James M. Broadus, "Seabed Materials," *Science,* Vol. 235, February 20, 1987, pp. 853–860. Allen L. Hammond, "Manganese Nodules (I): Mineral Resources on the Deep Seabed," *Science,* Vol. 183, February 8, 1974, pp. 502–503. Idyll, *Exploring,* pp. 170–175. Press and Siever, *Understanding Earth,* pp. 535–544.

6. Pasho, "Canada," p. 26. Broadus, "Seabed Materials," p. 858–859.

7. Evan Luard, *The Control of the Sea-bed: A New International Issue* (Taplinger, New York, 1974), pp. 81–193.

8. Allen L. Hammond, "Manganese Nodules (II): Prospects for Deep Sea Mining," *Science,* Vol. 183, February 15, 1974, pp. 644–646. Varner and Collier, *Matter of Risk,* pp. 58–68. William J. Broad, "Plan to Carve Up Ocean Floor Riches Nears Fruition," *The New York Times,* March 29, 1994, p. C1.

9. "Law-of-Sea Parley Opens in Venezuela with 5,000 Officials from 148 Lands," *The New York Times,* June 21, 1974, p. 12.

10. Jin S. Chung, "Advances in Manganese Nodule Mining Technology," *Marine Technology Society Journal,* Vol. 19, No. 4, 1985, pp. 39–44. Broadus, "Seabed Materials," pp. 857–858. National Oceanic and Atmospheric Administration, "Deep Seabed Mining: Report to Congress," U.S. Department of Commerce, NOAA, December 1993.

11. National Oceanic and Atmospheric Administration, "Deep Seabed," pp. 3–6. Letter to author, Karl Jugel, NOAA's Office of Ocean and Coastal Resource Management, March 22, 1994.

12. Judith Fenwick, *International Profiles on Marine Scientific Research* (Woods Hole Oceanographic Institution, Woods Hole, Mass., 1992), pp. v–vii, 33, 194. Thurman, *Introductory Oceanography,* pp. 458–460. For early history, see Luard, *The Control of the Sea-bed,* pp. 255–257. For the Cook Islands, see Taylor A. Pryor, "New Described Super-Nodule Resource," *Sea Technology,* September 1995, pp. 15–18.

13. Broadus, "Seabed Materials," pp. 857–859.

14. Alexander Malahoff, "A Comparison of the Massive Submarine Polymetallic Sulfides of the Galápagos Rift with Some Continental Deposits," *Marine Technology Society Journal,* Vol. 16, No. 3, 1982, pp. 39–45. Cone, *Fire Under the Sea,* pp. 99–100. Peter Lonsdale et al., "Metallogenesis at Seamounts on the East Pacific Rise," *Marine Technology Society*

Journal, Vol. 16, No. 3, 1982, pp. 54–60. Tom Burroughs, "Ocean Mining: Boom or Bust," *Technology Review,* April 1984, p. 54.

15. All the following articles are from *Marine Technology Society Journal,* Vol. 16, No. 3, 1982. Lonsdale et al., pp. 54–61. Alexander Malahoff, "The Ocean Floor, Our New Frontier: A Scientific Viewpoint," pp. 3–4. Malahoff, "A Comparison," pp. 39–45. Conrad G. Welling, "Polymetallic Sulfides: An Industry Viewpoint," pp. 5–7. David B. Duane, "Elements of a Proposed Five-Year Research Program on Polymetallic Sulfides," pp. 87–91.

16. Cone, *Fire,* pp. 152–156. William J. Broad, "Resource Wars: The Lure of South Africa," *Science,* Vol. 210, December 5, 1980, pp. 1099–1100. John Lehman, "Going for the High Ground in the Deep Seas," *Marine Technology Society Journal,* Vol. 16, No. 2, 1982, pp. 3–6.

17. William J. Broad, "Plan to Carve." For general EEZ diplomatic history, see Luard, *The Control,* pp. 143–168. For U.S. stance, see Cone, *Fire,* pp. 152–156. For state-by-state EEZ claims, see Fenwick, *International Profiles.*

18. For the EEZ declaration, see *United States Code Congressional and Administrative News,* 98th Congress, First Session, 1983, Vol. 3 (West Publishing, St. Paul, 1984), pp. A28–A29. For the policy statement, see John B. Smith, "Managing Nonenergy Marine Mineral Development—Genesis of a Program," *Proceedings, Oceans 85* (Marine Technology Society, Washington, 1985), pp. 339–351. Also see Cone, *Fire,* pp. 152–156, 256. Fenwick, *International Profiles,* pp. 159–160. For debates on EEZ area, see Millington Lockwood, "Memorandum for the Record," Joint Office for Mapping and Research, United States Geological Survey and National Oceanic and Atmospheric Administration, Reston, Va., April 30, 1993. For Hawaiian zone, see David M. Graham, "Hawaii Ocean R&D: Life on the 'Fast Track,' " *Sea Technology,* August 1994, pp. 26–38.

Some experts argue that the Reagan proclamation was mostly a symbolic act that carried little legal force not already spelled out in previous assertions of U.S. marine resource jurisdiction. Letter to author, Robert J. McManus, March 19, 1996. Mr. McManus at the time of the proclamation was NOAA's general counsel. Other experts disagree, saying no U.S. law spoke of the seabed's minerals up to two hundred miles from shore. My reading of the documents supports the latter view.

19. Jeffrey P. Zippin, "Draft Environmental Impact Statement: Proposed Polymetallic Sulfide Minerals Lease Offering, Gorda Ridge Area Offshore Oregon and Northern California," Minerals Management Service, U.S. Department of the Interior, Reston, Va., December 1983. Smith, "Managing," pp. 348–349. Cone, *Fire,* pp. 156–159.

20. Cone, *Fire,* pp. 159–165. Kaharl, *Water Baby,* p. 346. Broadus, "Seabed Materials," pp. 853–860. S. D. Scott, "Polymetallic Sulfide Riches from the Deep: Fact or Fallacy?" in K. J. Hsü and J. Thiede, eds., *Use and Misuse of the Seafloor* (Wiley, New York, 1992), pp. 87–115. For a good discussion of vent gold concentrations, see pp. 103–108 of this latter reference.

The vent episode was but one of several mineralogical dramas set in motion by the Reagan Administration, the final acts of which were often played out only after the protagonists left Washington. In its enthusiasm for EEZ mining, the Reagan Administration also targeted such places as titanium-rich deposits off southern Oregon, gold sands in the Norton Sound off Alaska, and cobalt-rich crusts around the Johnston and Hawaiian islands. In the latter case, these deposits were on the flanks of volcanic isles and seamounts stretching from the big island of Hawaii westward for hundreds of miles through the major and minor outposts of the archipelago. The deposits were judged to be vast—up to 2.6 million tons of cobalt, 1.6 million tons of nickel, and 81 million tons of manganese. But in all cases, despite some allure, no American company came forward with cash in hand ready to do any mining. Indeed, around Hawaii no one could dream up any cost-effective way of tapping deposits up to a mile and a half deep. By the early 1990s it became clear that the Reagan vision of a booming industry in EEZ mining was, at least for the moment, a mirage.

Roger V. Amato, "Pacific EEZ Marine Minerals Activities of the Minerals Management

Service," in *Proceedings of the 1991 Exclusive Economic Zone Symposium on Mapping and Research: Working Together in the Pacific EEZ* (U.S. Geological Survey, Denver, 1992), USGS Survey Circular 1092, pp. 152–157. Broadus, "Seabed Materials," p. 853.

21. Patrick E. Tyler, "China Revamps Forces with Eye to Sea Claims," *The New York Times,* January 2, 1995, section 1, p. 2.

"Geoeconomic aspects are becoming as important as geostrategic and geopolitical variables," Admiral Alfredo A. Ambrossiana, head of the Peruvian Navy, said of the new naval world after the cold war. "The Commanders Respond," U.S. Naval Institute *Proceedings,* March 1995, p. 28.

22. Babb et al., "Dual Use," p. 44.

23. NOAA, "Deep Seabed Mining: Report to Congress," pp. 7, 11–13. Ted Brockett and Caroline Z. Richards, "Deepsea Mining Simulator for Environmental Impact Studies," *Sea Technology,* August 1994, pp. 77–82.

24. NOAA, "Deep Seabed Mining: Final Programmatic Environmental Impact Statement," Vol. One, U.S. Department of Commerce, National Oceanic and Atmospheric Administration, Washington, D.C., September 1981, pp. 99–109.

25. Sagalevitch, " 'Mir' Submersibles," p. 242. For details on Malahoff's career, see Cone, *Fire,* pp. 97–100.

26. Interview, Shelley Lauzon, Woods Hole Oceanographic Institution, June 11, 1996.

27. Interview, Randy Koski, October 9, 1993. Broad, "A Voyage into the Abyss."

28. For Britain's lack of deep piloted submersibles, see Busby, *Undersea Vehicles Directory,* pp. 13–14.

It is worth noting that Britain excels in unmanned aspects of deep exploration and has an aggressive program to market its technologies abroad, especially to coastal states keen on assessing, developing, and defending their Exclusive Economic Zones. The British Department of Trade and Industry in the 1980s and '90s launched a number of development programs to help its industries perfect and sell maritime wares, including ones for mining manganese nodules and polymetallic sulfides. In 1993 it issued the *Exclusive Economic Zone Catalogue,* a 221-page compendium of marine goods and services. Its introduction is written in English, French, German, Italian, Portuguese, and Spanish.

Don E. Lennard, "The United Kingdom's Progress in Marine Technology," *Marine Technology Society Journal,* Vol. 19., No. 1, First Quarter, 1985, pp. 38–45. "Opening Up Davy Jones's Locker," *The Economist,* December 20, 1986, p. 117. Cliff Funnell, "Wealth from the Oceans: A U.K. Program," *Sea Technology,* October 1993, pp. 55–60. *Exclusive Economic Zone Catalogue* (Combined Service Publications, Ltd., Hampshire, Great Britain, 1993).

29. William J. Broad, "Weird Mound Offers Clues to Mysteries of the Deep," *The New York Times,* August 9, 1994, p. C1. Rona, "Deep-Sea Geysers," pp. 105–109. Scott, "Polymetallic Sulfide Riches," pp. 100–101. Woods Hole Oceanographic Institution, "The Hot Spot," "Stone Soup," "Ground Zero," *Woods Hole Currents,* Summer 1995, pp. 3–13.

30. For the gold concentration, see Kaharl, *Water Baby,* p. 346.

31. Susan Humphris et al., "Leg 158 Scientific Prospectus: Drilling an Active Hydrothermal System on a Sediment-Free Slow-Spreading Ridge," Ocean Drilling Program, Texas A&M University, March 1994. Broad, "Weird Mound." Woods Hole Oceanographic Institution, "Drilling into the Ocean Floor's Plumbing System May Provide New Insights into How Mineral Deposits Are Formed," WHOI two-page news release, July 27, 1994.

32. Interview, Susan Humphris, Woods Hole Oceanographic Institution, August 3, 1994. Gerrard Raven, "Scientists to Study Threatened Ocean Volcano Area," Reuters, August 26, 1994. Nick Nuttall, "Scientists Set Sail on Mission to the Bottom of the Sea," *The Times* (London), August 27, 1994. Sean Ryan, "Back from the Deep," *The Sunday Times* (London), October 9, 1994.

33. Interview, Peter A. Rona, Rutgers University, a TAG expert, June 24, 1996. "Active

Hydrothermal System Drilled at the Mid-Atlantic Ridge," *EOS, Transactions of the American Geophysical Union,* Vol. 76, No. 37, September 12, 1995, p. 361.

34. Letter to author, Ian R. Jonasson, *Geological Survey of Canada,* October 24, 1995. Horst U. Oebius, "Recent Advances, Future Needs and Collaboration in Ocean Technology in the Federal Republic of Germany," *Marine Technology Society Journal,* Vol. 24, No. 1, March 1990, pp. 22–27. Gage and Tyler, *Deep-Sea Biology,* p. 399. Peter Herzig et al., "Submarine Volcanism and Hydrothermal Venting Studied in Papua, New Guinea," *EOS,* Vol. 75, No. 44, November 1, 1994, pp. 513–516.

35. For examples of Japan's prowess and drive, see Clyde Haberman, "Japanese Fight Invading Sea for Priceless Speck of Land," *The New York Times,* January 4, 1988, p. A1. Hamilton-Paterson, *The Great Deep,* p. 78. Takahisa Nemoto, "The Ocean Research Institute, Tokyo," in Elisabeth Mann Borgese, ed., *Ocean Frontiers: Explorations by Oceanographers on Five Continents* (Abrams, New York, 1992), pp. 213–227. Noriyuki Nasu, "Japan and the Sea," *Oceanus,* Vol. 30, No. 1, Spring 1987, pp. 2–8. Japan Marine Science & Technology Center, "Jamstec Profile," *JAMSTEC,* Yokosuka, Kanagawa, November 1991. William J. Broad, "Japan Plans to Conquer Sea's Depths," *The New York Times,* October 18, 1994, p. C1.

36. Yojiro Ikeda, "Japan," *Mining Annual Review,* June 1988, p. 361. National Oceanic and Atmospheric Administration, "17th UJNR/Marine Facilities Panel Study Tour," 17th Meeting of the U.S.-Japan Marine Facilities Panel, *Conference Record,* U.S. Department of Commerce, NOAA, May 1991, p. 36. Yoshifumi Takaishi, "Fundamental Conception on Ocean Exploitation in Japan and Contribution of Shipbuilding Technology," 17th Meeting of the U.S.-Japan Marine Facilities Panel, *Conference Record,* U.S. Department of Commerce, NOAA, May 1991, pp. 69–74. Yoichi Kimoto, "Japan Mining Collector for Manganese Nodules," 18th Meeting of the U.S.-Japan Marine Facilities Panel, *1992 Conference Record,* U.S. Department of Commerce, NOAA, October-November 1992, pp. 1–3.

37. National Research Council, *Maximizing U.S. Interests in Science and Technology Relations with Japan* (National Academy Press, Washington, 1995), pp. 11–13. Tim Weiner et al., "C.I.A. Spent Millions to Support Japanese Right in the 50's and 60's," *The New York Times,* October 9, 1994, p. A1.

38. Interviews and observations, Woods Hole, July 27, 1994. Broad, "Japan Plans." Woods Hole Oceanographic Institution, "Woods Hole Oceanographic Institution–Japan Marine Science and Technology Center Relationship," WHOI two-page news release, July 1994. "An Emperor Comes to Call," *Newsletter,* Woods Hole Oceanographic Institution, December 1975, pp. 7–8. Hirosi Hotta, "Deep Sea Research Around the Japanese Islands," *Oceanus,* Vol. 30, No. 1, Spring 1987, pp. 32–33. Kenji Okamura, "Ocean Technology in Japan: Recent Advances, Future Needs and International Collaboration," *Marine Technology Society Journal,* Vol. 24, No. 1, March 1990, pp. 32–47. "Atlantis II Visits Japan," *Newsletter,* Woods Hole Oceanographic Institution, May-November 1987, pp. 1–4. Shinichi Takagawa, "Deep Submersible Project (6,500 m)," *Oceanus,* Vol. 30, No. 1, Spring 1987, pp. 29–32. Woods Hole Oceanographic Institution, "Scientists Aboard the Deep Diving Submersible, *Shinkai 6500,* Conduct a Unique Study of the Earth's Structure," WHOI two-page news release, July 27, 1994. Japan Marine Science & Technology Center, "Research Submersible Shinkai 6500 System," JAMSTEC brochure, undated.

39. Broad, "Weird Mound." Woods Hole Oceanographic Institution, "First American-Japanese Expedition to the Mid-Atlantic Ridge Increasing Knowledge of the Evolution of the Earth," WHOI, two-page news release, July 27, 1994. Japan Marine Science & Technology Center, "MODE '94: Mid Ocean Ridge Diving Expeditions," JAMSTEC brochure, undated.

40. Contrary to projections when the first wave of deep-mining assessment got underway, some experts now believe terrestrial mineral supplies are adequate for the next one hundred years. See Carol Ann Hodges, "Mineral Resources, Environmental Issues, and Land Use," *Science,* vol. 268, June 2, 1995, pp. 1305–1312.

41. Lloyd Williams, "World Law of the Sea, Long Opposed by US, Quietly Goes Into

Effect," Associated Press, November 16, 1994. For Boutros Boutros-Ghali quote, see Canute James, "Concerns Linger Over Seabed Settlement—Doubts About the Effectiveness of a New UN Authority," *Financial Times,* December 9, 1994, p. 29. For background, see Broad, "Plan to Carve."

42. "Indonesia Needs More Ships to Guard Eastern Waters," Associated Press, February 1, 1995. "Australia Extends Coastal Zone," *Facts on File World News Digest,* December 1, 1994, p. 902. David Miller, "Offshore Patrolling: New Responsibilities Demand Specialized Equipment," *International Defense Review,* January 1, 1995, p. 40. For positive EEZ aspects, see Kent Jeffreys, "Rescuing the Oceans," in Ronald Bailey, ed., *The True State of the Planet* (Free Press, New York, 1995) pp. 295–338. C. Barry Raleigh, "The Internationalism of Ocean Science vs. International Politics," *Marine Technology Society Journal,* Vol. 23, No. 1, March 1989, pp. 44–47.

43. Thurman, *Introductory Oceanography,* pp. 458–460. Raleigh, "The Internationalism," pp. 44–45. For Scholz quote, see James, "Concerns Linger." For an optimistic assessment of the treaty's political ramifications, see Borgese, "Ocean Mining and the Future of World Order," in Hsü and Thiede, *Use and Misuse,* pp. 117–126.

44. Broadus, "Seabed Materials." Broad, "Resource Wars." Milton R. Copulos, "The Lessons of History," *Heritage Foundation Reports,* January 16, 1989, p. 11.

45. Michael J. Cruickshank, "Pair of Important Meetings Laud Future of Non-U.S. Seabed Mining," *Sea Technology,* February 1995, p. 94. Lawrence Herman, "Law of the Sea Is Seen as a Bit of a Museum-piece: Canada's Non-ratification Shows Times Have Changed," *The Financial Post,* December 7, 1994, p. 17. For Owada quote, see James, "Concerns Linger." For more on Japanese mining program, see Seizo Nakao, "Deep-sea Mineral Activity in Japan," *Marine Technology Society Journal,* Vol. 29, No. 3, Fall 1995, pp. 74–78.

46. Pryor, "New Described Super-Nodule Resource." See also Michael J. Cruickshank, "Cook Island Nodule Mining Plans Closer to Reality," *Sea Technology,* February 1996, p. 105.

47. For a review of the UN's biological myopia and moves toward possible regulation, see Lyle Glowka, "The Deepest of Ironies: Genetic Resources, Marine Scientific Research and the International Deep Sea-bed Area," The World Conservation Union, Environmental Law Center, Bonn, Germany, August 1995.

48. Interviews, James F. Holden, November 11, 1993, and March 20, 1995.

49. Interview, John A. Baross, November 11, 1993.

50. Thomas D. Brock, "Life at High Temperatures in Water Environments," *Science,* Vol. 230, October 11, 1985, p. 132. Thomas D. Brock et al., *Biology of Microorganisms* (Prentice Hall, Englewood Cliffs, N.J., 1994), pp. 337–338. Michael Milstein, "Yellowstone Managers Eye Profits from Hot Microbes," *Science,* Vol. 264, April 29, 1994, p. 655.

51. Kary B. Mullis, "The Unusual Origin of the Polymerase Chain Reaction," *Scientific American,* April 1990, pp. 56–65. For a deeper look at the discovery episode, see Paul Rabinow, *Making PCR: A Story of Biotechnology* (University of Chicago, Chicago, 1996). Henry A. Erlich et al., "Recent Advances in the Polymerase Chain Reaction," *Science,* Vol. 252, June 21, 1991, p. 1643. Tim Appenzeller, "Chemistry: Laurels for a Late-Night Brainstorm," *Science,* Vol. 262, October 22, 1993, pp. 506–507. Frank Clifford, "Simpson Case Boosts Microbe Conservation," *The Los Angeles Times,* August 31, 1994, p. A1.

52. Interview, Holger W. Jannasch, November 10, 1993. Interview, Rebecca B. Kucera, New England Biolabs, November 9, 1993. William J. Broad, "Strange Oases in Sea Depths Offer Map to Riches," *The New York Times,* November 16, 1993, p. C1. Holger W. Jannasch, "Deep Sea Hydrothermal Vents: Underwater Oases," *The NEB Transcript of New England Biolabs,* September 1992, pp. 1–3. William E. Jack et al., "Biochemical Characterization of Vent and Deep Vent DNA Polymerases," *The NEB Transcript,* September 1992, pp. 4–5. Michael Milstein, "Who Has the Right to Own—and Lease—a Living Thing?" *The San Diego Union-Tribune,* June 1, 1994, p. E4.

53. Interview, Eric J. Mathur, November 12, 1993. Interview, Vincent Kazmer, Novem-

ber 11, 1993. Interview, Holger W. Jannasch, November 10, 1993. See also Broad, "Strange Oases." University of Maryland Biotechnology Institute, "Microbiology/Biotechnology at the Center of Marine Biotechnology," UMBI brochure, 1994, pp. 10–11. For background information see E. J. Mathur, "Applications of Thermostable DNA Polymerases in Molecular Biology," in Michael W. W. Adams and Robert M. Kelly, eds., *Biocatalysis at Extreme Temperatures* (American Chemical Society, Washington, 1992), ACS Symposium Series 498, pp. 189–206. See also Bruce L. Zamost et al., "Thermostable Enzymes for Industrial Applications," *Journal of Industrial Microbiology,* Vol. 8, 1991, pp. 71–81.

54. Sean Ryan, "Briton Explores the Mysteries of the Deep," *The Sunday Times* (London), August 21, 1994. Ryan, "Back from the Deep."

55. Kantaro Fujioka, "Quicklook Report on the TAG Hydrothermal Mound," *Shinkai 6500* dive number 216, August 4, 1994, and dive number 230, August 22, 1994. For these reports, my thanks to Woods Hole, where they were filed by fax from the Japanese mother ship *Yokosuka.*

56. Interview, Anna-Louise Reysenbach, August 2, 1994. Also see Broad, "Weird Mound."

57. Interview, Stephen R. Hammond, November 29, 1995.

58. Interview, John A. Baross, March 13, 1996. William J. Broad, "Volcanic Fury Rocks Pacific Sea Floor Where Inner Earth Meets Outer Earth," *The New York Times,* March 19, 1996, p. C1.

59. "Pfizer Tests Marine Microorganisms for Disease Fighting Potential," Pfizer news release, Groton, Conn., August 24, 1995. Michael Baum, "Commerce Department Announces 29 Advanced Technology Awards," U.S. Department of Commerce news release, November 4, 1993.

The cautiousness of the Clinton Administration in aiding such private endeavors rose markedly in November 1994 when the Republicans captured both houses of Congress and began a war to cut the Federal budget and Federal aid to American businesses.

60. Gregory S. Stone, "Japanese Ocean Research and Development," *Marine Technology Society Journal,* Vol. 26, No. 3, 1992, pp. 11–19. Anna Maria Gillis, "A Pressure-Filled Life: How Do Microorganisms Cope When They Live at Extremes?" *Bioscience,* Vol. 44, October 1994, p. 584.

61. Masanori Kyo and Ikuo Nakazaki, "Development of Collection, Isolation and Cultivation System for Deep-sea Microbes Study," *18th Meeting of the U.S.-Japan Marine Facilities Panel,* October–November 1992 Conference Record, National Oceanic and Atmospheric Administration, U.S. Department of Commerce. Frederick Shaw Myers and Alun Anderson, "Microbes from 20,000 Feet Under the Sea," *Science,* vol. 255, January 3, 1992, pp. 28–29. Mark Crawford, "ERATO Sparks Japan's Basic Research Effort," *New Technology Week,* December 18, 1995.

62. Eliot Marshall, "Superbugs in Waiting: Some Cautionary Tales," *Science,* vol. 255, January 3, 1992, p. 29. For Horikoshi quote, see Myers and Anderson, "Microbes from 20,000 Feet."

63. William J. Broad, "Creatures of the Deep Find Their Way to the Table," *The New York Times,* December 26, 1995, p. C1. Florence Fabricant, "A Motley Procession of Fish Finds Its Way to the Table," *The New York Times,* June 19, 1991, p. C1.

64. Carl Safina, "The World's Imperiled Fish," *Scientific American,* November 1995, pp. 46–53. Parfit, "Diminishing Returns." Broad, "Creatures of the Deep."

65. For estimate of growth in deep fishery, see Hans Ackefors, "Production of Fish and Other Animals in the Sea," *Ambio,* 1977, Vol. 6, No. 4, p. 192.

66. Interview, Jack Sobel, December 20, 1995. Interview, Bruce Morehead, December 20, 1995.

67. Interview, Peter J. Auster, December 18, 1995.

68. Press conference, Washington, D.C., Walter H. F. Smith, National Oceanic and Atmospheric Administration, and David T. Sandwell, Scripps Institution of Oceanography,

October 23, 1995. See also Broad, "Map Makes Ocean Floors as Knowable as Venus," and Carlowicz, "New Map of Seafloor."

69. John D. Isaacs and Richard A. Schwartzlose, "Active Animals of the Deep-Sea Floor," *Scientific American,* October 1975, pp. 85–91. For the camera project, also see Bascom, *The Crest,* pp. 247–249. For more on unusual animals discovered, see Gage and Tyler, *Deep-Sea Biology,* pp. 78–79. For similar frenzies with dead whales, see Craig R. Smith, "Whale Falls: Chemosynthesis on the Deep Seafloor," *Oceanus,* Vol. 35, No. 3, Fall 1992, pp. 74–78.

70. Scott and Scott, *Atlantic Fishes,* pp. 296–304. Idyll, *Abyss,* pp. 169–173, 334–338. Gage and Tyler, *Deep-Sea Biology,* pp. 394–396.

71. Gage and Tyler, *Deep-Sea Biology,* pp. 83–84, 395–396, 451, 484.

72. Ibid., p. 395. Fenwick, *International Profiles,* p. 107. "Orange Roughy Fact Sheet," Greenpeace International, New Zealand, February 23, 1994, kindly provided to author by Mike Hagler of Greenpeace. Interview, Mike Hagler, December 19, 1995. RICC Francis, "Use of Risk Analysis to Assess Fishery Management Strategies: A Case Study Using Orange Roughy *(Hoplostethus atlanticus)* on the Chatham Rise, New Zealand," *Canadian Journal of Fisheries & Aquatic Sciences,* Vol. 49, No. 5, May 1992, pp. 922–930.

73. Francis, "Use of Risk Analysis." "Orange Roughy Fact Sheet." "The Miraculous Roughy Fishery," *Liberation* (France), July 22, 1992, and "Looking for Deep Sea Fish," *Liberation* (France), September 29, 1992. My thanks to Mike Hagler for summaries of these articles.

74. Letter to author, Don Robertson, February 2, 1996. Robertson is a fisheries biologist at the successor agency to the Ministry of Agriculture and Fisheries, which is known as the National Institute of Water and Atmosphere. The TACC history for Chatham Rise, as cited by Robertson:

1981–82	23,000 tons
1982–83	23,000 tons
1983–84	30,000 tons
1984–85	30,000 tons
1985–86	29,865 tons
1986–87	38,065 tons
1987–88	38,065 tons
1988–89	38,300 tons
1989–90	32,787 tons
1990–91	23,787 tons
1991–92	23,787 tons
1992–93	14,000 tons
1993–94	14,000 tons
1994–95	8,000 tons

75. Interview, Mike Hagler, December 19, 1995.

76. Jay Harlow, "For Flavor and Value, Get a Little Hoki," *The San Francisco Chronicle,* June 3, 1992, food section, p. 6. "Ugly-looking Fish Bring Beautiful Bucks," *Quick Frozen Foods International,* January 1993, p. 56. Judith E. Foulke, "Cracking Down on Fresh Fish Fraud," *Consumers' Research Magazine,* October 1993, p. 25. Jeremy Lyle et al., "Oreos—an Underutilised Resource," *Australian Fisheries,* Vol. 51, No. 4, April 1992, pp. 12–15.

77. Interview, Clyde F. E. Roper, National Museum of Natural History at the Smithsonian Institution, Washington, D.C., August 10, 1994. Roper is an expert on the giant squid. William J. Broad, "Biologists Closing on Hidden Lair of Giant Squid," *The New York Times,* February 13, 1996, p. C1.

78. "Deepwater Fishing Grants Supported by the Northeast Region National Marine Fisheries Service, Fiscal Years 1994–95," two-page fact sheet, National Marine Fisheries Service, Gloucester, Mass., undated. "The Saltonstall-Kennedy Grant Program: Fisheries Research and Development. Report 1995," U.S. Department of Commerce, National Oceanic

and Atmospheric Administration, National Marine Fisheries Service, August 11, 1995, pp. 4, 5, 16, 53. For the campaign of the National Fisheries Institute, see Carole Sugarman, "Skating Away—Fish with Wings," *The Washington Post,* October 19, 1994, p. E1.

79. Interview, Kenneth L. Beal, December 18, 1995. Broad, "Creatures of the Deep."

80. Interview, Bill Bomster, December 19, 1995. Broad, "Creatures of the Deep."

Seven. TIDES

1. National Oceanic and Atmospheric Administration, "Preliminary Natural Resource Survey, Farallones Islands Radioactive Waste Dumps," NOAA report, March 29, 1990, pp. 3–7. NOAA, "Gulf of the Farallones National Marine Sanctuary," U.S. Department of Commerce, NOAA brochure, National Marine Sanctuary Program, undated.

2. NOAA, "Preliminary Natural Resource Survey," pp. 1, 7–9. Katherine Bishop, "U.S. to Determine Danger from Barrels of Atomic Waste in Pacific," *The New York Times,* January 20, 1991, p. A21. "Atomic Waste Reported Leaking in Ocean Sanctuary Off California," *The New York Times,* May 7, 1990, p. B12. Patrick E. Tyler, "The U.S., Too, Has Dumped Waste at Sea," *The New York Times,* May 4, 1992, p. A8. "Radioactive Waste at Sea Confirmed," *Los Angeles Times,* November 19, 1990, p. A24. "Atom Waste Termed Peril to West Coast," *The New York Times,* August 20, 1980, p. A13.

3. International Atomic Energy Agency, "Inventory of Radioactive Material Entering the Marine Environment: Sea Disposal of Radioactive Waste," IAEA, Vienna, Austria, March 1991, pp. 48–51.

4. Ibid. Polmar and Allen, *Rickover,* pp. 355–359. W. Jackson Davis et al., "Evaluation of Oceanic Radioactive Dumping Programs," report of the Environmental Studies Institute, University of California at Santa Cruz, July 1982.

5. International Atomic Energy Agency, "Inventory of Radioactive Material," pp. 23, 26, 28, 30, 32, 34, 36, 38, 40, 42, 45.

6. Davis, "Evaluation," pp. 13–15. NOAA, "Preliminary," pp. 7–9.

7. Davis, "Evaluation," pp. 46–47, 49–50, 88–91. NOAA, "Preliminary," pp. 7-11.

8. James M. Broadus and Raphael V. Vartanov, *The Oceans and Environmental Security: Shared U.S. and Russian Perspectives* (Island Press, Washington, 1994), pp. 112–128. Tyler, "The U.S., Too," p. A8.

The U.S. curtailment was codified in 1972 with the passage of the Marine Protection, Research, and Sanctuaries Act, often referred to as the ocean-dumping act. And because of the public's safety worries, the Navy in the 1980s decided to forgo any more scuttling of empty reactor chambers and hull sections from nuclear submarines, instead burying them in special dumps at the government's nuclear production site in Hanford, Washington.

9. International Atomic Energy Agency, "Inventory of Radioactive Material," pp. 9–11.

10. Aleksei V. Yablokov et al., "Facts and Problems Related to Radioactive Waste Disposal in Seas Adjacent to the Territory of the Russian Federation," Office of the President of the Russian Federation, Moscow, 1993, pp. 6–10, 16–34. The translation I worked from for this book was made public by the United States Geological Survey, based in Reston, Va., and was probably done by a Federal intelligence agency. An early translation was also done by the Washington, D.C., office of Greenpeace International; I worked from it to do the *Times* article cited here. William J. Broad, "Russians Describe Extensive Dumping of Nuclear Waste," *The New York Times,* April 27, 1993, p. A1. See also Patrick E. Tyler, "Soviets' Secret Nuclear Dumping Causes Worry for Arctic Waters," *The New York Times,* May 4, 1992, p. A1. Michael Dobbs, "In the Former Soviet Union, Paying the Nuclear Price," *The Washington Post,* September 7, 1993, p. A1.

11. Yablokov, "Facts and Problems," pp. 35–41.

12. Quoted in Broad, "Russians Describe Extensive Dumping."

13. William J. Broad, "No Global Risk Seen in Nuclear Waste in Oceans," *The New*

York Times, June 13, 1993, p. A6. *Nuclear Contamination in the Arctic: Addressing the Problem of Extensive Dumping of Radioactive Waste in the Arctic Ocean,* hearing before the subcommittee on oceanography, Gulf of Mexico, and the outer continental shelf of the Committee on Merchant Marine and Fisheries, House of Representatives, serial no. 103–81, September 30, 1993, pp. 24–30.

14. Mark E. Mount and Michael K. Sheaffer, "Estimated Inventory of Radionuclides in Former Soviet Union Naval Reactors Dumped in the Kara Sea," in *Nuclear Contamination in the Arctic,* pp. 103–160.

15. David E. Sanger, "Nuclear Material Dumped Off Japan," *The New York Times,* October 19, 1993, p. A1.

The episode began as a Russian Navy tanker, the *TNT-27,* was loaded with nine hundred tons of radioactive water and other material from a major submarine base in the Russian Far East and headed for a site due west of Hakodate, on Japan's northernmost island, Hokkaido. It was tracked by a Greenpeace ship, and eventually by Japanese television networks, which beamed pictures of the operation to amazed viewers around the world. Nuclear symbols were clearly discernible on the Russian ship's cargo, which was dumped from a large pipe in broad daylight directly into the sea, making the act seem wildly brazen. Later, in Tokyo, the Russian Ambassador was summoned to the Japanese Foreign Ministry as officials there issued a statement of "strong regret." The episode occurred just after a state visit to Japan by President Yeltsin, undercutting whatever good will the Russian President might have garnered.

16. David E. Pitt, "Nations Back Ban on Atomic Dumping," *The New York Times,* November 13, 1993, p. A7. Pitt, "Russia Is Pressed on Nuclear Waste Dumping," *The New York Times,* December 5, 1993, p. A6. Reuters, "Ban Is Now in Force on Nuclear Dumping," *The New York Times,* February 22, 1994, p. A8. Pearl Marshall, "U.K., China Agree to Abide by London Convention Sea Dump Ban," *Nucleonics Week,* February 24, 1994, p. 14.

17. Norwegian Radiation Protection Authority, "Radioactive Contamination at Dumping Sites for Nuclear Wastes in the Kara Sea," Joint Russian-Norwegian Experts Group for Investigation of Radioactive Contamination in the Northern Areas, Osteras, Norway, November 1994, pp. 7–8, 71–72. Associated Press, "Russia to Open Nuke Dump," August 22, 1994. Associated Press, "Radiation Leaks Found at Soviet Ocean Dump Site," December 7, 1994. Goldhawk Film and TV Productions, Ltd., "Russia's Deep Secrets," BBC Horizon, 1994. Copies available from Horizon, P.O. Box 7, London, W3 6XJ, United Kingdom.

18. *Nuclear Contamination in the Arctic,* pp. 5, 9, 12. "Arctic Seafood One Focus of FDA Radionuclides Monitoring," *Food Chemical News,* June 6, 1994. *Nuclear Wastes in the Arctic: An Analysis of Arctic and Other Regional Impacts from Soviet Nuclear Contamination* (Congressional Office of Technology Assessment, Washington, September 1995), OTA-ENV-623, p. 99.

19. Interview, Leonard G. Johnson, a scientist formerly with the Navy Office of Naval Research, June 8, 1995. Broad, "U.S. Navy's Attack Subs to Be Lent for Study." A. O. Deineko et al., "Deep-Water Scintillation Counter to Estimate Alpha-Active Substance Content in Natural Water," *Marine Technology Society Journal,* Vol. 30, No. 1, Spring 1996, pp. 51–54.

20. Anatoly M. Sagalevitch, "Results of Five Years of Operation with Deep Manned Submersibles 'Mir-1' and 'Mir-2' on Nuclear Submarine 'Komsomolets' Wreck," *Conference Proceedings, Oceans '95,* Marine Technology Society and Institute of Electrical and Electronic Engineers, October 1995, Vol. 1, pp. 6–10. William J. Broad, "Russians Seal Nuclear Sub on Sea Floor," *The New York Times,* September 8, 1994, p. A7. Broad, "Russians to Seal Sunken Nuclear Torpedoes," *The New York Times,* September 19, 1993, p. A19. Broad, "Hazard Is Doubted from Sunken Sub," *The New York Times,* September 5, 1993, p. A7.

21. Interviews, Ed Ueber, April 28, 1994, May 25, 1995. See also William J. Broad, "Sea Sanctuaries Expand in U.S., Offering Refuge to a Riot of Life," *The New York Times,* May 10, 1994, p. C1.

22. Quoted in Don Knapp, "Radioactive Wastes Threaten Pristine Farallones Islands," Cable News Network, September 17, 1994, transcript no. 236–3.

23. John J. Stegeman et al., "Monooxygenase Induction and Chlorobiphenyls in the Deep-Sea Fish *Coryphaenoides armatus*," *Science,* Vol. 231, March 14, 1986, pp. 1287–1289. Stegeman, "Detecting the Biological Effects of Deep-Sea Waste Disposal," *Oceanus,* Vol. 33, No. 2, Summer 1990, pp. 54–60. Stegeman's research was funded by the National Science Foundation, the National Oceanic and Atmospheric Administration, and the Richard King Mellon Foundation. For the ubiquity of *Coryphaenoides armatus,* see Gage and Tyler, *Deep-Sea Biology,* p. 395.

24. Stegeman, "Detecting the Biological Effects," pp. 54–56. For the ubiquity of the chemicals, see Michael J. Kennish, *Practical Handbook of Marine Science* (CRC Press, Boca Raton, Fla., 1994), pp. 416–417, 446–491. See also Thurman, *Introductory Oceanography,* pp. 490–494. For pollution overviews, see Anne E. Platt, "Dying Seas," *World Watch,* Vol. 8, No. 1, January/February 1995, pp. 10–19. Tim Benton, "Oceans of Garbage," in Peter Benchley and Judith Gradwohl, eds., *Ocean Planet: Writings and Images of the Sea* (Abrams, New York, 1995), pp. 156–158. Adam Markham, *A Brief History of Pollution* (St. Martin, New York, 1994), pp. 52–61. Michael L. Weber and Judith Gradwohl, *The Wealth of Oceans* (Norton, New York, 1995), pp. 21, 146–148.

25. Interview, John J. Stegeman, June 27, 1996. Stegeman, "Detecting the Biological Effects," pp. 54–60.

26. Sylvia A. Earle, *Sea Change: A Message of the Oceans* (Putnam, New York, 1995), pp. 198–200.

27. J. P. Barry et al., "Climate-Related, Long-Term Faunal Changes in a California Rocky Intertidal Community," *Science,* Vol. 267, February 3, 1995, pp. 672–675.

28. John McGowan and Dean Roemmich, "Climatic Warming and the Decline of Zooplankton in the California Current," *Science,* March 3, 1995, pp. 1324–1326. Kathy A. Svitil, "Collapse of a Food Chain," *Discover,* July 1995, pp. 36–37. Janet Howard, "Vanishing Act," *Explorations of the Scripps Institution,* Vol. 2, No. 2, Fall 1995, pp. 10–17.

29. Gregorio Parrilla et al., "Rising Temperatures in the Subtropical North Atlantic Ocean Over the Past 35 Years," *Nature,* Vol. 369, May 5, 1995, pp. 48–51. For Millard quote, see Richard Monastersky, "Temperatures on the Rise in Deep Atlantic," *Science News,* May 7, 1994, p. 295.

30. One critic dismissed the *Hesperides*'s findings of deep warming as "an indication that mid-ocean temperatures aren't as stable as was once thought. Period." See "Report Addresses Relation Between Deep Ocean Warming & Global Warming," *Global Warming Network Online Today,* July 27, 1994. For discussion of unexpected ocean temperature swings, see Thurman, *Introductory Oceanography,* p. 211.

31. Robert C. Balling, Jr., "Global Warming: Messy Models, Decent Data, and Pointless Policy," in Ronald Bailey, ed., *The True State of the Planet* (Free Press, New York, 1995), pp. 81–107. Easterbrook, *A Moment,* pp. 277–316. William K. Stevens, "Global Warming Experts Call Human Role Likely," *The New York Times,* September 10, 1995, p. A1. Stevens, "Richard S. Lindzen; A Skeptic Asks, Is It Getting Hotter, or Is It Just the Computer Model?" *The New York Times,* June 18, 1996, p. C1.

32. Thurman, *Introductory Oceanography,* pp. 144, 186–190, 356–358.

33. Interview, Walter Munk (ATOC principal investigator), March 31, 1994. Interview, David W. Hyde (ATOC project director), April 1, 1994. Walter Munk et al., "Acoustic Thermometry of Ocean Climate: Volume 1: Technical Proposal," University of California at San Diego report number 92–1321, June 1, 1992. William J. Broad, "2 Environmental Camps Feud Over Noisy Ocean Experiment," *The New York Times,* April 5, 1994, p. C4.

34. Interview, Lindy Weilgart, March 31, 1994. For Weilgart's whale research, see Hal Whitehead, "The Realm of the Elusive Sperm Whale," *National Geographic,* November 1995, pp. 56–73. Jocelyn Kaiser, "Of Whales and Ocean Warming," *Science News,* Vol. 147, June 3, 1995, pp. 350–351. Broad, "2 Environmental Camps." Neil Morgan, "Jury's In: Walter Munk Isn't a Whale Killer," *The San Diego Union-Tribune,* August 1, 1995, p. A2. Maria Goodavage and Sandra Sanchez, "Foes Sound the Alarm Over Ocean Noise Test,"

USA Today, May 13, 1994. For background on the experiment, see Jon Cohen, "Was Underwater 'Shot' Harmful to the Whales?" *Science,* Vol. 252, May 17, 1991, pp. 912–914.

35. Richard C. Paddock, "Undersea Noise Test Could Risk Making Whales Deaf," *The Los Angeles Times,* March 22, 1994, p. A1. Peter Monaghan, "Oceanographers in a Sea of Trouble," *The Chronicle of Higher Education,* June 1, 1994, p. A6. Broad, "2 Environmental Camps."

36. Letter, John A. Orcutt, Institute of Geophysics and Planetary Physics, Scripps, to *The Los Angeles Times,* March 28, 1994. Scripps Institution of Oceanography, "Acoustic Thermometry of Ocean Climate (ATOC) Media Advisory," Scripps, University of California at San Diego, April 11, 1994.

37. Interviews, Andrew Forbes (ATOC project manager), October 26, November 2, 1995. David Graham, "Ocean Sound Tests Get Go-ahead," *The San Diego Union-Tribune,* July 15, 1995, p. 1. Kaiser, "Of Whales." William J. Broad, "Ocean Study to Start After Big Changes and a Long Delay," *The New York Times,* November 7, 1995, p. C4.

38. Broad, "Ocean Study to Start." Broad, "Sea Sanctuaries Expand." National Oceanic and Atmospheric Administration, "An Overview of the Sanctuaries and Reserves Division," NOAA Congressional Briefing, March 1994, p. 75.

39. William J. Broad, "Test of Underwater Sound System Showing No Harm to Animals," *The New York Times,* January 21, 1996, p. A20.

40. Interview, Bradley Barr (manager of the Stellwagen Bank National Marine Sanctuary), October 20, 1995. I've taken the liberty of not noting individually most of the other interviews conducted during this one-day cruise of October 20, 1995. The names of the interviewees should be obvious in the text. For Stellwagen as a sanctuary, see NOAA, "An Overview of the Sanctuaries," p. 87, and Nathalie Ward, *Stellwagen Bank: A Guide to the Whales, Sea Birds, and Marine Life of the Stellwagen Bank National Marine Sanctuary* (Down East Books, Camden, Maine, 1995). For vanishing whales, see Usha Lee McFarling, "Biologists Document a Steady Humpback Exodus from Stellwagen," *The Boston Globe,* August 15, 1994, p. 25.

41. For the humpback deaths, see Donald M. Anderson, "Red Tides," *Scientific American,* August 1994, pp. 52–58. For polynuclear aromatic hydrocarbons, see Kennish, *Practical Handbook,* p. 465. For the radioactive dump, see International Atomic Energy Agency, "Inventory of Radioactive Material," pp. 48–51. For fishing problems, see McFarling, "Biologists Document."

42. James G. Bellingham et al., "A Small, Long-Range Autonomous Vehicle for Deep Ocean Exploration," Proceedings of the Second International Offshore and Polar Engineering Conference, 1992, pp. 461–467.

43. For an overview of the classes of undersea robots, see Dana R. Yoerger, "Robotic Undersea Technology," *Oceanus,* Vol. 34. No. 1, Spring 1991, pp. 32–37.

44. Interview, James G. Bellingham, October 20, 1995. For deep oxygenation, see Gage and Tyler, *Deep-Sea Biology,* pp. 12–18; and Thurman, *Introductory Oceanography,* pp. 200–205.

EPILOGUE

1. J. E. D. Williams, *From Sails to Satellites: The Origin and Development of Navigational Science* (Oxford, New York, 1992), pp. 21–126. Dava Sobel, *Longitude* (Walker, New York, 1995).

2. Walter A. McDougall, *Let the Sea Make a Noise: A History of the North Pacific from Magellan to MacArthur* (Basic, New York, 1993), pp. 56–57, 83–84. R. R. Palmer and Joel Colton, *A History of the Modern World* (Knopf, New York, 1965), pp. 90–94, 214–215.

3. Michael McCabe, "To the Deepest Spot on Earth: San Anselmo Inventor Designs New Class of Submarine," *San Francisco Chronicle,* October 29, 1996, p. A11.

4. Verena Tunnicliffe, "Coaxial Report: Beard Vent Fauna," draft report, January 25, 1994. Van Dover, *The Octopus's Garden,* pp. 115–126.

5. Interview, John A. Baross, December 23, 1993. Broad, "Life Springs Up." Broad, "A Voyage into the Abyss." Broad, "Strange Oases." For a hyperthermophile overview, see William J. Broad, "Clues to Fiery Origin of Life Sought in Hothouse Microbes," *The New York Times,* May 9, 1995, p. C1.

6. Interview, Verena Tunnicliffe, March 18, 1996. Julia Getsiv and Bob Embley, "The Study of Hydrothermal Venting Systems: Time Series, Trace Metals and Event Response: An A2/Alvin Expedition to the Juan de Fuca Ridge, June 17 to July 9, 1994," NOAA Pacific Marine Environmental Laboratory, Newport, Oregon. For a detailed overview of such research, including our site along the Juan de Fuca Ridge, see Susan E. Humphris et al., *Seafloor Hydrothermal Systems: Physical, Chemical, Biological and Geological Interactions* (American Geophysical Union, Washington, 1995), Geophysical Monograph 91.

7. Broad, "Anti-Sub Seabed Grid Thrown Open."

8. Interviews, Bradley A. Moran and Clyde F. E. Roper, January 10, 1997. For background, see Broad, "Biologists Closing on Hidden Lair."

9. Andrew C. Revkin, "Cold-War Laser Aids T.W.A. Hunt," *The New York Times,* August 1, 1996, p. B1. Don Van Natta, Jr., "Luggage Spotted in Debris Trail Suggests an Explosion to Experts," *The New York Times,* August 6, 1996, p. A1. Edward J. Saade and Drew Carey, "Laser Line Scanner: 'Highgrading' Search Targets," *Sea Technology,* October 1996, pp. 63–65.

10. William J. Broad, "Effort to Raise Part of Titanic Falters as Sea Keeps History," *The New York Times,* August 31, 1996, p. A6. Randy Kennedy, "With Ship's Hull Back on the Ocean Floor, Titanic Buffs Return to New York," *The New York Times,* September 2, 1996, p. B26.

11. For a discussion of the emerging gap, see George F. Bass and W. F. Searle, "Epilog," in George F. Bass, ed., *Ships and Shipwrecks of the Americas: A History Based on Underwater Archaeology* (Thames and Hudson, New York and London, 1988), pp. 251–259.

In 1996, treasure hunters founded the Deep Shipwreck Explorers' Association, Inc., a not-for-profit corporation that aims to raise industry standards. The association address is 3030 North Rocky Point Boulevard, Number 280, Tampa, Florida, 33607, and its president is Greg Stemm of Remarc International, Inc., in Tampa. At the other end of the spectrum, scholars and experts working for the United Nations in 1994 drew up a draft convention on the protection of underwater cultural heritage that would ban all commercial salvage and keep such patrimony in place beneath the sea "unless its removal is necessary for scientific or protective measures." At the moment, this so-called Buenos Aires draft convention has an uncertain political future.

12. Pryor, "New Described Super-Nodule Resource."

13. Malcolm W. Browne, "Fish That Dates Back to Age of Dinosaurs Is Verging on Extinction," *The New York Times,* April 18, 1995, p. C4. Keith Steward Thomson, "The Story of the Coelacanth," *Oceanus,* Vol. 34, No. 3, Fall 1991, pp. 38–43, and in the same issue Hans Fricke and Karen Hissmann, "Coelacanths . . . The Fate of a Famous Fish," pp. 44–45.

14. Letter to author from James D. Watkins, president of the Consortium for Oceanographic Research and Education, Washington, D.C., May 20, 1996. Jeffrey Mervis, "Hearing Highlights Hopes, Realities," *Science,* Vol. 271, February 2, 1996, pp. 591–592.

15. For global coastal sanctuaries, see Earle, *Sea Change,* pp. 295–317, 329–342.

16. Van Dover, *The Octopus's Garden,* p. 77.

17. National Research Council, *Understanding Marine Biodiversity* (National Academy Press, Washington, 1995), p. 9.

18. Interview, Paul R. Tidwell, July 13, 1995.

Bibliography

Adrião, Daniel. "Gomes Da Silva Becomes Part of the Treasure Hunt," *Seminario* (Lisbon), August 27, 1994, pp. 50–51.

All Lisbon and Its Surroundings (Editorial Escudo de Oro, Lisbon, 1994).

Alpern, David M. "CIA's Mission Impossible," *Newsweek,* March 31, 1975, pp. 24–32.

Amato, Ivan. "A Sub Surveillance Network Becomes a Window on Whales," *Science,* Vol. 261, July 30, 1993, pp. 549–550.

Amato, Roger V. "Pacific EEZ Marine Minerals Activities of the Minerals Management Service," in *Proceedings of the 1991 Exclusive Economic Zone Symposium on Mapping and Research: Working Together in the Pacific EEZ* (U.S. Geological Survey, Denver, 1992), USGS Survey Circular 1092, pp. 152–157.

Ambrossiana, Alfredo A. "The Commanders Respond," U.S. Naval Institute *Proceedings,* March 1995, p. 28.

Anderson, Donald M. "Red Tides," *Scientific American,* August 1994, pp. 52–58.

Appenzeller, Tim. "Chemistry: Laurels for a Late-Night Brainstorm," *Science,* Vol. 262, October 22, 1993, pp. 506–507.

Archaeonautica Studies Center. "Presentation," Lisbon, undated one-page description of the group and its activities.

——. *Archaeonautica Correspondence,* Lisbon, No. 2, 1st semester of 1995.

"Arctic Seafood One Focus of FDA Radionuclides Monitoring." *Food Chemical News,* June 6, 1994.

Arkin, William J., and Richard W. Fieldhouse. *Nuclear Battlefields* (Ballinger, Cambridge, Mass., 1985).

"ART LS-4096 Laser Line Scan System," Underwater News & Technology, July-August 1994, pp. 16–17.

Asimov, Isaac. "Haber, Fritz," *Asimov's Biographical Encyclopedia of Science and Technology* (Doubleday, Garden City, N.Y., 1982), pp. 624–625.

——. "Piccard, Jacques," *Asimov's Biographical Encyclopedia of Science and Technology* (Doubleday, Garden City, New York, 1982), pp. 694–695.

Associated Press. "Russia to Open Nuke Dump," August 22, 1994.

——. "Scientists Find Little Spread of Radiation in Arctic," September 16, 1994.

——. "Radiation Leaks Found at Soviet Ocean Dump Site," December 7, 1994.

——. "Indonesia Needs More Ships to Guard Eastern Waters," February 1, 1995.

"Atomic Waste Reported Leaking in Ocean Sanctuary Off California," *The New York Times,* May 7, 1990, p. B12.

"Atom Waste Termed Peril to West Coast," *The New York Times,* August 20, 1980, p. A13.

Babb, Ivar G., et al. "Dual Use of a Nuclear Powered Research Submersible: The U.S. Navy's NR-1," *Marine Technology Society Journal,* Vol. 27, No. 4, Winter 1993–94, pp. 39–48.

"Back to the Hunt," *Newsweek,* June 10, 1963, pp. 31–32.

Baker, Edward T. "Megaplumes," *Oceanus,* Winter 1991–92, p. 84.

Ballard, Robert D. "NR-1: The Navy's Inner-Space Shuttle," *National Geographic,* April 1985, pp. 450–459.

——. *Exploring Our Living Planet* (National Geographic Society, Washington, 1988).

——. "Riddle of the Lusitania," *National Geographic,* April 1994, pp. 68–85.

——, with Rick Archbold. *The Discovery of the Titanic* (Warner, New York, 1989).

——, with Rick Archbold. *The Lost Wreck of the Isis* (Madison, Toronto, 1990).

——, with Rick Archbold. *The Discovery of the Bismarck* (Warner, New York, 1990).

——, with Rick Archbold. *The Lost Ships of Guadalcanal* (Warner, New York, 1993).

——, with Malcolm McConnell. *Explorations: My Quest for Adventure and Discovery Under the Sea* (Hyperion, New York, 1995).

Balling, Robert C., Jr. "Global Warming: Messy Models, Decent Data, and Pointless Policy," in Ronald Bailey, ed., *The True State of the Planet* (Free Press, New York, 1995), pp. 81–107.

Barry, J. P., et al. "Climate-Related, Long-Term Faunal Changes in a California Rocky Intertidal Community," *Science,* Vol. 267, February 3, 1995, pp. 672–675.

Bascom, Willard. "Mining in the Sea," in Lewis M. Alexander, ed., *The Law of the Sea: Offshore Boundaries and Zones* (Ohio State University Press, Columbus, Ohio, 1967), pp. 160–171.

——. *The Crest of the Wave: Adventures in Oceanography* (Doubleday/Anchor, New York, 1990).

Bass, George F., and W. F. Searle. "Epilog," in George F. Bass, ed., *Ships and Shipwrecks of the Americas: A History Based on Underwater Archaeology* (Thames and Hudson, New York and London, 1988), pp. 251–259.

Baum, Michael. "Commerce Department Announces 29 Advanced Technology Awards," U.S. Department of Commerce News Release, November 4, 1993.

Beebe, William. *Half Mile Down* (Harcourt Brace, New York, 1934).

Behrman, Daniel. *The New World of the Oceans: Men and Oceanography* (Little-Brown, Boston, 1969).

Bellingham, James G., et al. "A Small, Long-Range Autonomous Vehicle for Deep Ocean Exploration," *Proceedings of the Second International Offshore and Polar Engineering Conference,* 1992, pp. 461–467.

Bennett, Vanora. "Japan, Americans Argue Over Undersea Treasure Trove," Reuters, July 27, 1995.

Benton, Tim. "Oceans of Garbage," in Peter Benchley and Judith Gradwohl, eds., *Ocean Planet: Writings and Images of the Sea* (Abrams, New York, 1995), pp. 156–158.

Berman, Robert P., and John C. Baker. *Soviet Strategic Forces* (Brookings Institution, Washington, 1982).

Bishop, Katherine. "U.S. to Determine Danger from Barrels of Atomic Waste in Pacific," *The New York Times,* January 20, 1991, p. A21.

Borgese, Elisabeth Mann. "Ocean Mining and the Future of World Order," in J. Thiede and K. J. Hsü, *Use and Misuse of the Seafloor* (Wiley, New York, 1992), pp. 117–126.

——, ed. *Ocean Frontiers: Explorations by Oceanographers on Five Continents* (Abrams, New York, 1992).

Boxer, Charles R. *Four Centuries of Portuguese Expansion, 1415–1825* (University of California, Berkeley, 1969).

Boyd, Carl. *The Japanese Submarine Force and World War II* (Naval Institute Press, Annapolis, 1995).

——. *American Command of the Sea through Carriers, Codes and the Silent Service: World War II and Beyond* (The Mariners' Museum, Newport News, Va., 1995).

British Broadcasting Company. "Russia's Deep Secrets," *BBC Horizon,* 1994.

Broad, William J. "Resource Wars: The Lure of South Africa," *Science,* Vol. 210, December 5, 1980, pp. 1099–1100.

——. "Wreckage of Titanic Reported Discovered 12,000 Feet Down," *The New York Times,* September 3, 1985, p. A1.

——. "Titanic Wreck Was Surprise Yield of Underwater Tests for Military," *The New York Times,* September 8, 1985, p. A1.

——. "Airhorns Blare as Titanic Researchers Sail In," *The New York Times,* September 10, 1985, p. A1.

——. "Finder of Titanic Aims to Capitalize," *The New York Times,* September 11, 1985, p. D26.

——. "U.S. Designs Spy Satellites to Be More Secret Than Ever," *The New York Times,* November 3, 1987, p. C1.

——. "Undersea Robots Open a New Age of Exploration," *The New York Times,* November 13, 1990, p. C1.

——. "Deepest Wrecks Now Visible to Undersea Cameras," *The New York Times,* February 2, 1993, p. C1.

——. "Into the Abyss: New Robots Probe the Deep," *The New York Times,* March 9, 1993, p. C1.

——. "Russians Describe Extensive Dumping of Nuclear Waste," *The New York Times,* April 27, 1993, p. A1.

——. "No Global Risk Seen in Nuclear Waste in Oceans," *The New York Times,* June 13, 1993, p. A6.

——. "Russia Says U.S. Got Sub's Atom Arms," *The New York Times,* June 20, 1993, p. A14.

——. "Items on Titanic Brought to U.S.," *The New York Times,* July 7, 1993, p. A11.

——. "Deep-Sea Scavenging Gets Easier," *The New York Times,* July 18, 1993, section 4, p. 6.

——. "Graham Hawkes: Racing to the Bottom of the Deep, Black Sea," *The New York Times,* August 3, 1993, p. C1.

——. "Long-Secret Navy Devices Allow Monitoring of Ocean Eruption," *The New York Times,* August 20, 1993, p. A1.

——. "Navy Listening System Opening World of Whales," *The New York Times,* August 23, 1993, p. A12.

——. "Hazard Is Doubted from Sunken Sub," *The New York Times,* September 5, 1993, p. A7.

——. "New Idea on Titanic Sinking Faults Steel as Main Culprit," *The New York Times,* September 16, 1993, p A14.

——. "Russians to Seal Sunken Nuclear Torpedoes," *The New York Times,* September 19, 1993, p. A19.

——. "Life Springs Up in Ocean's Volcanic Vents, Deep Divers Find," *The New York Times,* October 19, 1993, p. C4.

——. "A Voyage into the Abyss: Gloom, Gold and Godzilla," *The New York Times,* November 2, 1993, p. C1.

——. "Navy Says 2 Subs Pose No Hazards," *The New York Times,* November 7, 1993, p. A30.

——. "Strange Oases in Sea Depths Offer Map to Riches," *The New York Times,* November 16, 1993, p. C1.

——. "Navy Has Long Had Secret Subs for Deep-Sea Spying, Experts Say," *The New York Times,* February 7, 1994, p. A1.

——. "Sunken Soviet Sub Leaks Radioactivity in Atlantic," *The New York Times,* February 8, 1994, p. A13.

——. "Plan to Carve Up Ocean Floor Riches Nears Fruition," *The New York Times,* March 29, 1994, p. C1.

——. "2 Environmental Camps Feud Over Noisy Ocean Experiment," *The New York Times,* April 5, 1994, p. C4.

——. "Sea Sanctuaries Expand in U.S., Offering Refuge to a Riot of Life," *The New York Times,* May 10, 1994, p. C1.

——. "Fantastic World Found in Sea's Middle Depths," *The New York Times,* July 26, 1994, p. C1.

——. "Weird Mound Offers Clues to Mysteries of the Deep," *The New York Times,* August 9, 1994, p. C1.

——. "Squids Emerge as Smart, Elusive Hunters of Mid-Sea," *The New York Times,* August 30, 1994, p. C1.

——. "Russians Seal Nuclear Sub on Sea Floor," *The New York Times,* September 8, 1994, p. A7.

——. "Japan Plans to Conquer Sea's Depths," *The New York Times,* October 18, 1994, p. C1.

——. "In Ocean Depths, Scientists Find Fast Growth," *The New York Times,* October 20, 1994, p. B13.

——. "Secret Sub to Scan Sea Floor for Roman Wrecks," *The New York Times,* February 7, 1995, p. C1.

——. "U.S. Navy's Attack Subs to Be Lent for Study of Arctic Icecap," *The New York Times,* February 21, 1995, p. C1.

——. "Clues to Fiery Origin of Life Sought in Hothouse Microbes," *The New York Times,* May 9, 1995, p. C1.

——. "Watery Grave of the Azores to Yield Shipwrecked Riches," *The New York Times,* June 6, 1995, p. C1.

——. "Lost Japanese Sub with 2 Tons of Axis Gold Found on Floor of Atlantic," *The New York Times,* July 18, 1995, p. C1.

——. "Calming Stormy Seas with New Kinds of Ship Hulls," *The New York Times,* August 15, 1995, p. C1.

——. "The World's Deep, Cold Sea Floors Harbor a Riotous Diversity of Life," *The New York Times,* October 17, 1995, p. C1.

——. "Map Makes Ocean Floors as Knowable as Venus," *The New York Times,* October 24, 1995, p. C1.

——. "Ocean Study to Start After Big Changes and a Long Delay," *The New York Times,* November 7, 1995, p. C4.

——. "Navy Is Releasing Treasure of Secret Data on World's Oceans," *The New York Times,* November 28, 1995, p. C1.

——. "Test of Underwater Sound System Showing No Harm to Animals," *The New York Times,* January 21, 1996, p. A20.

——. "Biologists Closing on Hidden Lair of Giant Squid," *The New York Times,* February 13, 1996, p. C1.

——. "Volcanic Fury Rocks Pacific Sea Floor Where Inner Earth Meets Outer Earth," *The New York Times,* March 19, 1996, p. C1.

——. "Anti-Sub Seabed Grid Thrown Open to Research Uses," *The New York Times,* July 2, 1996, p. C1.

——. "Titanic Relics Open Era in Deep-Sea Commerce," *The New York Times,* August 25, 1996, p. A24.

——. "Effort to Raise Part of Titanic Falters as Sea Keeps History," *The New York Times,* August 31, 1996, p. A6.

Broadus, James M. "Seabed Materials," *Science,* Vol. 235, February 20, 1987, pp. 853–860.

Brock, Thomas D. "Life at High Temperatures in Water Environments," *Science,* Vol. 230, October 11, 1985, p. 132.

——, et al. *Biology of Microorganisms* (Prentice Hall, Englewood Cliffs, N.J., 1994).

Brockett, Ted, and Caroline Z. Richards. "Deepsea Mining Simulator for Environmental Impact Studies," *Sea Technology,* August 1994, pp. 77–82.

Brodie, John, and Andrew Hindes. "Fox, Par Board 'Titanic,' " *Variety,* July 22, 1996, p. 17.

Brown, Lester, and Hal Kane. *Full House: Reassessing the Earth's Population Carrying Capacity* (Norton, New York, 1994), pp. 80–81.

Browne, Malcolm W. "Fish That Dates Back to Age of Dinosaurs Is Verging on Extinction," *The New York Times,* April 18, 1995, p. C4.

Brusca, Richard C., and Gary J. Brusca. *Invertebrates* (Sinauer, Sunderland, Mass., 1990).

Bruun, Anton F., et al. *The Galathea Deep Sea Expedition: 1950–1952* (Macmillan, New York, 1956).

Buchsbaum, Ralph, et al. *Animals Without Backbones* (University of Chicago, Chicago, 1987).

Burroughs, Tom. "Ocean Mining: Boom or Bust," *Technology Review,* April 1984, p. 54.

Busby, Frank. *Undersea Vehicles Directory—1990–91* (Busby Associates, Arlington, Va., 1990).

——, and Joseph R. Vadus. "Autonomous Underwater Vehicle R&D Trends," *Sea Technology,* May 1990, pp. 65–71.

Carlowicz, Mark. "New Map of Seafloor Mirrors Surface," *EOS,* Vol. 76, No. 44, October 31, 1995, p. 441.

Carpenter, Scott. "Escape from the Deep," *Popular Science,* September 1969, pp. 79–81, 214.

Carson, Rachel L. *The Sea Around Us* (Oxford, New York, 1951).

Central Intelligence Agency. "Glomar Explorer: Recovery and Burial of Soviet Sailors," CIA statement, Washington, November 12, 1993.

Childress, James J. "Oceanic Biology: Lost in Space?" in P. Brewer, ed., *Oceanography: The Present and Future* (Springer-Verlag, New York, 1983), pp. 127–135.

Chung, Jin S. "Advances in Manganese Nodule Mining Technology," *Marine Technology Society Journal,* Vol. 19, No. 4, 1985, pp. 39–44.

Clancy, Tom. *Submarine: A Guided Tour Inside a Nuclear Warship* (Berkley, New York, 1993).

Clifford, Frank. "Simpson Case Boosts Microbe Conservation," *Los Angeles Times,* August 31, 1994, p. A1.

Cochran, Thomas B., et al. *Nuclear Weapons Databook: Volume Two, U.S. Nuclear Warheads Production* (Ballinger, Cambridge, Mass., 1987).

Codevilla, Angelo. *Informing Statecraft: Intelligence for a New Century* (Free Press, New York, 1992).

Cohen, Daniel. *A Modern Look at Monsters* (Dodd Mead, New York, 1970).

——. *Encyclopedia of Monsters* (Dodd Mead, New York, 1982).

Cohen, Jon. "Was Underwater 'Shot' Harmful to the Whales?" *Science,* Vol. 252, May 17, 1991, pp. 912–914.

Cone, Joseph. *Fire Under the Sea: The Discovery of the Most Extraordinary Environment on Earth—Volcanic Hot Springs on the Ocean Floor* (Morrow, New York, 1991).

Connor, Judith, and Charles Baxter. "Kelp Forests." Monterey Bay Aquarium Foundation, 1989.

Connor, Judith, and Nora L. Deans. "David Packard and Julie Packard: Monterey Bay Profiles in Depth," *Oceanus,* Summer 1993, pp. 74–79.

Copulos, Milton R. "The Lessons of History," *Heritage Foundation Reports,* January 16, 1989, p. 11.

Corliss, John B., et al. "Submarine Thermal Springs on the Galápagos Rift," *Science,* March 16, 1979, pp. 1073–1082.

——, J. A. Baross, and S. E. Hoffman. "An Hypothesis Concerning the Relationship Between Submarine Hot Springs and the Origin of Life on Earth," *Oceanologica Acta,* supplement, 1981, pp. 59–69.

Cowell, Alan. "Italian Obsession: Was Airliner Shot Down?" *The New York Times,* February 10, 1992, p. A7.

Craven, John P. "Sea Power and the Sea Bed," U.S. Naval Institute *Proceedings,* April 1966, pp. 36–51.

——. "The Geopolitical Significance of a Deepest Ocean Presence," *Marine Technology Society Journal,* June 1990, Vol. 24, No. 2, pp. 13–15.

Crawford, Mark. "ERATO Sparks Japan's Basic Research Effort," *New Technology Week,* December 18, 1995.

Cruickshank, Michael J. "Pair of Important Meetings Laud Future of Non-U.S. Seabed Mining," *Sea Technology,* February 1995, p. 94.

——. "Cook Island Nodule Mining Plans Closer to Reality," *Sea Technology,* February 1996, p. 105.

Dahlburg, John-Thor. "CIA's Raising of Soviet Sub Told," *The Los Angeles Times,* October 17, 1992, p. A2.

Dalton, John H., et al. "Forward . . . from the Sea," U.S. Naval Institute *Proceedings,* December 1994, pp. 46–49.

Darwin, Charles. *The Origin of Species* and *The Descent of Man* (Modern Library, New York, 1948).

Davis, W. Jackson, et al. "Evaluation of Oceanic Radioactive Dumping Programs," Environmental Studies Institute, University of California at Santa Cruz, July 1982.

Deacon, Margaret. *Scientists and the Sea, 1650–1900: A Study of Marine Science* (Academic Press, New York, 1971).

de Duve, Christian. *Vital Dust: Life as a Cosmic Imperative* (Basic, New York, 1995).

Delaney, John R. "An Interdisciplinary Response to the Earthquake Swarms and Submarine Eruption, CoAxial Segment of the Juan de Fuca Ridge," University of Washington Proposal to the National Science Foundation, August 20, 1993.

Deming, Jody W., and John A. Baross. "Deep-sea Smokers: Windows to a Subsurface Biosphere?" *Geochimica et Cosmochimica Acta,* 1993, Vol. 57, pp. 3219–3230.

Dibner, Bern. "Communications," in Melvin Kranzberg and Carroll W. Pursell, Jr., *Technology in Western Civilization,* Vol. I (Oxford, New York, 1967), pp. 454–460.

Dietz, Robert S. "The Underwater Landscape," in *Exploring the Ocean World* (Crowell, New York, 1969).

Dobbs, Michael. "In the Former Soviet Union, Paying the Nuclear Price," *The Washington Post,* September 7, 1993, p. A1.

Drew, Christopher, et al. "A Risky Game of Cloak-and-Dagger Under the Sea," *The Chicago Tribune,* January 7, 1991, p. 1.

Duane, David B. "Elements of a Proposed Five-Year Research Program on Polymetallic Sulfides," *Marine Technology Society Journal,* Vol. 16, No. 3, 1982, pp. 87–91.

Duncan, T. Bentley. *Atlantic Islands: Madeira, the Azores and the Cape Verdes in Seventeenth-Century Commerce and Navigation* (University of Chicago, Chicago, 1972).

Dunham, Roger C. *Spy Sub: A Top Secret Mission to the Bottom of the Pacific* (Naval Institute Press, Annapolis, 1996).

Dybas, Cheryl Lyn. "The Oarfish," *Oceanus,* Vol. 36, No. 1, Spring 1993, pp. 98–99.

——. "Beautiful, Ethereal Larvaceans Play a Central Role in Ocean Ecology," *Oceanus,* Vol. 36, No. 2, Summer 1993, pp. 84–86.

——. "Comb Jelly: A Jewel of a Creature," *Sea Frontiers,* December 1994, pp. 12–13, 55.

Earle, Sylvia A. "Ocean Everest—An Idea Whose Time Has Come," *Marine Technology Society Journal,* June 1990, Vol. 24, No. 2, pp. 9–12.

——. *Sea Change: A Message of the Oceans* (Putnam, New York, 1995).

Easterbrook, Gregg. *A Moment in Time: The Coming Age of Environmental Optimism* (Viking, New York, 1995).

Eaton, John P. "Descent to History," *Voyage,* No. 18, Autumn 1994, pp. 66–77.

——, and Charles A. Haas. *Titanic: Triumph and Tragedy* (Norton, New York, 1995).

Edmond, John M., and Karen L. Von Damm. "Hydrothermal Activity in the Deep Sea," *Oceanus,* Spring 1992, p. 76.

——. "Hot Springs on the Ocean Floor," *Scientific American,* April 1983, p. 78.

Egerton, Frank N., 3rd. "Forbes, Edward, Jr.," *Dictionary of Scientific Biography,* Vol. 5 (Scribner's, New York, 1972), pp. 66–68.

Ellis, Richard. *Monsters of the Sea* (Knopf, New York, 1994).

Emerson, Tony, and Hideko Takayama. "Down to the Bottom," *Newsweek,* July 5, 1993, pp. 60–64. A longer version of this article by the same authors appeared in "Deep Dreams," *Newsweek,* international edition, July 5, 1993, pp. 34–40.

Erlich, Henry A., et al. "Recent Advances in the Polymerase Chain Reaction," *Science,* Vol. 252, June 21, 1991, p. 1643.

Facts on File World News Digest. "Australia Extends Coastal Zone," December 1, 1994, p. 902.

Fenwick, Judith. *International Profiles on Marine Scientific Research* (Woods Hole Oceanographic Institution, Woods Hole, Mass., 1992).

Fernández-Armesto, Felipe, ed. *The Times Atlas of World Exploration* (HarperCollins, New York, 1991).

Fisher, Lawrence M. "David Packard, 83, Pioneer of Silicon Valley, Is Dead," *The New York Times,* March 27, 1996, p. D20.

Fricke, Hans, and Karen Hissmann. "Coelacanths . . . the Fate of a Famous Fish," *Oceanus,* Vol. 34, No. 3, Fall 1991, pp. 44–45.

Fujioka, Kantaro. "Quicklook Report on the TAG Hydrothermal Mound," filed to Woods Hole Oceanographic Institution, *Shinkai 6500* dive number 216, August 4, 1994, and dive number 230, August 22, 1994.

Funnell, Cliff. "Wealth from the Oceans: A U.K. Program," *Sea Technology,* October 1993, pp. 55–60.

Gage, J. D., and P. A. Tyler. *Deep-Sea Biology* (Cambridge University Press, New York and Cambridge, 1991).

Gates, Robert M. *From the Shadows: The Ultimate Insider's Story of Five Presidents and How They Won the Cold War* (Simon & Schuster, New York, 1996).

Georgetown University. "The Soviet Military Technological Challenge," The Center for Strategic Studies, Washington, Special Report No. 6, September 1967.

Getsiv, Julia, and Bob Embley. "The Study of Hydrothermal Venting Systems: Time Series, Trace Metals and Event Response: An A2/Alvin Expedition to the Juan de Fuca Ridge," June 17 to July 9, 1994, NOAA Pacific Marine Environmental Laboratory, Newport, Oregon.

Gibbs, Nancy. "The Race for Sunken Treasure: The Ocean Gold Rush," *Time,* international edition, October 25, 1993, pp. 48–55.

Gillis, Anna Maria. "A Pressure-Filled Life: How Do Microorganisms Cope When They Live at Extremes?" *Bioscience,* Vol. 44, October 1994, p. 584.

Given, Deam. "Underwater Technology in the USSR," *Oceanus,* Vol. 34, No. 1, Spring 1993, pp. 67–73.

Global Warming Network Online Today. "Report Addresses Relation Between Deep Ocean Warming & Global Warming," July 27, 1994.

Glowka, Lyle. "The Deepest of Ironies: Genetic Resources, Marine Scientific Research and the International Deep Sea-bed Area," The World Conservation Union, Environmental Law Center, Bonn, Germany, August 1995.

Gold, Thomas. "The Deep, Hot Biosphere," *The Proceedings of the National Academy of Sciences,* Vol. 89, July 1992, pp. 6045–6049.

Goldberg, Carey. "With Fortune Built, Packard Heirs Look to Build a Legacy," *The New York Times,* May 6, 1996, p. A1.

Goodavage, Maria, and Sandra Sanchez. "Foes Sound the Alarm Over Ocean Noise Test," *USA Today,* May 13, 1994.

Goran, Morris. "Haber, Fritz," *Dictionary of Scientific Biography,* Vol. Five (Scribner's, New York, 1972), pp. 620–623.

Gordon, Alan. "Underwater Laser Line Scan Technology," *Underwater Intervention '93 Conference Proceedings* (Marine Technology Society, Washington, 1993), pp. 164–170.

Gore, Rick. "Between Monterey Tides," *National Geographic,* February 1990, pp. 2–43.

Graham, David. "Ocean Sound Tests Get Go-ahead," *The San Diego Union-Tribune,* July 15, 1995, p. 1.

Graham, David M. "Seven-Vessel Fleet Marks Russian CGGE," *Sea Technology,* June 1994, pp. 67–69.

——. "Hawaii Ocean R&D: Life on the 'Fast Track,' " *Sea Technology,* August 1994, pp. 26–38.

Graham, Winston. *The Spanish Armadas* (Doubleday, Garden City, N.Y., 1972).

Grassle, J. Frederick, and Nancy J. Maciolek. "Deep-Sea Species Richness: Regional and Local Diversity Estimates from Quantitative Bottom Samples," *The American Naturalist,* Vol. 139, No. 2, February 1992, pp. 313–341.

Greenberg, Daniel S. *The Politics of Pure Science* (New American Library, New York, 1967).

Griggs, Gary B. "Monterey Bay, National Center for Marine Science," *Sea Technology,* May 1995, pp. 43–53.

Grzimek, Bernhard. *Grzimek's Animal Life Encyclopedia: Mollusks* (Van Nostrand, New York, 1974).

Guidera, Mark. "High-Tech Hunters Close in on Lost Sub," *The Sun* (Baltimore), July 23, 1995, p. 1B.

Haas, Charles A. "Destination, North Atlantic," "Dives 1 and 2: Precision in Preparation," "A Voyage for Knowledge," *Voyage,* No. 18, Autumn 1994, pp. 40–65.

——. ed. *Voyage,* No. 22, Winter 1996, pp. 51–83.

Haberman, Clyde. "Japanese Fight Invading Sea for Priceless Speck of Land," *The New York Times,* January 4, 1988, p. A1.

Hall, Stephen S. *Mapping the Next Millennium* (Vintage, New York, 1993).

Hamilton-Paterson, James. *The Great Deep: The Sea and Its Thresholds* (Random House, New York, 1992).

Hammond, Allen L. "Manganese Nodules (I): Mineral Resources on the Deep Seabed," *Science,* Vol. 183, February 8, 1974, pp. 502–503.

——. "Manganese Nodules (II): Prospects for Deep Sea Mining," *Science,* Vol. 183, February 15, 1974, pp. 644–646.

Handler, Joshua, et al. "Naval Safety 1989: The Year of the Accident," Greenpeace, Washington, D.C., April 1990.

Hansen, Chuck. *U.S. Nuclear Weapons: The Secret History* (Orion, New York, 1988).

Harbison, G. Richard. "The Gelatinous Inhabitants of the Ocean Interior," *Oceanus,* Fall 1992, pp. 18–23.

Hawker, Kenneth E., Jr., ed. "Scientific Utility of Naval Environmental Data: A MEDEA Special Task Force Report," MEDEA, McLean, Va., June 1995.

Hawkes, Graham S. "The Future of Atmospheric Diving Systems and Associated Manipulator Technology with Special Reference to the New Omads, Deep Rover," *Marine Technology Society Journal,* Vol. 17, No. 3, Fall 1983, pp. 51–60.

——, and Philip J. Ballou. "The Ocean Everest Concept: A Versatile Manned Submersible for Full Ocean Depth," *Marine Technology Society Journal,* June 1990, Vol. 24, No. 2, pp. 79–86.

Haymon, R. A., et al. "Volcanic Eruption of the Mid-ocean Ridge Along the East Pacific Rise," *Earth and Planetary Science Letters,* 1993, Vol. 119, pp. 85–101.

Hayward, Tony. "Introduction," *ROV Review* (Windate Enterprises, Spring Valley, Calif., 1988), pp. 3–4.

Helm, Thomas. *Dangerous Sea Creatures: A Complete Guide to Hazardous Marine Life* (Funk & Wagnalls, New York, 1976).

Hendrickson, Robert. *The Ocean Almanac* (Doubleday, New York, 1984).

Herman, Lawrence. "Law of the Sea Is Seen as a Bit of a Museum-piece: Canada's Non-ratification Shows Times Have Changed," *The Financial Post,* December 7, 1994, p. 17.

Herold, Robert Cameron. *The Politics of Decision-Making in the Defense Establishment: A Case Study* (Ph.D. Thesis, George Washington University, September 1969).

——, and Shane E. Mahoney. "Military Hardware Procurement," *Comparative Politics,* Vol. 6, No. 4, July 1974, pp. 571–599.

Hersh, Seymour M. "C.I.A. Salvage Ship Brought Up Part of Soviet Sub Lost in 1968, Failed to Raise Atom Missiles," *The New York Times,* March 19, 1975, p. 1.

——. "Human Error Is Cited in '74 Glomar Failure," *The New York Times,* December 9, 1976, p. 1.

Herzig, Peter, et al. "Submarine Volcanism and Hydrothermal Venting Studied in Papua, New Guinea," *EOS,* Vol. 75, No. 44, November 1, 1994, pp. 513–516.

Hickman, Cleveland P. *Biology of the Invertebrates* (Mosby, St. Louis, 1973).

Hodges, Carol Ann. "Mineral Resources, Environmental Issues, and Land Use," *Science,* Vol. 268, June 2, 1995, pp. 1305–1312.

Hollaway, D. M. A., Roy D. Haul, and Robertson P. Dinsmore. "Waves and the Future: Swath for Ocean Science," *MTS 94 Conference Proceedings* (Marine Technology Society, Washington, 1994), pp. 548–551.

Hotta, Hirosi. "Deep Sea Research Around the Japanese Islands," *Oceanus,* Vol. 30, No. 1, Spring 1987, pp. 32–33.

Howard, Janet. "Vanishing Act," *Explorations* (Scripps), Vol. 2, No. 2, Fall 1995, pp. 10–17.

Hsü, Kenneth J. *Challenger at Sea: A Ship That Revolutionized Earth Science* (Princeton University Press, Princeton, N.J., 1992).

Humphris, Susan E., et al. "Leg 158 Scientific Prospectus: Drilling an Active Hydrothermal System on a Sediment-Free Slow-Spreading Ridge," Ocean Drilling Program, Texas A&M University, March 1994.

——. *Seafloor Hydrothermal Systems: Physical, Chemical, Biological and Geological Interactions* (American Geophysical Union, Washington, 1995), Geophysical Monograph 91.

Hutchinson, Gillian. *The Wreck of the Titanic* (National Maritime Museum, Greenwich, England, 1994).

Idyll, C. P. *Abyss: The Deep Sea and the Creatures That Live in It* (Crowell, New York, 1963).

——, ed. *Exploring the Ocean World: A History of Oceanography* (Crowell, New York, 1969).

Ikeda, Yojiro. "Japan," *Mining Annual Review,* June 1988, p. 361.

International Atomic Energy Agency. "Inventory of Radioactive Material Entering the Marine Environment: Sea Disposal of Radioactive Waste," IAEA, Vienna, Austria, March 1991, pp. 48–51.

Isaacs, John D., and Richard A. Schwartzlose. "Active Animals of the Deep-Sea Floor," *Scientific American,* October 1975, pp. 85–91.

Izvestia, as translated from the Russian by the Joint Publications Research Service of the National Technical Information Service, Springfield, Va., July 22, 1992, pp. 8–9, and July 29, 1992, pp. 12–15.

Jack, William E., et al. "Biochemical Characterization of Vent and Deep Vent DNA Polymerases," *The NEB Transcript* of New England Biolabs, September 1992, pp. 4–5.

James, Canute. "Concerns Linger Over Seabed Settlement—Doubts About the Effectiveness of a New UN Authority," *Financial Times,* December 9, 1994, p. 29.

Jane's Fighting Ships (McGraw-Hill, New York, and Jane's Information Group, Alexandria, Va.).

Jannasch, Holger W. "Deep Sea Hydrothermal Vents: Underwater Oases," *The NEB Transcript* of New England Biolabs, September 1992, pp. 1–3.

Japan Marine Science & Technology Center (JAMSTEC). "Research Submersible Shinkai 6500 System" (brochure), JAMSTEC, Yokosuka, Kanagawa, Japan, undated.

——. "MODE '94: Mid Ocean Ridge Diving Expeditions" (brochure), JAMSTEC, Yokosuka, Kanagawa, Japan, undated.

——. "JAMSTEC Profile," JAMSTEC, Yokosuka, Kanagawa, Japan, November 1991.

Jeffreys, Kent. "Rescuing the Oceans," in Ronald Bailey, ed., *The True State of the Planet* (Free Press, New York, 1995) pp. 295–338.

Johnson, Nicholas L. *Soviet Space Programs: 1980–1985* (American Astronautical Society, San Diego, 1987).

——. *The Soviet Year in Space: 1986* (Teledyne Brown, Colorado Springs, 1987).

Jourdan, David. "Searching for the I-52," *Meridian Passages,* Vol. 2, No. 2, Spring 1995, pp. 1–3.

Kaharl, Victoria A. *Water Baby: The Story of Alvin* (Oxford University Press, New York, 1990).

Kaiser, Jocelyn. "Of Whales and Ocean Warming," *Science News,* Vol. 147, June 3, 1995, pp. 350–351.

Keach, Donald L. "Down to Thresher by Bathyscaph," *National Geographic,* June 1964, pp. 764–777.

Kennedy, Randy. "With Ship's Hull Back on the Ocean Floor, Titanic Buffs Return to New York," *The New York Times,* September 2, 1996, p. B26.

Kennish, Michael J. *Practical Handbook of Marine Science* (CRC Press, Boca Raton, 1994).

Kimoto, Yoichi. "Japan Mining Collector for Manganese Nodules," *18th Meeting of the U.S.-Japan Marine Facilities Panel, 1992 Conference Record,* U.S. Department of Commerce, National Oceanic and Atmospheric Administration, October-November 1992, pp. 1–3.

King, Dean, et al. *A Sea of Words* (Henry Holt, New York, 1995).

Knapp, Don. "Radioactive Wastes Threaten Pristine Farallones Islands," Cable News Network, September 17, 1994, transcript no. 236–3.

Kurkchubasche, Ramon R. "Underwater Applications for Structural Ceramics," *Underwater Intervention '94 Conference Proceedings* (Marine Technology Society, Washington, 1994), pp. 357–363.

Kyo, Masanori, and Ikuo Nakazaki. "Development of Collection, Isolation and Cultivation System for Deep-sea Microbes Study," *18th Meeting of the U.S.-Japan Marine Facili-*

ties Panel, October-November 1992 Conference Record, National Oceanic and Atmospheric Administration, U.S. Department of Commerce.

LaFond, Charles. "Hope for Disabled Undersea Vessels," *Science News,* September 12, 1970, pp. 231–233.

Lambshead, P. John D. "Recent Developments in Marine Benthic Biodiversity Research," *Oceanis,* Vol. 19, No. 6, 1993, pp. 5–24.

Langseth, Marcus, et al. "SCICEX-93: Arctic Cruise of the U.S. Navy Nuclear Powered Submarine USS Pargo," *Marine Technology Society Journal,* Vol. 27, No. 4, Winter 1993–94, pp. 4–12.

Leatham, James, and Bryan W. Coles. "Use of Laser Sensors for Search & Survey," *Underwater Intervention '93 Conference Proceedings* (Marine Technology Society, Washington, 1993), pp. 171–186.

Lehman, John. "Going for the High Ground in the Deep Seas," *Marine Technology Society Journal,* Vol. 16, No. 2, 1982, pp. 3–6.

Lennard, Don E. "The United Kingdom's Progress in Marine Technology," *Marine Technology Society Journal,* Vol. 19, No. 1, First Quarter, 1985, pp. 38–45.

Lewis, Flora. *One of Our H-Bombs Is Missing* (McGraw-Hill, New York, 1967).

Link, Edwin A. "Tomorrow on the Deep Frontier," *National Geographic,* June 1964, pp. 778–801.

Logsdon, John M. "The Space Shuttle Program: A Policy Failure?" *Science,* Vol. 232, May 30, 1986, pp. 1099–1105.

Lonsdale, Peter, et al. "Metallogenesis at Seamounts on the East Pacific Rise," *Marine Technology Society Journal,* Vol. 16, No. 3, 1982, pp. 54–61.

Love, Robert W., Jr. *History of the U.S. Navy,* Vol. 2, 1942–1991 (Stackpole, Harrisburg, Pa., 1992).

Luard, Evan. *The Control of the Sea-bed: A New International Issue* (Taplinger, New York, 1974).

Lutz, Richard A., and Rachel M. Haymon. "Rebirth of a Deep-Sea Vent," *National Geographic,* November 1994, pp. 114–126.

Lynch, Don, and Ken Marschall. *Titanic: An Illustrated History* (Hyperion, New York, 1992).

MacInnis, Joseph. *Titanic in a New Light* (Thomasson-Grant, Charlottesville, 1992).

——, ed., *Saving the Oceans* (Key Porter, Toronto, 1992).

Malahoff, Alexander. "The Ocean Floor, Our New Frontier: A Scientific Viewpoint," *Marine Technology Society Journal,* Vol. 16, No. 3, 1982, pp. 3–4.

——. "A Comparison of the Massive Submarine Polymetallic Sulfides of the Galápagos Rift with Some Continental Deposits," *Marine Technology Society Journal,* Vol. 16, No. 3, 1982, pp. 39–45.

Marchette, Victor, and John D. Marks. *The CIA and the Cult of Intelligence* (Dell, New York, 1975).

Markham, Adam. *A Brief History of Pollution* (St. Martin, New York, 1994).

Marques, A. H. de Oliveira. *History of Portugal* (Casa da Moeda, Lisbon, 1991).

Marshall, Eliot. "Superbugs in Waiting: Some Cautionary Tales," *Science,* Vol. 255, January 3, 1992, p. 29.

Marshall, N. B. *Aspects of Deep Sea Biology* (Philosophical Library, New York, 1954).

Marshall, Pearl. "U.K., China Agree to Abide by London Convention Sea Dump Ban," *Nucleonics Week,* February 24, 1994, p. 14.

Marx, Robert F. "Phoenician Explorations, Ltd., Azorean Underwater Archaeological Project, Plan of Operations," September 25, 1994.

——, and Jenifer Marx. *The Search for Sunken Treasure: Exploring the World's Great Shipwrecks* (Key Porter, Toronto, 1993).

——, and Jenifer Marx. *New World Shipwrecks 1492–1825: A Comprehensive Guide* (Ram, Dallas, 1994).

Massachusetts Institute of Technology Sea Grant College Program. "Exploring Underwater Volcanoes on the Cheap (and Fast)," *Quarterly Report,* Summer 1995, p. 3.

Mathur, E. J. "Applications of Thermostable DNA Polymerases in Molecular Biology," in Michael W. W. Adams and Robert M. Kelly, eds., *Biocatalysis at Extreme Temperatures* (American Chemical Society, Washington, 1992), ACS Symposium Series 498, pp. 189–206.

McCabe, Michael. "To the Deepest Spot on Earth: San Anselmo Inventor Designs New Class of Submarine," *San Francisco Chronicle,* October 29, 1996, p. A11.

McCann, Anna Marguerite, and Joann Freed. "Deep Water Archaeology: A Late-Roman Ship from Carthage and an Ancient Trade Route Near Skerki Bank off Northwest Sicily," *The Journal of Roman Archaeology,* Supplementary Series No. 13 (Ann Arbor, Mich., 1994).

McDonald, Ian R., et al. "Deep-Ocean Use of the SM2000 Laser Line Scanner on Submarine NR-1 Demonstrates System Potential for Industry and Basic Science," *Conference Proceedings, Oceans '95,* Marine Technology Society and Institute of Electrical and Electronic Engineers, October 1995, Vol. 1, pp. 555–565.

McDougall, Walter A. *Let the Sea Make a Noise: A History of the North Pacific from Magellan to MacArthur* (Basic, New York, 1993).

McFarling, Usha Lee. "Biologists Document a Steady Humpback Exodus from Stellwagen," *The Boston Globe,* August 15, 1994, p. 25.

McGowan, John, and Dean Roemmich. "Climatic Warming and the Decline of Zooplankton in the California Current," *Science,* March 3, 1995, pp. 1324–1326.

Menard, H. W. *The Ocean of Truth: A Personal History of Global Tectonics* (Princeton University Press, Princeton, 1986).

Mervis, Jeffrey. "Hearing Highlights Hopes, Realities," *Science,* Vol. 271, February 2, 1996, pp. 591–592.

Miller, David. "Offshore Patrolling: New Responsibilities Demand Specialized Equipment," *International Defense Review,* January 1, 1995, p. 40.

Miller, Walter James, and Frederick Paul Walter, trans. *Jules Verne's Twenty Thousand Leagues Under the Sea: The Definitive, Unabridged Edition Based on the Original French Texts* (Naval Institute Press, Annapolis, 1993).

Mills, Eric L., and Jacqueline Carpine-Lancre. "The Oceanographic Museum of Monaco," in Elisabeth Mann Borgese, ed., *Ocean Frontiers: Explorations by Oceanographers on Five Continents* (Abrams, New York, 1992), pp. 121–134.

Milstein, Michael. "Yellowstone Managers Eye Profits from Hot Microbes," *Science,* Vol. 264, April 29, 1994, p. 655.

——. "Who Has the Right to Own—and Lease—a Living Thing?" *The San Diego Union-Tribune,* June 1, 1994, p. E4.

Monaghan, Peter. "Oceanographers in a Sea of Trouble," *The Chronicle of Higher Education,* June 1, 1994, p. A6.

Monastersky, Richard. "Temperatures on the Rise in Deep Atlantic," *Science News,* May 7, 1994, p. 295.

Monterey Bay Aquarium Foundation. "Monterey Bay Aquarium," Monterey, Calif., 1992.

Moore, Valerie. "Mysteries of the Deep," *Sky* magazine of Delta Airlines, Vol. 22, No. 8, August 1993, pp. 30–40.

Morgan, Neil. "Jury's in: Walter Munk Isn't a Whale Killer," *The San Diego Union-Tribune,* August 1, 1995, p. A2.

Morris, Christopher. *The Day They Lost the H-Bomb* (Coward-McCann, New York, 1966).

Mount, Mark E., and Michael K. Sheaffer. "Estimated Inventory of Radionuclides in Former Soviet Union Naval Reactors Dumped in the Kara Sea," in *Nuclear Contamination in the Arctic: Addressing the Problem of Extensive Dumping of Radioactive Waste in the Arctic Ocean,* hearing before the Subcommittee on Oceanography, Gulf of Mexico, and the Outer Continental Shelf of the Committee on Merchant Marine and Fisheries, House of Representatives, serial no. 103-81, pp. 103–160.

Mukerji, Chandra. *A Fragile Power: Scientists and the State* (Princeton University, Princeton, 1989).

Mullen, Craig. "Eastport International Solves the M/V Lucona Sinking Mystery," *Waves,* September-October 1991, pp. 18–19.

Mullis, Kary B. "The Unusual Origin of the Polymerase Chain Reaction," *Scientific American,* April 1990, pp. 56–65.

Munk, Walter, et al. "Acoustic Thermometry of Ocean Climate. Volume 1: Technical Proposal," University of California at San Diego report number 92-1321, June 1, 1992.

Myers, Frederick Shaw, and Alun Anderson. "Microbes from 20,000 Feet Under the Sea," *Science,* Vol. 255, January 3, 1992, pp. 28–29.

Nakao, Seizo. "Deep-sea Mineral Activity in Japan," *Marine Technology Society Journal,* Vol. 29, No. 3, Fall 1995, pp. 74–78.

Nasu, Noriyuki. "Japan and the Sea," *Oceanus,* Vol. 30, No. 1, Spring 1987, pp. 2–8.

National Aeronautics and Space Administration. *Aeronautics and Space Report of the President: Fiscal Year 1992 Activities* (NASA, Washington, 1993).

National Oceanic and Atmospheric Administration. "Gulf of the Farallones National Marine Sanctuary," U.S. Department of Commerce, NOAA brochure, National Marine Sanctuary Program, undated.

——. "Monterey Bay National Marine Sanctuary," U.S. Department of Commerce, NOAA, Washington, undated.

——. "Deep Seabed Mining: Final Programmatic Environmental Impact Statement, Volume One," U.S. Department of Commerce, NOAA, September 1981.

——. "Preliminary Natural Resource Survey, Farallones Islands Radioactive Waste Dumps," NOAA report, March 29, 1990.

——. "17th UJNR/Marine Facilities Panel Study Tour," *17th Meeting of the U.S.-Japan Marine Facilities Panel, Conference Record,* U.S. Department of Commerce, NOAA, May 1991.

——. "The Monterey Bay Connection: a Prospectus for Scientific Excellence in the 1990's," U.S. Department of Commerce, NOAA, June 1992.

——. "Deep Seabed Mining: Report to Congress," U.S. Department of Commerce, NOAA, December 1993.

——. "An Overview of the Sanctuaries and Reserves Division," NOAA Congressional Briefing, March 1994.

——. "Deepwater Fishing Grants Supported by the Northeast Region National Marine Fisheries Service, Fiscal Years 1994–95," two-page fact sheet, National Marine Fisheries Service, Gloucester, Mass.

——. "The Saltonstall-Kennedy Grant Program: Fisheries Research and Development; Report 1995," U.S. Department of Commerce, NOAA, National Marine Fisheries Service, August 11, 1995.

National Research Council. *Oceanography in the Next Decade: Building New Partnerships* (National Academy Press, Washington, 1992).

——. *Understanding Marine Biodiversity* (National Academy Press, Washington, 1995).

——. *Maximizing U.S. Interests in Science and Technology Relations with Japan* (National Academy Press, Washington, 1995).

——. *Undersea Vehicles and National Needs* (National Academy Press, Washington, 1996).

"Naval Undersea Museum," *Ocean News & Technology,* September/October 1995, p. 34.

"Navy Plumbs the Ocean's Depths," *Business Week,* July 6, 1968, pp. 58–59.

Nemoto, Takahisa. "The Ocean Research Institute, Tokyo," in Elisabeth Mann Borgese, ed., *Ocean Frontiers: Explorations by Oceanographers on Five Continents* (Abrams, New York, 1992), pp. 213–227.

Newman, James B., and Bruce H. Robison. "Development of a Dedicated ROV for Ocean Science," *MTS Journal* (Marine Technology Society), Vol. 26, No. 4, pp. 46–53.

The New York Times Index 1963 (The New York Times Company, New York, 1964), pp. 707–708.

Nitze, Paul H. "Harnessing the Ocean Depths," *Vital Speeches,* October 1, 1964, pp. 749–751.

Noonan, Tim. "The Greatest Treasure Ever Found," *Life,* March 1992, pp. 32–42.

Norse, Elliott A., ed. *Global Marine Biological Diversity* (Island Press, Washington, 1993).

Norwegian Radiation Protection Authority. "Radioactive Contamination at Dumping Sites for Nuclear Wastes in the Kara Sea," report of Joint Russian-Norwegian Experts Group for Investigation of Radioactive Contamination in the Northern Areas, Osteras, Norway, November 1994.

Nuttall, Nick. "Scientists Set Sail on Mission to the Bottom of the Sea," *The Times* (London), August 27, 1994.

Oceaneering Technologies. "*Magellan 725,*" product brochure, Oceaneering Technologies (Upper Marlboro, Md., undated).

"Oceanography: Work Beneath the Waves," *Time,* January 19, 1968, pp. 68–75.

Oebius, Horst U. "Recent Advances, Future Needs and Collaboration in Ocean Technology in the Federal Republic of Germany," *Marine Technology Society Journal,* Vol. 24, No. 1, March 1990, pp. 22–27.

Offley, Ed. "Secret Navy Sub Finds New Home at Bangor Base," *Seattle Post-Intelligencer,* November 23, 1994, p. A1.

Okamura, Kenji. "Ocean Technology in Japan: Recent Advances, Future Needs and International Collaboration," *Marine Technology Society Journal,* Vol. 24, No. 1, March 1990, pp. 32–47.

O'Keefe, Sean, et al. ". . . From the Sea," U.S. Naval Institute *Proceedings.* November 1992, pp. 93–96.

"Opening Up Davy Jones's Locker," *Economist,* December 20, 1986, p. 117.

Oswald, Robert B. "Dual Use Technology: Using Department of Defense Technology to Support Marine Research," *Marine Technology Society Journal,* Vol. 27, No. 4, Winter 1993–94, pp. 49–51.

Packard, David. *The HP Way: How Bill Hewlett and I Built Our Company* (Harper-Collins, New York, 1995).

Paddock, Richard C. "Undersea Noise Test Could Risk Making Whales Deaf," *The Los Angeles Times,* March 22, 1994, p. A1.

Palmer, R. R., and Joel Colton. *A History of the Modern World* (Knopf, New York, 1965).

Palowitch, A. W. "Lasers in Underwater Imaging," *Underwater News & Technology,* Vol. 1, No. 1, July-August 1993, pp. 4–6.

Parfit, Michael. "Diminishing Returns," *National Geographic,* November 1995, pp. 2–37.

Parrilla, Gregorio, et al. "Rising Temperatures in the Subtropical North Atlantic Ocean Over the Past 35 Years," *Nature,* Vol. 369, May 5, 1994, pp. 48–51.

Pasho, David W. "Canada and Ocean Mining," *Marine Technology Society Journal,* Vol. 19, No. 4, 1985, pp. 26–30.

"Pfizer Tests Marine Microorganisms for Disease Fighting Potential," Pfizer news release, Groton, Conn., August 24, 1995.

Piccard, Jacques, and Robert S. Dietz. *Seven Miles Down: The Story of the Bathyscaph Trieste* (Putnam, New York, 1961).

Pickford, Nigel. *The Atlas of Shipwrecks and Treasure: The History, Location, and Treasures of Ships Lost at Sea* (Dorling Kindersley, New York, 1994).

Pitt, David E. "Nations Back Ban on Atomic Dumping," *The New York Times,* November 13, 1993, p. A7.

——. "Russia Is Pressed on Nuclear Waste Dumping," *The New York Times,* December 5, 1993, p. A6.

Platt, Anne E. "Dying Seas," *World Watch,* Vol. 8, No. 1, January-February 1995, pp. 10–19.

Polmar, Norman. *Death of the Thresher* (Chilton, New York, 1964).

——. "The Deep Submergence Vehicle Fleet," U.S. Naval Institute *Proceedings,* June 1986, pp. 119–120.

——. "How Many Spy Subs . . . ?" U.S. Naval Institute *Proceedings,* December 1996, pp. 87–88.

——, and Thomas B. Allen. *Rickover: Controversy and Genius* (Simon & Schuster, New York, 1982).

Poore, Gary C. B., and George D. F. Wilson. "Marine Species Richness," *Nature,* Vol. 361, February 18, 1993, pp. 597–598.

Pope, Gregory T. "Deep Flight," *Popular Mechanics,* April 1990, pp. 70–72.

Portman, Jamie. "Cameron Can't Avoid the Big Picture: Director Embarks on Another Blockbuster—This Time About Titanic," *The Gazette* (Montreal), March 5, 1996, p. C9.

Portugal Regional Tourism Board. "Terceira," Angra do Heroismo, Azores, undated.

Portuguese Commission for the Lisbon World Exposition. "Expo '98, The Last Exposition of the 20th Century," Lisbon, 1995.

Press, Frank, and Raymond Siever. *Understanding Earth* (Freeman, New York, 1994).

Pryor, Taylor A. "New Described Super-Nodule Resource," *Sea Technology,* September 1995, pp. 15–18.

Rabinow, Paul. *Making PCR: A Story of Biotechnology* (University of Chicago, Chicago, 1996).

"Radioactive Waste at Sea Confirmed," *Los Angeles Times,* November 19, 1990, p. A24.

Raleigh, C. Barry. "The Internationalism of Ocean Science vs. International Politics," *Marine Technology Society Journal,* Vol. 23, No. 1, March 1989, pp. 44–47.

Raven, Gerrard. "Scientists to Study Threatened Ocean Volcano Area," Reuters, August 26, 1994.

Reuters. "Ban Is Now in Force on Nuclear Dumping," *The New York Times,* February 22, 1994, p. A8.

Revkin, Andrew C. "Cold-War Laser Aids T.W.A. Hunt," *The New York Times,* August 1, 1996, p. B1.

Richardson, Henry R., and Lawrence D. Stone. "Operations Analysis During the Underwater Search for Scorpion," *Naval Research Logistics Quarterly,* Vol. 18, No. 2, June 1971, pp. 141–157.

Richelson, Jeffrey T. *The U.S. Intelligence Community* (Ballinger, Cambridge, Mass., 1985).

——. *American Espionage and the Soviet Target* (Morrow, New York, 1987).

Riding, Alan. "1,800 Objects from the Titanic: Any Claims?" *The New York Times,* December 16, 1992, p. A17.

Ringle, Ken. "New Depths for Titanic Promoter? Cruise to Ship Site Reopens Controversy," *The Washington Post,* August 6, 1996, p. B1.

Robigou, Veronique, and John R. Delaney. "Large Massive Sulfide Deposits in a Newly Discovered Active Hydrothermal System, the High-Rise Field, Endeavour Segment, Juan de Fuca Ridge," *Geophysical Research Letters,* September 3, 1993, Vol. 20, No. 17, pp. 1887–1890.

Robison, Bruce H. "Midwater Biological Research with the WASP ADS," *Marine Technology Society Journal,* Vol. 17, No. 3, Fall 1983, pp. 21–27.

——. "Light in the Ocean's Midwaters," *Scientific American,* July 1995, pp. 60–64.

——, et al. "Cold Seep Research Program," annual research plan, Monterey Bay Aquarium Research Institute, 1993.

Rolland, Daniel, and Thomas Demas. "Oceanographic Research Ships of the Twenty-First Century," *MTS 94 Conference Proceedings* (Marine Technology Society, Washington, 1994), pp. 556–562.

Romer, A. S. "Darwin and the Fossil Record," in Philip Appleman, ed., *Darwin: A Norton Critical Edition* (Norton, New York, 1970), pp. 374–375.

Rona, Peter A. "The Dynamic Abyss," in Joseph MacInnis, ed., *Saving the Oceans* (Key Porter, Toronto, 1992), pp. 99–116.

——. "Deep-Sea Geysers of the Atlantic," *National Geographic,* October 1992, pp. 105–109.

Roper, Clyde F. E., and Kenneth J. Boss. "The Giant Squid," *Scientific American,* Vol. 246, No. 4, April 1982, pp. 96–105.

Rule, Margaret. *The Mary Rose: The Excavation and Raising of Henry VIII's Flagship* (Conway Maritime Press, London, 1983).

Ryan, Sean. "Back from the Deep," *The Sunday Times* (London), October 9, 1994.

Saade, Edward J., and Drew Carey. "Laser Line Scanner: 'Highgrading' Search Targets," *Sea Technology,* October 1996, pp. 63–65.

Sagalevitch, Anatoly M. " 'Mir' Submersibles in Science and Underwater Technology," *MTS 94 Conference Proceedings* (Marine Technology Society, Washington, 1994), pp. 241–244.

——. "Results of Five Years of Operation with Deep Manned Submersibles 'Mir-1' and 'Mir-2' on Nuclear Submarine 'Komsomolets' Wreck," *Conference Proceedings, Oceans '95* (Marine Technology Society and Institute of Electrical and Electronics Engineers, October 1995), Vol. 1, pp. 6–10.

"The Salvage Controversy," *USA Today,* the magazine, March 1995, p. 68.

Sanger, David E. "Nuclear Material Dumped Off Japan," *The New York Times,* October 19, 1993, p. A1.

Santos, Pedro Marta. "Fruits of the Sea," Vida supplement of *O Independente* (Lisbon), April 28, 1995, pp. 14–16.

Schlee, Susan. *The Edge of an Unfamiliar World: A History of Oceanography* (Dutton, New York, 1973).

Scott, S. D. "Polymetallic Sulfide Riches from the Deep: Fact or Fallacy?" in K. J. Hsü and J. Thiede, eds., *Use and Misuse of the Seafloor* (Wiley, New York, 1992), pp. 87–115.

Scott, W. B., and M. G. Scott. *Atlantic Fishes of Canada* (University of Toronto, Toronto, 1988).

Scripps Institution of Oceanography. "Acoustic Thermometry of Ocean Climate (ATOC) Media Advisory," Scripps, University of California at San Diego, April 11, 1994.

Smith, Craig R. "Whale Falls: Chemosynthesis on the Deep Seafloor," *Oceanus,* Vol. 35, No. 3, Fall 1992, pp. 74–78.

Smith, John B. "Managing Nonenergy Marine Mineral Development—Genesis of a Program," *Conference Proceedings, Oceans '85* (Marine Technology Society, Washington, 1985), pp. 339–351.

Smith, N. Eugene. "DSV Sea Cliff—1992 Hawaiian Science Cruise," *Underwater Intervention '93 Conference Proceedings* (Marine Technology Society, Washington, 1993).

——. "Civilian Scientists Use the Navy's Deep Submergence Vehicles," *Marine Technology Society Journal,* Vol. 27, No. 4, Winter 1993, pp. 64–69.

Sobel, Dava. *Longitude* (Walker, New York, 1995).

Sparck, Ragnar. "Background and Origin of the Expedition," in Anton F. Bruun et al., *The Galathea Deep Sea Expedition: 1950–1952* (Macmillan, New York, 1956), pp. 11–17.

Stegeman, John J. "Detecting the Biological Effects of Deep-Sea Waste Disposal," *Oceanus,* Vol. 33, No. 2, Summer 1990, pp. 54–60.

——, et al. "Monooxygenase Induction and Chlorobiphenyls in the Deep-Sea Fish *Coryphaenoides armatus,*" *Science,* Vol. 231, March 14, 1986, pp. 1287–1289.

Stevens, William K. "Feeding a Booming Population Without Destroying the Planet," *The New York Times,* April 5, 1994, p. C1.

——. "Emissions Must Be Cut to Avert Shift in Climate, Panel Says," *The New York Times,* September 20, 1994, p. C4.

——. "Global Warming Experts Call Human Role Likely," *The New York Times,* September 10, 1995, p. A1.

——. "Richard S. Lindzen; A Skeptic Asks, Is It Getting Hotter, or Is It Just the Computer Model?" *The New York Times,* June 18, 1996, p. C1.

Stone, Gregory. "Dual Use of Military Assets for Environmental Research," *Marine Technology Society Journal,* Vol. 27, No. 4, Winter 1993–94, p. 3.

Stone, Gregory S. "Japanese Ocean Research and Development," *Marine Technology Society Journal,* Vol. 26, No. 3, 1992, pp. 11–19.

Stone, Lawrence D. "Search for the SS Central America: Mathematical Treasure Hunting," *Interfaces,* January-February 1992, pp. 32–54.

Sugarman, Carole. "Skating Away—Fish with Wings," *The Washington Post,* October 19, 1994, p. E1.

Svitil, Kathy A. "Collapse of a Food Chain," *Discover,* July 1995, pp. 36–37.

Takagawa, Shinichi. "Deep Submersible Project (6,500 m)," *Oceanus,* Vol. 30, No. 1, Spring 1987, pp. 29–32.

——. "Advanced Technology Used in Shinkai 6500 and Full Ocean Depth ROV Kaiko," *Marine Technology Society Journal,* Vol. 29, No. 3, Fall 1995, pp. 15–25.

Takaishi, Yoshifumi. "Fundamental Conception on Ocean Exploitation in Japan and Contribution of Shipbuilding Technology," *17th Meeting of the U.S.–Japan Marine Facilities Panel, Conference Record,* U.S. Department of Commerce, NOAA, May 1991, pp. 69–74.

Thiede, J., and K. J. Hsü, eds. *Use and Misuse of the Seafloor* (Wiley, New York, 1992).

——. "The Future of Ocean Resources," in J. Thiede and K. J. Hsü, eds., *Use and Misuse of the Seafloor* (Wiley, New York, 1992).

Thomas, Robert McG., Jr. "Eva Hart, 91, a Last Survivor with Memory of Titanic, Dies," *The New York Times,* February 16, 1996, p. A30.

Thomas, William. "Memphis Salvagers Plan Trip to Titanic," *The Commercial Appeal* (Memphis), August 15, 1992, p. A1.

Thomson, C. Wyville. *The Depths of the Sea* (Macmillan, London, 1874).

Thomson, Keith Steward. "The Story of the Coelacanth," *Oceanus,* Vol. 34, No. 3, Fall 1991, pp. 38–43.

Thurman, Harold V. *Introductory Oceanography* (Macmillan, New York, 1994).

Tidwell, Paul R. "Operation Rising Sun," set of reproductions of archival documents, undated.

Travis, John. "Invader Threatens Black, Azov Seas," *Science,* November 26, 1993, pp. 1366–1367.

Tunnicliffe, Verena. "Hydrothermal-Vent Communities of the Deep Sea," *American Scientist,* July-August 1992, pp. 336–349.

——. "Coaxial Report: Beard Vent Fauna," draft report, January 25, 1994.

Tyce, Robert, et al. "NECOR Sea Beam Data Collection and Processing Development," *Marine Technology Society Journal,* Vol. 21, No. 3, June 1987, pp. 80–92.

Tyler, Patrick E. *Running Critical* (Harper & Row, New York, 1986).

——. "The U.S., Too, Has Dumped Waste at Sea," *The New York Times,* May 4, 1992, p. A8.

——. "China Revamps Forces with Eye to Sea Claims," *The New York Times,* January 2, 1995, section 1, p. 2.

United Kingdom. *Exclusive Economic Zone Catalogue* (Combined Service Publications Ltd., Hampshire, Great Britain, 1993).

United Press International. "In Search of Sunken Treasure—Copper," March 28, 1982.

——. "Robot That Spotted Titanic Also Found Sunken Submarine," *The New York Times,* September 17, 1985, p. A23.

——. "Texan Loses Bid to Salvage Titanic," October 3, 1992.

United States Code Congressional and Administrative News, 98th Congress, First Session, 1983, Vol. 3 (West Publishing, St. Paul, 1984).

United States Congress. *Nuclear Contamination in the Arctic: Addressing the Problem of Extensive Dumping of Radioactive Waste in the Arctic Ocean.* Hearing Before the Subcommittee on Oceanography, Gulf of Mexico, and the Outer Continental Shelf of the Committee on Merchant Marine and Fisheries, House of Representatives, serial no. 103-81.

United States Congressional Office of Technology Assessment. *Nuclear Wastes in the Arctic: An Analysis of Arctic and Other Regional Impacts from Soviet Nuclear Contamination* (OTA, Washington, September 1995), OTA-ENV-623.

United States Navy. "Naval Ocean Systems Center Underwater Vehicle History," Technical Document 1530, Naval Ocean Systems Center, San Diego, April 1989.

——. "Advanced Unmanned Search System," Document No. 2348, Naval Command, Control and Ocean Surveillance Center, RDT&E Division, San Diego, December 1992.

——. First Court of Inquiry report on the *Scorpion,* which was declassified in October 1993.

——. Second Court of Inquiry report on the *Scorpion,* which was declassified in October 1993.

University of Maryland Biotechnology Institute. "Microbiology/Biotechnology at the Center of Marine Biotechnology," UMBI brochure, 1994, pp. 10–11.

Van Dover, Cindy Lee. "Do 'Eyeless' Shrimp See the Light of Glowing Deep-Sea Vents?" *Oceanus,* Winter 1988–89, Vol. 31, No. 4, pp. 47–52.

——. *The Octopus's Garden: Hydrothermal Vents and Other Mysteries of the Deep Sea* (Addison, New York, 1996).

Van Natta, Don, Jr. "Luggage Spotted in Debris Trail Suggests an Explosion to Experts," *The New York Times,* August 6, 1996, p. A1.

Varner, Roy, and Wayne Collier. *A Matter of Risk: The Inside Story of the CIA's Hughes Glomar Explorer Mission to Raise a Russian Submarine* (Random House, New York, 1978).

Wade, Nicholas. "Do Moles Matter?" *The New York Times Magazine,* March 20, 1994, p. 22.

Wakelin, James H., Jr. "Thresher: Lesson and Challenge," *National Geographic,* June 1964, pp. 759–763.

Waldrop, M. Mitchell. "Goodbye to the Warm Little Pond?" *Science,* November 23, 1990, pp. 1078–1080.

——. "The Golden Crystal of Life," *Science,* November 23, 1990, p. 1080.

Walsh, Don. "Looking Backwards at the Future," *Naval Institute Proceedings,* January 1985, pp. 103–104.

——. "Thirty Thousand Feet and Thirty Years Later: Some Thoughts on the Deepest Presence Concept," *Marine Technology Society Journal,* June 1990, Vol. 24, No. 2, pp. 7–8.

Walton, James, et al. "Advanced Unmanned Search System," *Underwater Intervention '93 Conference Proceedings* (Marine Technology Society, Washington, 1993), pp. 243–249.

Ward, Nathalie. *Stellwagen Bank: A Guide to the Whales, Sea Birds, and Marine Life of the Stellwagen Bank National Marine Sanctuary* (Down East Books, Camden, Maine, 1995).

Weber, Michael L., and Judith A. Gradwohl. *The Wealth of Oceans* (Norton, New York, 1995).

Weiner, Tim, et al. "C.I.A. Spent Millions to Support Japanese Right in the 50's and 60's," *The New York Times,* October 9, 1994, p. A1.

Welling, Conrad G. "Polymetallic Sulfides: An Industry Viewpoint," *Marine Technology Society Journal,* Vol. 16, No. 3, 1982, pp. 5–7.

Whitehead, Hal. "The Realm of the Elusive Sperm Whale," *National Geographic,* November 1995, pp. 56–73.

Whitman, Edward C. "Defense Conversion in Marine Technology—Past, Present, and Potential Possibilities," *MTS 94 Conference Proceedings* (Marine Technology Society, Washington, 1994), pp. 121–126.

——. "Defense Conversion in Marine Technology," *Sea Technology,* November 1994, pp. 21–25.

Widder, E. A., et al. "Bioluminescence in the Monterey Submarine Canyon: Image Analysis of Video Recordings from a Midwater Submersible," *Marine Biology,* Vol. 100, 1989, pp. 541–551.

Williams, J. E. D. *From Sails to Satellites: The Origin and Development of Navigational Science* (Oxford, New York, 1992).

Williams, Lloyd. "World Law of the Sea, Long Opposed by US, Quietly Goes into Effect," Associated Press, November 16, 1994.

Wilson, Edward O. *The Diversity of Life* (Norton, New York, 1992).

Winston, Judith E. "Systematics and Marine Conservation," in Niles Eldredge, ed., *Systematics, Ecology and the Biodiversity Crisis* (Columbia University Press, New York, 1992).

Woods Hole Oceanographic Institution. "An Emperor Comes to Call," *Newsletter,* Woods Hole Oceanographic Institution, December 1975, pp. 7–8.

——. "Atlantis II Visits Japan," *Newsletter,* Woods Hole Oceanographic Institution, May-November 1987, pp. 1–4.

——. "Woods Hole Oceanographic Institution–Japan Marine Science and Technology Center Relationship," WHOI two-page news release, July 1994.

——. "Scientists Aboard the Deep Diving Submersible, Shinkai 6500, Conduct a Unique Study of the Earth's Structure," WHOI two-page news release, July 27, 1994.

——. "First American-Japanese Expedition to the Mid-Atlantic Ridge Increasing Knowledge of the Evolution of the Earth," WHOI two-page news release, July 27, 1994.

——. "Drilling into the Ocean Floor's Plumbing System May Provide New Insights into How Mineral Deposits Are Formed," WHOI two-page news release, July 27, 1994.

——. "The Hot Spot," "Stone Soup," "Ground Zero," *Woods Hole Currents,* Summer 1995, pp. 3–13.

Woodward, Bob. *Veil: The Secret Wars of the CIA, 1981–1987* (Simon & Schuster, New York, 1987).

World Technology Evaluation Center. "WTEC Panel Report on Research Submersibles and Undersea Technologies," Loyola College, Baltimore, June 1994.

——. "WTEC Panel Report on Submersibles and Marine Technologies in Russia's Far East and Siberia," Loyola College, Baltimore, August 1996.

Yablokov, Aleksei V., et al. "Facts and Problems Related to Radioactive Waste Disposal in Seas Adjacent to the Territory of the Russian Federation," Office of the President of the Russian Federation, Moscow, 1993.

Yoerger, Dana R. "Robotic Undersea Technology," *Oceanus,* Vol. 34, No. 1, Spring 1991, pp. 32–37.

Zamost, Bruce L., et al. "Thermostable Enzymes for Industrial Applications," *Journal of Industrial Microbiology,* Vol. 8, 1991, pp. 71–81.

Zippin, Jeffrey P. "Draft Environmental Impact Statement: Proposed Polymetallic Sulfide Minerals Lease Offering, Gorda Ridge Area Offshore Oregon and Northern California," Minerals Management Service, U.S. Department of the Interior, Reston, Va., December 1983.

Acknowledgments

FIRST AND FOREMOST, my thanks go to the explorers, scientists, and technologists who graciously took time to describe their work and answer my questions, both for this book and for my articles in *The New York Times,* which helped lay its foundations. Many of these experts are cited in this book and its notes or in the *Times* stories. But many are not. Conversations that took place at a trade show or on a ship's fantail may have escaped my notebook and laptop computer but nonetheless aided my education, at times significantly. All these individuals, cited or not, have my gratitude. I'd also like to thank the people who spoke to me about military secrets on the condition of anonymity. Far from posing a danger to national security, their disclosures promise to strengthen its foundations by informing the electorate and stimulating public discussion. They are patriots in an unassuming way.

Many groups and individuals supplied me with written materials, background information, and other aid. My thanks to Francisco J. S. Alves of Portugal's National Museum of Archaeology; Roger V. Amato of the United States Interior Department; William M. Arkin; Susan Artigiani of the Naval Institute Press; Peter J. Auster of the University of Connecticut; Robert D. Ballard; Doug Bandow of the Cato Institute; John A. Baross and John R. Delaney of the University of Washington; James G. Bellingham of the Massachusetts Institute of Technology; Bruce D. Berkowitz; Carl Boyd of Old Dominion University; Ted Brockett and Caroline Richards of Sound Ocean Systems; Jeff Burns and David W. Jourdan of Meridian Sciences; James J. Childress of the University of California at Santa Barbara; Duane A. Cox of the Scientific Environmental Research Foundation; John P. Craven; D. Roy Cullimore; Clifton Curtis, Mike Hagler, and Joshua Handler of Greenpeace; Roger C. Dunham; Cheryl Lyn Dybas of the National Science Foundation; Amos S. Eno of the National Fish and Wildlife Foundation; Alex Foley and George Tulloch of RMS *Titanic;* Joseph Fromm of the International Institute for Strategic Studies; William H. Garzke, Jr.; Thomas Gold of Cornell University; Stanley Goldberg; J. Frederick Grassle and Peter A. Rona of Rutgers University; Graham S. Hawkes; Karen Hawkes; Dave Herasimchuk of Global Marine; Daniel Hirsch of the Committee to Bridge the Gap; Geoff Holdridge of the World Technology Evaluation Center; J. Timothy

Hudson of Marex International; Randolph A. Koski and Janet L. Morton of the United States Geological Survey; Emory Kristof, Barbara Moffet, and Melissa A. Montefiore of the National Geographic Society; Laurence Lippsett and Faye Yates of the Lamont-Doherty Earth Observatory of Columbia University; F. Michael Lorz of the Columbus-America Discovery Group; Anna McCann; Alfred S. McLaren of the Explorers Club; Robert F. Marx; Eric J. Mathur; David L. Mearns of Blue Water Recoveries; Jonathan Medalia of the Congressional Research Service; Shane Merz of Senator Daniel K. Akaka's office; Henry Ng of the J. M. Kaplan Fund; Robert S. Norris of the Natural Resources Defense Council; James E. Oberg; Ed Offley of the *Seattle Post-Intelligencer;* Charles Pellegrino; Lyle D. Perrigo of the Arctic Research Commission; John E. Pike of the Federation of American Scientists; Norman Polmar; Clyde F. E. Roper of the Smithsonian Institution; William F. Searle, Jr.; Jack Sobel of the Center for Marine Conservation; Rick Spinrad and James D. Watkins of the Consortium for Ocean Research and Education; Lawrence D. Stone of Metron; Per Strand of the Norwegian Radiation Protection Authority; David W. Thomas of the American Geophysical Union; Paul R. Tidwell; Verena Tunnicliffe of the University of Victoria; Lindy Weilgart; and Aaron H. Woods of the Ocean Drilling Program at Texas A&M University.

At the United States Navy, my thanks to Gail Cleere, Dennis M. Conlon, Commander David B. Knox, Thomas J. LaPuzza, Commander Stephen Pietropaoli, Lieutenant Commander Kenneth B. Ross, John Sanders, Bob Wernli, Edward C. Whitman, and many other individuals at various branches and news offices.

At the National Oceanic and Atmospheric Administration, my thanks to Edward T. Baker, Bradley Barr, Ken Beal, Robert W. Embley, Christopher G. Fox, Eliot Hurwitz, M. Karl Jugel, Justin Kenney, Dane Konop, Jeanne G. Kouhestan, Marilyn Mayo, Betty S. Rosser, N. Eugene Smith, Walter H. F. Smith, Ed Ueber, and Joseph R. Vadus, among many other helpful individuals.

In the Monterey Bay region, my thanks to Peter G. Brewer, Judith L. Connor, Annette Gough, James Hunt, Khosrow Lashkari, and Bruce H. Robison of the Aquarium Research Institute; to Ken Peterson of the Aquarium; and to Gretchen Dennis of David Packard's office.

At the Woods Hole Oceanographic Institution, my thanks to James M. Broadus, Judith Fenwick, Charles D. Hollister, Shelley M. Lauzon, Hugh D. Livingston, Cathy Offinger, Richard F. Pittenger, David A. Ross, and John J. Stegeman.

At the Scripps Institution of Oceanography, my thanks to Cindy L. Clark, Chuck Colgan, David W. Hyde, Walter H. Munk, Susie Pike, Cindy Rogers, and David T. Sandwell.

My education over the years was aided by two periodicals, *Sea Technology* and the *Marine Technology Society Journal,* whose editors explored many of the germane issues. So, too, my thanks go to the Marine Technology Society, whose staff was quick to help in all kinds of ways, especially in retrieving back issues of the *Journal* and back papers and proceedings of society conferences.

For scuba guidance, thanks to my instructors at Pan Aqua Diving in Manhattan, and their instructors at International PADI, Inc.

My colleagues and editors at *The New York Times* were most indulgent of my deep-sea interests. Nicholas Wade, my coauthor from a previous book and head of the Science Department during the book's genesis and writing, was generous in encouraging my curiosity and played a significant role in shaping my stories and ideas about the undersea world. Thanks also go to William K. Stevens, whose environmental reporting was a continual education. Walter Sullivan, a senior colleague who recently passed away, kindly opened his files to me for this book and was an inspiration in his long and distinguished career of going wherever he needed to go to get the story.

Most generally, I am indebted to *The New York Times* as an institution for giving me the freedom to follow my journalistic instincts in writing dozens of articles about deep-sea science and exploration. More than anything else, that privilege put me in the right place at the right time.

For research assistance on this book, my thanks to Rayan Feris for her library searches and to Lawrence M. La Fountain–Stokes for his Portuguese translations.

In London, the people at the Oxford Television Company stimulated me to clarify important issues as they worked on a spinoff documentary. My thanks to Will Aslett, Steve Davis, Jeremy Hall, Ned Johnston, Nicholas Kent, and Alex Marengo.

My agent, Peter Matson of Sterling Lord Literistic, initially spurred me on with his enthusiasm and offered much encouragement along the way. My thanks also to Jody Hotchkiss for support over the long haul.

A number of experts read through various parts or drafts of the manuscript, making thoughtful comments and catching a number of errors of fact and interpretation. I am very grateful and take full responsibility for any mistakes that remain. It is important to note that in cases where a reader was also an interview subject at an earlier stage, such interviewing was never done on the condition or promise of manuscript review. Only after the fact did I seek out expert advice in an attempt to make this book as solid as possible. Many thanks to Peter J. Auster of the University of Connecticut, Angelo M. Codevilla of Boston University, Kathleen Crane of the Naval Research Laboratory, Kevin J. Crisman of the Institute of Nautical Archaeology at Texas A&M University, Michael J. Cruickshank of the University of Hawaii, Sylvia A. Earle of Deep Ocean Exploration and Research, James F. Holden of the University of Georgia, Ian R. Jonasson of the Geological Survey of Canada, David W. Jourdan of Meridian Sciences, John E. Pike of the Federation of American Scientists, Bruce H. Robison of the Monterey Bay Aquarium Research Institute, Russell Seitz of the John M. Olin Institute for Strategic Studies at Harvard University, and Cindy Lee Van Dover of the University of Alaska.

Dimitry Schidlovsky, who did the illustrations, has my heartfelt thanks for his hard work and magical touch, which turned my disorganized ramblings into revealing and often gripping works of art.

At Simon & Schuster, my thanks go first and foremost to Alice E. Mayhew. It was she who persuaded me of the merits of a general book on deep-sea exploration, who gave me unfailing encouragement over the years, and, as rough drafts came in, who applied her penetrating simplicity to cut to the heart of the matter, over and over. She made it a book. Many thanks also to her colleagues Elizabeth Stein and Lisa Weisman for their deft assaults on the manuscript, which greatly improved its organization and artfully removed all kinds of excess baggage. Also at Simon & Schuster, my thanks to Michael Accordino, Bob Asahina, Bob Bender, Margaret Cheney, Steve Messina, Marcia B. Paul, Eric Rayman, and Becky Saletan.

My parents in Milwaukee and my siblings and wider family cheered me on with steady enthusiasm, which was indispensable. Thanks also to my wife's family and to brother-in-law Jarl Mohn of Los Angeles for a raised eyebrow that ended up restructuring a chapter. The encouragement of friends and neighbors is gratefully acknowledged, with special thanks to John Risner for putting me onto Patrick O'Brian's sea novels, which were a wonderful antidote to deep pressure.

My wife, Tanya, has my greatest thanks. She showed loving forbearance over the years of writerly toil, graciously putting up with delays and distractions, lost weeks and weekends, planned escapes and vacations that failed to materialize. When research promised to take me far from home, she was quick to give her unqualified support, and do so year after year, despite the hardships. Moreover, throughout the long writing process she used her sharp eye and judgment to improve the book in all kinds of important ways, which is no small accomplishment for a person taking care of three small children as well as starting to do her own work again. No words can express my gratitude. Without her, and her love, this book would not be.

Last, I would like to thank my children, Max, Isabelle, and Juliana, also known as Bug, Izzie, and Nana, to whom this book is dedicated. Their curiosity about all sorts of things, including the deep sea, has been an education and a source of pride. They have also strengthened my hope. The next century will doubtless be more unsettled than we can imagine. Yet thinking about their lives through the perspective of this book helped me see that they and their peers will face not only social uncertainty but all kinds of new frontiers that will enrich their own voyages of discovery.

WILLIAM J. BROAD
Larchmont, New York
January 1997

Index

Page numbers in *italics* refer to illustrations.